AEROSPACE MARKETING MANAGEMENT

AEROSPACE MARKETING MANAGEMENT

*Manufacturers • OEM • Airlines • Airports
• Satellites • Launchers*

by

PHILIPPE MALAVAL

Toulouse Business School (ESC Toulouse), France

and

CHRISTOPHE BÉNAROYA

Business Conseil, France

*KLUWER ACADEMIC PUBLISHERS
Boston/Dordrecht/London*

Distributors for North America:
Kluwer Academic Publishers
101 Philip Drive
Assinippi Park
Norwell, Massachusetts, 02061 USA
Telephone (781) 871-6600
Fax (781) 871-6528
E-Mail <kluwer@wkap.com>

Distributors for all other countries:
Kluwer Academic Publishers Group
Distribution Centre
Post Office Box 322
3300 AH Dordrecht, THE NETHERLANDS
Telephone 31 78 6392 392
Fax 31 78 6546 474
E-Mail <orderdept@wkap.nl>

Electronic Services <http://www.wkap.nl>

Library of Congress Cataloging-in-Publication Data
Malaval, Philippe.
Bénaroya, Christophe.
 [Marketing aéronautique et spatial. English]
 Aerospace Marketing Management. Manufacturers - OEM - Airlines - Airports - Satellites - Launchers / by Philippe Malaval and Christophe Bénaroya
 p.cm.
 Translation of: Marketing aéronautique et spatial.
 Includes bibliographical references and index.
 ISBN

Copyright © 2002 by Kluwer Academic Publishers

All rights reserved. No part of this publication may be reproduced, stored in a retrieval system or transmitted in any form or by any means, mechanical, photocopying, recording, or otherwise, without the prior written permission of the publisher, Kluwer Academic Publishers, 101 Philip Drive, Assinippi Park, Norwell, Massachusetts 02061

Printed on acid-free paper.

Printed in the United States of America.

This book was first published in France in October 2001, Pearson Education France, Paris. Translated and published with permission of Pearson Education France.

CONTENTS

Preface	*xv*
Chapter 1	
Marketing in the Aeronautics and Space Industry	***1***
1. The aeronautics and space sector environment	***1***
1.1 Characteristics of the aeronautics and space sector	1
1.2 Market deregulation	2
1.3 The role of regulatory organizations: FAA, DGAC, CAA, IATA	3
The organizations which control air transport	***5***
2. The marketing approach	***9***
2.1 The rise of marketing	9
2.2 The two facets of marketing	11
2.3 An essential part of companies and organizations	12
2.4 A set of methods with wide-ranging applications	13
3. The role of marketing in the aeronautics and space industry	***14***
3.1 The growing importance of marketing in the aeronautics and space sector	14
3.2 Different types of marketing	15
3.3 Marketing in the aeronautics and space supply chain	18
3.4 Recent marketing trends in aeronautics	20
Chapter 2	
The Individual and Organizational Purchase	***23***
1. The individual purchase	***23***
1.1 Factors influencing the buying behavior	23
The child as influencer	***25***
Senior marketing	***26***
1.2 The buying process	30
2. The organizational purchase	***34***

2.1 The buying center	34
2.2 Buying phases	37

"Make or Buy ?", the example of the Super Transporter: from Super Guppy to A300-600ST — *40*

2.3 Different situations	42
2.4 The behavior of professional buyers	44
2.5 Bidding	47
2.6 E-procurement and the development of the marketplaces	49
3. Case study: the aircraft constructor's approach to the airline's buying center	*51*
3.1 First action level (1): the customer airline	52
3.2 Second action level (2): air traffic regulatory bodies	55
3.3 Third action level (3): airports	56
3.4 Fourth action level (4): passengers and citizen groups	56
4. Case study: Airing's approach to the buying center of the Sultan of M	*58*
5. Purchase marketing	*60*
5.1 The different conceptions of purchase marketing	61
5.2 Purchase marketing objectives	61
5.3 Means of action	62
Purchase marketing at EADS	*66*

Chapter 3
Business Marketing Intelligence — 75
1. The information system — *75*

Dassault Aviation: Increasing Falcon customer satisfaction by using an efficient information system — *76*

2. Market surveillance: active listening — *78*

2.1 The different types of surveillance	79
2.2 Setting up surveillance	80

Docland: Aerospatiale-Matra's information and documentation center (EADS) — *82*

3. Information sources — *82*

3.1 Main information sources	83
3.2 Information protection	87

Market studies made by Boeing and Airbus — *88*

4. The main types of studies — *95*

4.1 Qualitative studies	96
4.2 Quantitative studies	97

The international tourism study: a key analysis for airline destination strategy — *98*

4.3 Permanent and ad hoc studies	101

4.4 The other objectives for studies	106

From the Airbus A3XX project to the A380: the primordial role of studies — *109*

Chapter 4
Market Segmentation and Positioning — *115*
1. Segmenting a market — *115*
    1.1 The objectives of segmentation — 115

The segmentation of the satellite market — *116*

    1.2 The main segmentation methods — 117

An example of "top-down" segmentation applied to catering — *119*

"Bottom-up" segmentation: the example of the business jet market — *124*

    1.3 Other segmentation methods used in B to B — 128

The segmentation of the freight market — *131*

2. Positioning — *133*
    2.1 Positioning objectives — 133

Examples of positioning: AeroMexico and Thai Airways — *134*

Four examples of positioning: Singapore Airlines, Air France, Swissair, American Airlines — *136*

Two further examples: Virgin Atlantic and easyJet — *138*

    2.2 Setting up positioning — 139

Lufthansa Technik: positioning based on customer success — *143*

Chapter 5
Marketing and Sales Action Plan — *147*
1. The marketing plan — *147*
    1.1 Part one: the analysis — 149
    1.2 The objectives — 152
    1.3 The means or action plan — 154

2. The sales action plan — *156*
    2.1 The objectives of the sales action plan — 157
    2.2 How the sales action plan is carried out — 158
    2.3 An example of the contents of a sales action plan — 159

Chapter 6
Innovation and Product Management — *161*
1. The learning curve — *161*
2. The product life cycle — *163*
    2.1 Phases of the life cycle — 164
    2.2 Applying the life cycle concept — 165
    2.3 The characteristics of each life cycle phase — 167
3. Managing the product portfolio — *168*

3.1 The BCG model	169
An analysis of the Airbus and Boeing product portfolios	***172***
3.2 The McKinsey model	174
3.3 The Little model	175
3.4 Marketing and management of the product portfolios	176
The Bombardier niche strategy	***177***
4. Managing the product range	**178**
4.1 The characteristics of the range	178
The CFM International engine range	***180***
5. Managing innovation	**182**
Latécoère's innovation: on-board video systems	***182***
The different generations of a product: military aircraft	***184***
5.1 In-house innovations: the "push" strategy	186
Weber Aircraft: innovation at the heart of the offer	***187***
5.2 Innovations from outside: the "pull" strategy	189
The evolving range of Rolls-Royce Trent engines	***191***
5.3 Product development phases	191
5.4 The conditions for successful development	194
Innovation at Spot Image: Spot Thema	***196***
6. Innovation, the key to development of the A380 wide-body jet	**198**
6.1 Analysis of the market and the range	198
6.2 Taking into account customer's expectations	199
6.3 Producing a solution	200
6.4 Innovation in the cockpit	201
6.5 Production innovation to reduce costs	202
6.6 The involvement of industrial partners	203
6.7 Cabin fittings	205
6.8 Future versions of the A380	207
Chapter 7	
Marketing of Services	***211***
1. The characteristics of services	**211**
1.1 Intangibility	211
On-line information: a service to increase maintenance efficiency	***212***
1.2 "Perishability" and stock-impossibility	213
1.3 Inseparability	214
1.4 Variability	214
Spot Image: a service with multiple applications	***216***
2. Different categories of services	**218**
3. Professional services	**220**
3.1 Services which are required by law and regulations	221
3.2 More general services linked to management and strategy	222

	3.3 The aeronautical "marketplaces": a new type of service from MyAircraft to AirNewco	222
	3.4 Services linked to the production process	227
	3.5 Sales related services	227
	3.6 Technical and commercial, global services	227

Servair: a global service fulfilling the expectations of airlines and their passengers — 228

4. Consumer services: transport and tourism — 230

Qualiflyer, airlines serving customers — 231

Taking care of unaccompanied minors — 232

An example of a health service: MEDjet International — 233

The exemplary service quality of Thai Airways — 235

5. Focus on the freight market — 238

    5.1 The evolution of the freight market: expansion of the "integrated" carriers — 240

    5.2 The reaction of the cargo airline companies — 243

Chapter 8
Pricing Policy — 245

1. Factors involved in pricing — 245

    1.1 External constraints — 246

    1.2 Internal constraints — 250

The business plan or forecasting financial profitability — 251

2. Pricing approaches — 254

    2.1 Cost-based pricing — 254

Example of a modulated pricing policy for domestic Air France flights — 256

    2.2 Value-based pricing — 256

Helicopter engine manufacturers: taking into consideration the market and after sales in the pricing policy — 259

    2.3 Bidding — 262

3. Pricing strategies — 263

    3.1 The skimming strategy — 263

    3.2 The penetration strategy — 264

Charter on-line business aircraft — 265

    3.3 Flexibility strategies — 266

    3.4 Yield management — 266

Bold easyJet: Internet, yield management and non-conformism — 268

    3.5 The development of the "gray market" — 270

4. Price-adjustment policy — 270

    4.1 Adjusting the conditions of sale — 270

    4.2 The leasing — 272

    4.3 The development of fractional ownership on the business aircraft market — 274

The NetJets (from Executive Jet) fractional ownership program — 274

Chapter 9
Selecting Distribution Channels and Sales Team Management — 277
1. Logistics — 277
2. Choosing a distribution system — 280
    2.1 Choosing an external solution — 281
    2.2 Choosing multi-brand or exclusive distributors — 281
Breitling's selective distribution — 282
    2.3 Selecting partners and managing the network — 282
The development of e-ticketing — 284
3. Managing the sales point: adjusting supply to demand — 286
    3.1 The basis of the merchandising approach — 286
Travel agency merchandising — 287
    3.2 Merchandising objectives — 289
A special application of merchandising: managing spare parts — 290
4. Direct channel: the role of the sales representative — 291
    4.1 Communication — 291
    4.2 Pre-sales: prospecting — 292
    4.3 Sales presentation and negotiation — 293
    4.4 After-sales: the follow-up — 294
The recent evolution of sales representatives — 294
    4.5 Information feedback — 295
The sale of aircraft — 295
5. Managing the sales team — 296
    5.1 Defining objectives — 297
    5.2 Choosing the structure — 298
    5.3 The size of the sales team — 299
    5.4 Recruiting sales representatives — 299
    5.5 Supervising the team — 300
    5.6 Remunerating sales representatives — 301
    5.7 Sales representatives: motivation, training and career management — 304

Chapter 10
Project Marketing — 307
1. The specific nature of project marketing — 307
    1.1 High financial stakes — 307
The Beijing Capital International Airport — 308
    1.2 A "one-off" project — 309
    1.3 Generally predefined buying procedures — 310
    1.4 A generally discontinuous supplier-customer relationship — 312
2. Building demand — 313
    2.1 Identifying the customer's latent demand — 313

2.2 Helping to formulate dissatisfaction	313
2.3 Developing a solution	314
An example of constructing demand: building industry and airports	***314***
2.4 Drafting a solution	316
3. Customer intimacy	**317**
3.1 The depth of the interaction	317
3.2 The extent of the interaction	318
4. Influencing specifications	**320**
4.1 Intervening upstream of the deal	320
Industrial partnerships to win bids: the tender concerning missiles for the British Eurofighter	***321***
4.2 Intervening in the deal	322
Services at the very heart of the creative bid strategy: application to the military market	***324***

Chapter 11
Communication Policy — 327

1. Different types of communication	**327**
1.1 The objectives of communication	328
1.2 The four main types of communication	330
2. The communication plan	**337**
2.1 Determination of targets and budgets	337
2.2 Setting up the communication plan	341

Chapter 12
Selecting Media — 347

1. Trade shows	**347**
1.1 The specific nature of trade shows	347
1.2 Exhibiting at a show	352
The Paris Air Show – Le Bourget: the leading international aeronautics and space show	***354***
1.3 The different stages of participating in a show	356
2. The trade press	**358**
2.1 Main characteristics	358
Aviation Week & Space Technology	***358***
2.2 Resources and tools	364
3. The Internet	**366**
Boeing and the Internet	***369***
4. Direct marketing	**371**
4.1 The objectives of direct marketing	371
4.2 The different tools	372
5. Television, billboards and radio	**373**

	5.1 Television	373
An original operation: Airbus Beluga "Delacroix"		*374*
	5.2 Radio	375
	5.3 Billboards	376
6.	**Lobbying**	**377**
Lobbying the regulatory organizations		*378*
Lobbying for the Airbus Military Company A400M project		*379*
7.	**Public relations and sponsoring**	**388**
	7.1 Public relations	388
	7.2 Sponsoring	390
8.	**Sales promotion**	**391**
Frequent Flyer Programs (FFP)		*392*
The example of Spot Image's promotion for the launch of SpotView		*396*

Chapter 13
Brand Management — 399
1. Brand foundation — **399**
    1.1 Brand mechanisms — 400
    1.2 Brand functions for the company — 404
    1.3 Brand functions for the customer — 406
Aeronautics and space brands and performance facilitation — *413*
2. Special characteristics of the industrial brand — **413**
    2.1 "Purchaseability" levels of the industrial brand — 414
    2.2 The visibility strategy — 415
    2.3 Airbus: "Setting the Standards" — 418
3. Industrial brands classification — **424**
    3.1 According to the use of goods — 424
    3.2 According to international brand policy — 424
    3.3 According to brand origins — 425
4. Visual identity code, logos and slogans — **436**
    4.1 Logos — 436
    4.2 Slogans — 439
    4.3 Jingles — 441
    4.4 Visual identity code — 441
5. Latécoère: technical partnership and its own products — **442**
    5.1 The rise of Latécoère — 442
    5.2 Latécoère, technical performance facilitator brand — 444
    5.3 Sales performance facilitation — 447
6. Zodiac: managing a brand portfolio by sector — **447**
    6.1 History — 447
    6.2 Zodiac today — 449
    6.3 Brand policy — 450

Chapter 14
Building loyalty: Maintenance, Customer Training and Offsets — 455
1. Maintenance — 455
A key to aircraft safety — 461
    1.1 The different forms of maintenance — 462
    1.2 Maintenance: a tool for the marketing-mix — 466
    1.3 Maintenance: a tool for the marketing information system — 467
Airbus' after-sales marketing function: 4 main objectives — 469
2. Customer training — 472
Pilot training: a major and essential expense — 472
    2.1 Different training objectives — 473
    2.2 The contents of training — 474
Training, an essential part in the Airbus strategy — 475
    2.3 The main types of training — 481
GDTA: cutting-edge training — 482
3. Offset, a business tool — 484
    3.1 Offset: a means of payment — 484
    3.2 A business argument — 487

Chapter 15
Alliance Strategies — 491
1. Traditional forms of company development — 491
2. Specific objectives of alliances — 492
    2.1 Financial objectives — 493
    2.2 Marketing and sales objectives — 494
    2.3 The international political stakes — 498
3. Different forms of alliances — 498
    3.1 Tactical alliances — 501
    3.2 Strategic alliances while maintaining the company's initial identity — 502
Star Alliance: a worldwide air network — 505
SkyTeam: a new worldwide air network — 507
    3.3 Strategic alliances with creation of a specific structure — 512
The CFM International alliance — 512
Starsem: an alliance for a reliable and competitive space transport system — 516
Sea Launch: from the sea into space — 521
    3.4 From alliance to merger — 522
From GIE Airbus Industrie to the integrated company EADS — 523
Eurocopter: once a joint-venture, today a subsidiary — 525

Bibliography — 529

Subject index — 533

PREFACE

This book presents an overall picture of both business to business and business to consumer marketing in the aeronautics and space sector. Designed to be a practical handbook, it deals with all of the basic marketing concepts and tools, both from an operational and strategic point of view.

Aeronautics and Space Management is the first marketing manual entirely illustrated with examples taken from the aerospace sector: parts suppliers, aircraft builders, airlines, helicopter manufacturers, aeronautics service providers, airports, defense and military companies, satellites integrators, and space launchers, among others.

Organized so as to allow the reader to select those chapters that are most useful to him or her, it does not require prior in-depth marketing knowledge. While it is not exhaustive, particularly concerning the buying process or segmentation, it does address the most recent marketing trends (e-marketing, one-to-one marketing, customer orientation, project marketing…). Accessible to a wide public, this book was written in particular for executives working in the field as well as for students from Engineering and Business Schools.

The main marketing tools are discussed from purchase marketing, market surveillance and surveys, quality management and norms on the one hand, to trade shows, the trade press, direct marketing, public relations, lobbying, maintenance, customer training and offsets on the other. There is also a chapter focusing on branding strategy and brand management, in particular their role in facilitating customer performance, as well as a chapter on the project marketing approach. Last but not least, Chapter 15 presents the alliance strategies of the leading aeronautics manufacturers, part suppliers and airlines.

For their encouragement, trust and the information they have graciously provided, we would like to thank the companies mentioned in the book and in particular:
- Aeromexico: José Diaz de la Rivera
- Aéroports de Paris: Jacques Reder, Isabelle de Villeneuve, Catherine Bénétreau, Corine Bokobza-Servadio
- Aéroport de Toulouse-Blagnac: Michel Gay
- Air & Cosmos: Jean-Pierre Casamayou
- Air & Space Europe: Jean-Pierre Sanfourche
- Air Cargo News: Geoffrey Arend
- Air France: Laurent Clabé-Navarre, Benoît Guizard, Georges Rochas, Pierre-Yves Reville, François Roppe
- Airbus / EADS: Michel Guérard, Roland Sanguinetti, Claire Labedaix, Gilles Meric, Bruno Piquet, David Jennings, Richard Carcaillet, Jean-François Lasmezas, Bernard Bousquet, Philippe de Saint-Aulaire, Philippe Girard, Didier Lenormand, Françoise Soulier, Jérôme Verzat, Jean-Bernard Oresve, Anne Carrere, Daniel Baubil, Jean-Marc Thomas, Patrick Roger, Alain Vilanove
- Alcatel Space Industries: Thierry Deloye, Dominique Murat, Christiane Grimal, Rémy Sergent
- Alcatel: Caroline Mille
- AlliedSignal Aerospace: Bettina Pei-Ching Li
- American Airlines: Michael Gunn, Talcott J. Franklin, Phyllis J. Sawyers
- ATR: Carlo Logli, Bernard Fondo, Jacques Dronnet
- Aviapartner: Hervé Bonnan
- Aviation International News: Charles Alcock
- Aviation Sales Company: Dale Baker
- Aviation Week & Space Technology: Mark A. Lipowicz, Pierre Sparaco
- BAe Systems: Richard H. Evans, Nina Collins
- Bell (Textron): Dave Wyatt, Jenny Jackson, Robert N. Kohn
- BFGoodrich Aerospace: David L. Burner, Marshall O. Larsen
- Boeing: Philip M. Condit, Dan Kays
- Bombardier / Canadair: Yvan Allaire, Catherine Chase, Leo Knaapen
- Breitling: Nathalie Frésard
- Cathay Pacific: Nicholas Reynolds
- CFM International / Snecma: Pierre Bry, Sylvie Béamonte, Olivier Laroche, Nadia Zaïdi, Jean-Paul Rouot, M. Hucher, Jacques Geneste

  Civil Aviation Authoriy, CAA: Malcom Field, Douglas Andrews
- Dassault Aviation: Nicolas Renard

- Direction Générale de l'Aviation Civile, DGAC: Jacques Girerd, Brun-Potard, Philippe Beghelli
- easyJet: Toby Nicol, Samatha Day
- École Nationale Supérieure de l'Aéronautique et de l'Espace (Supaero): Pierre Jeanblanc, Chantal Gauthier, Joël Daste
- Eros: Denis Taieb
- Euresas: Sabine Kramer, William Gibson
- Eurocopter: Alain Journeau, Jean-Louis Boireau
- Euromissile: Jean Thyrard
- Explorer Consulting Company: Laurent Husson
- Farnborough: Maureen Symester, Linda Lloyd, Lindsey Hart
- Federal Aviation Administration, FAA: Norm Simenson
- Fedex: David J. Bronczek
- Foster and Partners: Norman Foster, Katy Harris
- GDTA: Jean-Luc Bessis
- GEIA: Dan Heinemeier, Jim Serafin, Mary Petitt, Michael Dooner
- General Electric: John F. Welch, David L. Calhoun
- Gore-Tex: Édouard Frignet des Préaux
- Institut Aéronautique et Spatial: Hervé Schwindenhammer, Michel Chauvin, Alain Riesen
- Jane's Airport Review: Alan Condron
- Latécoère: François Junca, Jean-Pierre Robert, Philippe Martin, Raphaël Bolzan
- Lockheed Martin: James Fetig, Lee Whitney
- Lucas Aerospace (TRW): Joseph T. Gorman, William K. Maciver
- Lufthansa Technik: Aage Duenhaupt, Angelika Albert-Graffert, Claudia Ungeheuer
- MartinBaker America: Bill Harrison, B.A. Miller
- Medjet: Jeffrey Tolbert, Sam Jackson, Stan Bradley, Debbie L. Ebner
- Messier-Dowty: Philippe Merle
- Motorola: Christian Ollivry
- Pechiney Aerospace: Caroline Lenglin, Jean-Claude Nicolas
- Rockwell-Collins: Jack Barbieux
- Rolls-Royce: John Purcell, John Crocker, Richard Deeks
- Royal Air Maroc: Asmâa Oudghiri, Nezha Alaoui, Karina Marmech
- Sea Launch: Terrance Scott, Glenn Anderson
- Sennheiser: Karlheinz Koinzack
- Servair: Jean-Luc Neirynck, Nathalie Hedin, Sandrine Audivert,
- Smiths Industrie: Jennifer Villarreal
- Spot Image: Alain Hirschfeld
- Starsem: Claire Coulbeaux, Romain Lavault
- Swissair: Michael Eggenschwiler

- Thai Airways: Pascale Baret
- Thalès Detexis (ex-Thomson-CSF): Gérard Bouy, Jean-Paul Perrier, Jean-Luc Pomier, Carole Caubet
- Turbomeca: Guillaume Giscard d'Estaing, Jean-Marc Soulier
- Virgin Atlantic: Charlotte Tidball
- Weber Aircraft: William Chase
- Zodiac: Dominique Puig.

We would like to thank the "Delegates" of the Aerospace MBA of the Toulouse Business School (ESC Toulouse) and our students who have indirectly contributed to this book thanks to their remarks and suggestions, and in particular:

- Nathalie Brun, Jun Cao, Robin Calot, Nicolas Coma, Olivier Courrèges, Thierry Debergé, Vincent Ducamin, Esmeralda Herrera, Julien Joncquiert, J. C. Jong, Thomas Lelièvre, Benoît Machefert, Olivier Malavallon, Gaël Maugendre, Hakim Messoussa, Laurence Milhade, Guillaume Pellegrino, Thomas Pelon, Göknur Pilli, Olivier Royer-Manoha, Xavier-Alexandre Scalbert, Vincent Varnier.

We would also like to thank of course Toulouse Business School (ESC Toulouse) for its support and in particular:

- Hervé Passeron, Andres Atenza, Jean-Luc Guiraud, Jonathan Winterton, Denis Lacoste,
- Florence Lévy and the whole Documentation Center,
- Jean-Luc Bouchot and his team.

Thanks to Geoffrey Staines of Pearson/Publi-Union as well as Monika Neumann for authorizing this American edition. Thanks to Marika Seletti and Adrian Pavely for the translating of the English version.

We are deeply grateful to our parents and our respective families. Thanks in particular to Christine for her kind understanding and support. Thanks to Isabelle and to my children Guilhem, Bertrand, Robin and Marion.

Chapter 1

MARKETING IN THE AERONAUTICS AND SPACE INDUSTRY

Over the last few years the aeronautics and space industry has undergone an evolution which has fundamentally modified managerial practice within the sector. From being a product based industry, there has been a progressive move towards a customer focus one, representing a turning point in marketing for an industrial sector with many particularities.

1. THE AERONAUTICS AND SPACE SECTOR ENVIRONMENT

1.1 Characteristics of the aeronautics and space sector

There are certain specificities in the aeronautics and space sector regardless of country or period of observation:

- *A high level of technology* with a multitude of special skills for the design of highly complex aircraft and satellites. Research is the driving force behind progress in the aeronautics and space market.
- *A never-ending need for capital*, necessary for developing and applying new technologies in response to customer requirements. Research and development centers around the software, miniaturization, calculations and alloys sectors.
- *A high strategic importance for the country*: outside civil applications, the aeronautics and space sector is militarily highly sensitive.
- *Governmental authorities are heavily involved in new programs*. In fact this is a stumbling block between the United States and the European Union who are accused of lending support to their aeronautics and space sector. The truth is that audits made

by English speaking companies since the 1960's, clearly show that the aid given to American companies is far higher than the total accorded by Europe and the European States. An important difference lies in the mechanisms chosen: the European programs work using reimbursable, short-term loans (17 years maximum) which cannot exceed 33% of the program's total cost (1992 *Bilateral Agreement* between the United States and the European Union). American aid is much less clear cut because it is mainly through the Department of Defense (DoD) which finances technological improvement programs, sufficiently wide to cover not only the military but also the civil version of a new aircraft (Lawrence and Braddon, 1998). The EELV program (*Evolutive Expandable Launch Vehicle*) which was started in the mid 1990's, applied to a family of new generation launchers, Delta 4 (Boeing) and Atlas 5 (Lockheed Martin). It relied on an annual finance packet of a billion dollars from NASA and the American Air Force, plus a captive market of about twenty military launches per year.

- *A high concentration of constructors and integrators*, as much in the States as in Europe. The 1990's were marked by an increase in the grouping together of companies whether by alliances or mergers. The alliances are sometimes one-off agreements for a particular project such as a new engine or aircraft, however they can develop into the setting up of a long-term organization[1]. For the moment there are about a dozen main companies in the aeronautics sector with the Boeing/Airbus dipole for large capacity civil aircraft.

- *A high concentration of customers* within the airline companies and even more so in freight transport, with a total of less than 300 potential customers on the civil market.

- *The progressive disengagement of the State* from within the capitalistic structure of firms in the sector, culminating in the total or partial privatization of national companies.

1.2   Market deregulation

Airlines

At the end of the Second World War, aviation became the symbol of modern life. A national airline company was a must for a country's image. Indeed some time later, with the fall of the Soviet Empire, we saw the political importance attached to the existence of a national airline company, announced in this case by the Baltic States just after their new flag and national anthem.

For a long time the aeronautics sector has been characterized by a highly regulated structure. In this way, from the 1950's to the 1980's, airline companies were protected from competitors by their trustees, regardless of the economic regime in force. Indeed in 1944, the *Chicago Convention* laid down a system of governmental regulation. However, although each national company obviously benefited from a monopoly on its own territory, it encountered serious difficulties when trying to set up abroad.

Following the deregulation policy started by the United States at the beginning of the 1980's (*Deregulation Act*), most countries opted for an end to protectionism in air transport. Today, companies work in an " open sky, open market " context. Here they have to compete with others at a local level but can on the other hand function in the world market. The end of monopolies in air transport, has thus allowed companies within the aeronautics sector to gradually take part in an open market where marketing has come into its own.

Aeronautics and space manufacturers

By making aeronautic and space activities strategically less sensitive, the fall of the Soviet Empire has allowed national industries to develop alliances. In addition, the 1990's has seen this liberalization gradually reach the airline' suppliers, i.e. the manufacturers. The restructuring of aeronautics and space companies in the United States as well as in the European Union, has subsequently opened up this liberalization to companies within the space sector.

1.3 The role of regulatory organizations: FAA, DGAC, CAA, IATA

National organization

One of the specificities of the aeronautics sector arises from the risks taken by users, whether they be passengers or salaried staff. This is why the main countries concerned have sought to prevent risks by instigating their own certification system for the different equipment involved. These bodies play a predominant role in the launch phase of a new aircraft which must be approved before using the country's airspace. The three main bodies are:
- the FAA, Federal Aviation Administration in the United States,
- the DGAC, Direction Générale de l'Aviation Civile in France,
- the CAA, Civil Aviation Authority in the United Kingdom.

These organizations intervene as soon as a modification is envisaged on an aircraft. Manufacturers wishing to make such a change which will apply to several aircraft, are bound to engage in a costly procedure, the Service Bulletin, with these administrative bodies. When a company wishes to make specific changes to one or two aircraft, bought second hand for example, it is the equipment manufacturers and companies which install cables that must engage in a procedure known as STC, Supplementary Trade Certificate ; these modifications must be presented country by country. Equivalents allowing a large number of transfers exist, notably between the American FAA and the European JAA. It follows that an aircraft not certified by either the European or the American organizations can be refused the right to fly over or land on the territory concerned.

Figure 1-1. Organization regulating the aeronautics market

As in other domains, decisions made by these organizations act as guidelines for equivalent administrations in other countries. In terms of trade communication, the aeronautics and space sector companies do not hesitate to feature certifications from these administrations (Figure 1.2). This approval, thus corresponds to a seal of quality and so becomes a sales argument.

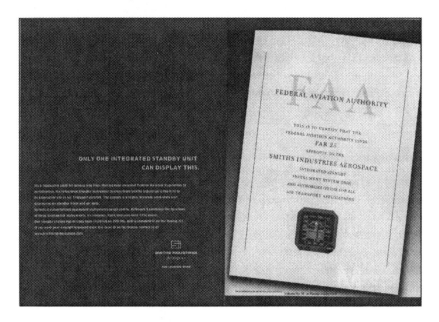

Figure 1-2. An example of a trade communication campaign highlighting approval which has been obtained: Smiths Industries

International organizations

- The IATA (International Aviation Transport Authority) is the central organization which establishes the international air transport rules. Because of this, IATA listens to the market:

 - it must answer a wide variety of questions coming from the airports, in particular because of traffic saturation in certain capitals ;

 - it is open to suggestions from parts manufacturers and constructors concerning the introduction of new norms which will allow an improvement in the security and or efficiency of air traffic.

A good example is the integration of air control at the European level. Faced with the saturation of certain German, British, Belgian or French airports the idea was gradually adopted whereby the information concerning the various European flights would be centralized in Brussels.

- The JAA (European Joint Aviation Authority) is the pan-European organization joining together the countries of the European Union as well as about fifteen other European states. Its aim is to harmonize European airspace in terms of traffic and also aircraft design norms, in-flight safety, maintenance and crew training. Nevertheless the JAA does not have any real regulatory authority which is why a European security agency (European Aviation Safety Authority) is presently being established.

THE ORGANIZATIONS WHICH CONTROL AIR TRANSPORT

The development of air transport has brought people together and favored the growth of economic activity in different parts of the world. Players in the aeronautics field, innovate within what has become a highly competitive market following deregulation. In order to provide long-term safety and assurance for this mode of transport, highly restrictive international regulations are necessary. This implies vigilance right from the design phase through to crew and air traffic control training.

- *In the United States: the Federal Aviation Administration (FAA)* [2]

The Federal Aviation Administration is entrusted by the American government to oversee civil aviation safety. The FAA plays a major role in the United States and in the world in the domain of air regulations.

Figure 1-3. Meticulous verification of aircraft

Its main functions are:

- Safety: the mission of the FAA is to establish and enforce the regulations and minimum standards for aircraft manufacture, use and maintenance. It is in charge of pilot evaluation and certification as well as that of commercial airports. The certification of a plane starts with the production site being visited by FAA experts. If everything is positive, the new model will obtain a type certificate followed by a production authorization one. The engines and other components of the aircraft are followed up in a like manner. In the event of problems during the life of the plane, the FAA will set out how to correct them. Any person involved in the use or maintenance of the aircraft must be in possession of an FAA certificate. The FAA controls a civil aviation safety program and enforces, via the Hazardous Materials Transportation Act, air cargo transport regulations.

Figure 1-4. Increased vigilance to guarantee passenger safety

Safety is also about security and equipment checks, and about being vigilant concerning the detection of arms or explosives when boarding as well as in baggage holds. The FAA is endeavoring to reduce the risks and vulnerability in airports and their associated infrastructures.

- Air traffic management: the FAA's main objective is to ensure the safe and efficient use of navigable airspace. This means managing the traffic in the best possible fashion using a network of airport control towers (450), air traffic control centers (21) and air service structures. Complex equipment and highly qualified staff are necessary in order to offer a flexible system, which reduces waiting and delays. The FAA establishes the air traffic rules, assigns airspace and ensures the control of air traffic security in conformity with the interests of national defense.

- The setting up of air navigation equipment: the FAA is in charge of the manufacture and installation of navigational visual aids as well as their maintenance, use and quality control. The other systems used cover voice data equipment, radar installations, computer systems and visual equipment within air service structures.

- Civil aviation abroad: In conformity with its legal structure, the FAA is also committed to air security abroad. This mainly entails exchanging aeronautical information with other foreign regulatory bodies. The FAA must also ensure certification of foreign repair centers, pilots, mechanics, help with training, negotiation of bilateral navigation agreements and a technical presence at international conferences.

1. Marketing in the Aeronautics and Space Industry

- Commercial space flights: the FAA regulates and helps the American space transport industry. It authorizes commercial space launch structures, e.g. the Sea Launch ocean platform from where the Zenit 3SL rockets are launched[3].

- Research, engineering and development: the FAA participates alongside universities and companies, in the research and development of systems and procedures which improve the efficiency and security of air navigation and traffic control. It takes part in the aeromedical research domain and contributes to improving aircraft, engines and equipment used. The FAA carries out tests and evaluations on aviation systems, security devices, materials and procedures.

- Information management

The FAA is authorized to maintain a system recording all aircraft, their documents and components. It manages an aeronautical insurance program and develops the specifications for aeronautical guidelines. It publishes information concerning airways and airport services as well as on technical subjects related to aeronautics.

On its website, the FAA facilitates the exchange of information between professionals and the administration as well as between the latter and the passengers. This could be meteorological data, security data (aircraft, component, airport, etc.), explanations concerning the regulations (Federal Aviation Regulations, FARs), certification of pilots, advice and information for passengers (risks, directions, cabin security, baggage control, etc.).

- *In France: the DGAC*[4]

For France, the Direction Générale de l'Aviation Civile (DGAC) carries out an essential role in terms of regulations, provision of services, risk prevention and industrial partnership. One of the roles of the DGAC, is to watch over the functioning of the market to ensure that the increased competition benefits above all the passenger consumer in terms of price, service quality and security. This surveillance is carried out in close collaboration with other European and international partners. The DGAC manages that part of the government's budget concerned with civil aviation programs. This money is destined to support wide-ranging aircraft, helicopter, engine and equipment programs as well as technological research activities. The DGAC has five essential missions.

- Responsibility for civil aviation security : The DGAC monitors the application and adaptation of the civil aviation laws. It establishes regulations and transposes European or international rules. In order to maintain the highest level of security, the DGAC ensures permanent vigilance in terms of certification, technical verification, approval of transport and maintenance companies, the inspection and control of all the different airlines and their staff, and feedback concerning experience after incidents or accidents.

- Providing operational services : The DGAC provides airline companies with the services necessary for security and regularity in air navigation. It manages the necessary material, human, technical and financial resources, and adapts its staff and organization to meet changes in air traffic. As a specialist in airport engineering, it is the project manager for the planning of infrastructures, territorial development factors and local development. An air transport equating fund has been set up to allow airline companies to achieve a financial equilibrium when serving destinations kept open in the interests of the community.

- Training quality : The DGAC monitors the quality of teaching within all the training structures. It has its own advanced level training institute with 8 centers. In addition, the École Nationale de l'Aviation Civile carries out the initial and in-service training of air traffic controllers.

- Partnerships with both small and large aeronautical companies. The DGAC works with all the companies within the aeronautics sector: manufacturers, engine builders, airlines, airport owners, service suppliers attached to the Ministry/Department of Defense, administrations and local organizations.

- Prevention : The DGAC monitors human safety but also the preservation of the environment by fighting against nuisances generated by air transport, noise, air and water pollution.

Figure 1-5. Air traffic management: complex equipment and qualified staff[5]

- ***In the United Kingdom: the Civil Aviation Authority (CAA)*** [6]

The Civil Aviation Authority possesses a body of aeronautical know-how and expertise which is one of the best in the world. Its responsibilities stretch from security of the airways to economic regulations via defense of the consumer. In addition, the CAA advises the British government on changes in aeronautics, represents the interests of the end customer, carries out scientific and economic research, produces statistical data and supplies specialist services.

- Economic regulations : The CAA acts via the Economic Regulation Group to control airlines and airports. In its role as expert, it advises the government on areas concerning the companies and the airports. It gathers together, analyses and publishes statistical data on the people within aeronautics. Finally, it collects the taxes paid by the airlines at certain airports (Heathrow, Gatwick Stansted and Manchester).

- Security : Through its specialized body the Safety Regulation Group, the CAA establishes security standards. In order for these to be reached, a process must be followed such that pilots are competent, correctly trained and physically capable, and the aircraft are designed, tested, certified and maintained according to these norms. The airline companies as well as the airports must be reliable and competent. Only the highest standards of safety are accepted by the CAA and these necessitate research into aircraft design and also production and technical maintenance. Although the SRG is in theory only concerned with British territory, it exchanges a

1. Marketing in the Aeronautics and Space Industry 9

maximum amount of information with other organizations and adopts world standards. The CAA collaborates internationally with the ICAO (International Civil Aviation Organization), the ECAC (European Civil Aviation Conference) and notably with the JAA (European Joint Aviation Authority).

In this context, the CAA has played an essential role in the constitution of the JAA which endeavors to establish common rules (JAR) and enforce them across the 30 European member states (of which 15 belong to the European Union). An important number of European rules have been ratified concerning aircraft, helicopter and engine design codes, aircraft certification procedures, all weather operations, production and maintenance organizations as well as those dealing with aeronautic maintenance engineer qualification, etc. The JAA works with the American FAA in an effort to harmonize security demands at a truly international level.

The CAA is part of a European GIE (Economic Interest Group) called Air EuroSafe, which brings together Air EuroSafe, Germany, Bureau Veritas, France and Sensa in Spain. These four partners, in charge of regulations within their own countries, combine their expertise in an effort to improve air safety. They intervene in order to help the airline system, and to develop air safety systems and in-staff training.

- Consumer protection : The CAA protects consumers who have bought package holidays, charter flights and reduced price tickets against the consequences and shortcomings of the tour operators. They grant licenses to airline companies which fulfill the national and European requirements in terms of financial resources, reliability and insurance.

---

2. *THE MARKETING APPROACH*

2.1 *The rise of marketing*

As a general rule, marketing develops when offer outstrips demand, in other words when the products or services that companies produce exceed what customers can buy (Table 1.6). Economists generally distinguish four types of economic environment: the production economy, the distribution economy, the market economy and the environment economy.

The production economy: This is the situation where companies are not producing enough to satisfy customer needs. This phase is characterized by not enough supply to meet demand and under these conditions, there is no real need for marketing. The company basically needs to produce while at the same time being economically efficient and controlling the cost price. The sales function becomes a question of rationalizing distribution: how to disseminate the product over the chosen territory ? As the end customers are waiting for the goods produced, the companies are not over inclined to dress up their offer by adding sophistication or lowering the selling price. The production economy is characterized by low competitive pressure since the companies are sure to be able to place what they produce. This sort of economy was a characteristic, regardless of country, of centuries past. The same sort of conditions are to be found when there is a shortage of goods following disturbances, conflicts or natural catastrophes.

The distribution economy: This corresponds to an ideal economic equilibrium where supply and demand are balanced. Certain economists situate this phase in the 1970's, however today it tends to be regarded as an idyllic vision of the economy.

The market economy: This corresponds to an economy where supply sometimes greatly exceeds the demand. Here, it is impossible for customers to buy all the goods on offer and so obviously marketing becomes all important. It is no longer good enough to merely produce, even efficiently, and then distribute the goods with an effective sales force. It now becomes essential to analyze the demand in order to know the customer's motives for satisfaction or dissatisfaction. The products or services offered can then be improved in anticipation of expectations, thereby outshining competitors. The market economy is thus characterized by a very intense level of competition.

The environment economy: This is closely related to the market economy and follows on from it by taking into account those non-economic organizations or people taking part indirectly in the transactions. These could be government organizations, particularly for certification, but also international bodies who establish rules and regulations for trade and technical norms (Federal Aviation Administration FAA, Direction Générale de l'Aviation Civile DGAC, Civil Aviation Authority CAA, etc.). It could also be the media, whose greatly increased importance affects the efficiency of companies: journalists who are constantly in search of new information have now become influencers and should be considered by companies as new targets.

Production economy	Supply	<	Demand
Distribution economy	Supply	≡	Demand
Market economy	Supply	>	Demand
Environment economy	Supply	>>	Demand

Figure 1-6. The appearance of marketing as a result of the economic turnaround

Remarks

1. Whatever the country, the marketing department should be in tune with its market, however management and shareholders do not always share this vision. This is not obvious when the company is run by engineers whose main objective is to extol the virtues of the research department; in fact when the company is far too product oriented rather than customer oriented. Nor is it generally the case within state organizations.

1. Marketing in the Aeronautics and Space Industry

2. There are also those companies who still fantasize about returning to the golden age of a production economy in which call all the shots. Normally, research and development helps the company to innovate, producing new products and services. If the latter are successful, demand will increase, putting the company temporarily into a production economy situation. Similar cases can still be found today among suppliers for the space industry.

2.2 The two facets of marketing

To define marketing we can look at it from a professional's point of view or from that of a person with no particular training:

- when asked about marketing, the non professional will cite adverts and other forms of promotion. Encouraged to explain what these techniques consist of, he will talk about the influence exerted by companies on people, inciting them to purchase one or another product or service. Marketing is thus perceived essentially as a tool which influences ;

- asked the same question, the professional will on the contrary explain that marketing consists of getting " into the customer's head " in order to know what he wants today and in the future. Once this has been answered, two more questions must be asked: Can the company produce the goods wanted ? Is the project profitable ? When the answer is 'yes' to both of these, the company can put together an offer which corresponds to the predefined demand. Marketing is thus perceived here as an essentially analytical tool.

In fact the real definition of marketing incorporates both of these viewpoints: it is a tool which influences as well as being an analytical one (Figure 1.7), although obviously it is imperative for the study phase to precede the influencing phase.

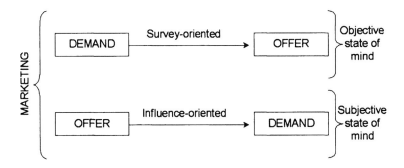

Figure 1-7. The two facets of marketing

This double-barreled approach dispenses with any ambiguity as to honesty or objectivity in marketing. It is essential to be objective in the analytical phase in order to be effective when defining the offer. This is not a moral position, just a technical requirement. On the other hand, influence marketing may contain a certain subjectivity. This involves winning customers over to convince them to buy the company's product or service.

Just like other companies, those in the aeronautics and space sector are concerned by both these aspects of marketing:

- " Survey-oriented Marketing " focuses on mastering market studies whether they be quantitative or qualitative[7]. It is also essential to know how to segment the market, i.e. divide it into homogenous sub groups[8]. This will allow an offer to be produced which is more attractive for each of the segments or targets. These subjects are dealt with in the chapters of the first part of the book: " Understanding the market ".

- "Influence-oriented Marketing" can be found in pricing policy (see chapter 8), distribution policy (see chapter 9) and communication policy (see chapters 11 & 12). The tools and techniques necessary are presented in the second part: " Implementing marketing ".

2.3 An essential part of companies and organizations

At the outset in the 70's, sales was one of the important areas alongside production, administration, finance and personnel management. Gradually this became enriched by market studies upstream and communication downstream to the point where it was called marketing-sales. Bit by bit, as the role of the customer was seen as increasingly important, this area grew to the point of sometimes becoming, as in a caricature, the central pillar of the company. The marketing-sales department is not fundamentally more important than others. However it has become an essential part of the organization by playing the " customer's ambassador " role relative to the other departments, in particular production, administration and finance (Figure 1.8).

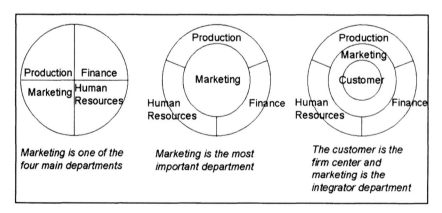

Figure 1-8. The evolution in the conceptions of marketing relative to its role within the company[9]

In just the same way, inspectors in airline companies sometimes play the "mystery passenger"; by putting themselves in the customer's shoes they are better able to detect anomalies in cabin service, whether this be a steward's attitude or something not right with a meal tray. The aim is to make each person aware of the importance of every single detail, including the presentation of the finished product. As far as the administrative and finance departments are concerned, the marketing department has got to show them the importance of, for example, presenting an invoice or an estimate which is customer-friendly. Thus the document is written with the customer first of all in mind rather than applying the company's standard in-house procedures.

One important point is that the word "marketing" does not carry the same meaning from one continent to the other. In European countries, the marketing director normally takes care of studies and communication but not sales. There is thus a sales director. In the United States, on the other hand, the marketing manager or director covers communication and sales. The area concerned with studies normally comes under the development department which brings together both technical studies (research and development) and market studies (Table 1.9).

Neither of these two solutions is ideal. When marketing and sales are juxtaposed they risk overlapping, thus fueling internal rivalry. Conversely this pairing stimulates self criticism from within the steering committee resulting in more frequent calling into question of actions. The American solution with a single marketing-sales director does away with any internal rivalry but risks producing an over powerful department relative to the others.

Marketing director	Studies	Communication	Sales
Europe	▓▓▓▓▓	▓▓▓▓▓	-----
United States	-----	▓▓▓▓▓	▓▓▓▓▓

Table 1-9. The different boundaries of marketing directors

2.4 A set of methods with wide-ranging applications

Marketing is based on a set of methods and techniques for measuring and analyzing information, making choices, planning, communicating and influencing the market, customers, etc. These methods can be found in the marketing plan[10] and in the sales actions plan. The latter allows the marketing plan to be implemented operationally and broken up in time.

The word marketing comes from the word "market" which implies some form of trade. However the past few years have witnessed the extension of marketing to new sectors outside of the industry and sales such as public organizations addressing citizens' groups in order to explain the wisdom of decisions taken and investments made (creation or extension of an airport, roads, etc.) or non-profit humanitarian organizations seeking funds such as Médecins sans Frontières, the Red Crescent, the Red Cross or the World Wildlife Fund. Finally it could also be a one-off association created after a natural catastrophe, such as in Guatemala in 1999 or in Salvador in 2001.

3. THE ROLE OF MARKETING IN THE AERONAUTICS AND SPACE INDUSTRY

3.1 The growing importance of marketing in the aeronautics and space sector

The aeronautics and space sector has been characterized right from the outset by a high level of innovation and technology. When engineers manage to design and make products or services which are more effective than previous ones, the company gets ahead of its competitors. If the product is much better, the company can make less or no extra sales effort either in terms of distribution strategy, price and/or communication. The customers are in effect already waiting for the new product. Under these conditions, marketing would seem to be of secondary importance if not superfluous among the priorities. This is one of the main reasons why it took a long time for marketing to be integrated into the general policy of aerospace companies.

In this way aeronautical companies have for a long time been characterized by the pre-eminence of the product to the detriment of the other marketing variables (distribution, price, communication). This 'engineering' based reasoning often results in an obsession with product excellence which in turn engenders a non-stop race to innovate and improve. Taken to an extreme, the product-oriented process leads to excessive sophistication without really taking into account what price the customer is ready to pay. The move to a customer-oriented policy, means taking account of the marketing mix which corresponds to the group of tools which the company has at its disposal to achieve its objectives in the target market. This mix is made up of four elements which allow the company to construct a coherent personality for its product, with a view to distinguishing it from the competitors. The marketing mix is the keystone of positioning[11]. The four pillars of the mix are: the product policy, the pricing policy, the distribution policy, the communication policy.

It is up to the marketing department to make coherent choices between these four policies. The product should no longer be defined uniquely by its technical description and by its intrinsic characteristics but rather be an entity (price, communication, distribution). For example a differential in the choice of price or communication can be just as important as a differential based on product characteristics.

Figure 1-10. The components of the marketing-mix

Furthermore, in the aerospace sector, engineers have long reacted on an ad hoc basis to the demands of customer companies, often defined by putting out to tender. Scientifically and technically dynamic, the suppliers have yet remained commercially passive for a long time, to the point of sometimes being sub-contractors to their economic partners.

Unlike product orientation, customer orientation consists of developing products once the potential market has been systematically verified. In this way, marketing corresponds to a state of mind, giving priority to the demand rather than the offer. For companies within the aeronautics sector, this customer orientation has become even more necessary since they are now in a truly competitive situation owing to market deregulation.

3.2 Different types of marketing

In the commercial sector the two main types of marketing can be separated according to their targets:
- Consumer goods marketing, or business to consumer (B to C) when it is selling to the end consumer,
- Business to business marketing (B to B) when it is selling to a company or an organization.

The distinction between product marketing and service marketing

Marketing was initially developed in the 1950's around consumer goods for the general public. Little by little, companies in the service sector tried to apply the same rules and techniques. Since then, marketing has been adapted to the specific nature of services (intangible, perishable, indivisible, variable)[12]. In the 1970's marketing split into product marketing and service marketing.

The distinction between consumer goods marketing and business to business marketing

The need to put the focus on the interests of the customer has opened up wide perspectives for marketing in technological companies from sectors as different as automotive parts, electronics, aeronautics and space, etc. Here again, the specificities of the targets has forced experts to offer better adapted analytical tools and influence marketing techniques. The expression, " marketing business to business " applies where the supplier company sells products, services or a mix of both, to other companies or organizations. Several characteristics illustrate the distinction between business to business marketing and consumer goods marketing (Table 1.11).

	Consumer goods marketing	Business to business marketing
Target	- (Very) large number of small customers (consumers)	- (Very) small number of large professional customers
Product	- Mass production	- Limited series production or one off's
Demand	- Variable, heterogeneous	- Derived, specialized, very heterogeneous, demanding
Main media	- TV, posters, radio, national press, - direct marketing	- Trade shows, trade press, direct marketing
Customer	- Passive - Buying process usually individual	- Active - Group buying
Marketing organization	- Usually by category of product - Service marketing and sales, - separate	- Usually by type of market or main customer - Integrated service and sales marketing
Type of approach	- Global by segment - Development of one to one	- Specific by key accounts, one to one
Use of Internet	- Frequent for communication - Variable for sales (e-commerce)	- Very frequent, Intranet and Extranet, e-communication, - e-commerce and e-partners

Table 1-11. The main differences between consumer goods and B-to-B marketing

1. Marketing in the Aeronautics and Space Industry

The special nature of project marketing

Project marketing, one branch of business to business marketing, involves very technical products[13]. It can be used by companies offering something unique, non repetitive and complex. With very sophisticated know-how and high financial stakes, project marketing needs to be specially adapted. For example, the buying procedures are generally pre-defined, by putting out to tender and selecting on the basis of the best or the lowest bid. The marketing strategy consists of creating a relationship upstream of the business in order to be able to advise the customer company on the very definition of specifications. Project marketing can be found just as easily in the building sector (airport development, airport extensions, office furniture, construction of factories, large buildings, sports stadiums, schools, head offices, etc.) with companies such as Bouygues or Vinci, as in the consulting sector (management or organizational consulting, service engineering) with companies like Cap Gemini-Ernst&Young, Altran, Peat & Marwick, PricewaterhouseCoopers and KPMG.

Project marketing, which is comparatively more sophisticated, is developing at full speed relative to traditional industrial marketing and consumer goods marketing (Figure 1.12).

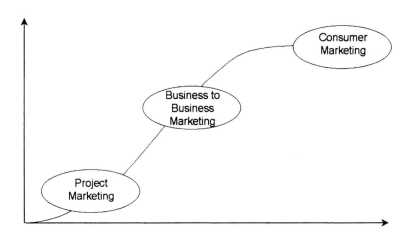

Figure 1-12. The degrees of maturity of marketing relative to domain of application

It must be emphasized that the same product can call for project marketing as well as traditional industrial marketing. For example the design and development of a new aircraft or some new equipment comes under project marketing. That same aircraft or equipment, once launched, is no longer a project. It has become the object of traditional, repetitive, transactions with suppliers and sales communication with customers.

3.3 Marketing in aeronautics and space supply chain

The supply chain concept must be linked to one of the main specificities of business to business marketing: the existence of derived demand. Within related industries the demand from companies downstream, determines the level of activity of those upstream. This " industrial chain " can be represented vertically to give an overall view of production, from the basic materials upstream to the final product downstream. This clearly shows that any manufacturer within such a chain depends on his customer who in turn depends on his customer.

In the aeronautical sector

A chain can be drawn to show relative values, from the aeronautical parts manufacturer down to the airline passenger or end customer, passing through the leasing companies such as ILFC (International Lease Finance Corp.), GECAS, GATX, Flightlease or express transporters like FedEx or UPS. Eventually we can see that companies in this sector are concerned by all types of marketing except for that which applies to the sale of consumer goods (Figure 1.13).

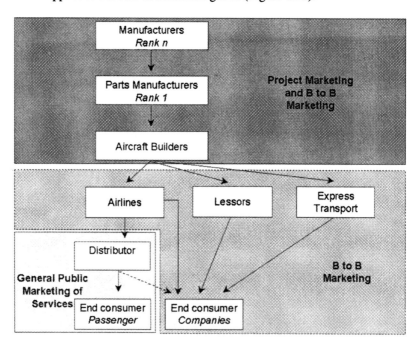

Figure 1-13. Diagram of aeronautics-related industries and type of marketing practiced

- The relationships between parts manufacturers and aircraft builders are determined by:
    - project marketing in the design and development phases ;

1. Marketing in the Aeronautics and Space Industry

- business to business marketing as soon as there is series production.
• The relationships between aircraft builders and their customers, the airlines or the rental firms, are characterized by b-to-b marketing.
• The relationships between airlines and their customers are concerned by:
- b-to-b marketing of services when the customers are companies,
- general public marketing of services when they are private individuals.

In the space sector

All transactions within the space sector involve business to business marketing between the different parts or sub-assembly manufacturers, the integrators and their customers (Figure 1.14).

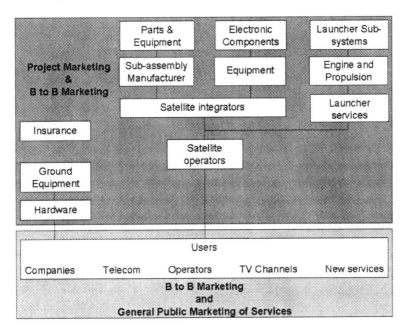

Figure 1-14. Diagram of related industries in the space sector[14]

Only the concluding transaction concerns the consumer goods market: final applications are aimed at consumers, like a private individual taking out a subscription to a cluster of satellite TV channels, or it could be a subscription to a weather, navigation or telecom (telephone, Internet) service.

General public applications	Professional applications
Telecom	Telecom
Internet	Internet
Television, radio	Television, radio
Weather	Weather
Navigation, guidance, search	Navigation, guidance, search
-	Earth observation

Figure 1-15. Main satellite applications, general public/ professionals

In fact, numerous functions can be carried out by the same satellite system thanks to the digitalization of electrical signals: direct TV, data transmission, telecommunications. The other satellite applications are linked to earth observation whether civil or military (Table 1.15).

3.4 Recent marketing trends in aeronautics

From EDI to e-business

The Electronic Data Interchange has rationalized the production and supply process and has modified inter-company transactions making them faster, able to react quicker and giving improved production management. But at present, the biggest upheaval in marketing comes from the expansion of information technology and Internet offering new possibilities of communication, for selling and working from day to day with partner companies throughout the world. Internet improves efficiency as much in Business to Business as in Business to Consumer relationships e.g. information, reservation and purchasing of airline e-tickets[15].

One-to-one marketing

One-to-one marketing is characterized by a commitment to addressing each customer within a company, individually and in a personalized fashion. This has been presented as something really new, which effectively it is as far as consumer goods marketing is concerned thanks to the considerable improvements made in IT tools and data processing. However, in Business to Business marketing and Project marketing, one-to-one is in fact quite old since Business to Business is based on a specific approach per key account. Still, one-to-one marketing is a step forward for sales of products or services to private individuals. It can be thought of as a democratization of the made-to-measure concept, which was up to now reserved for personalities and high income customers. In the airline sector, one-to-one applies particularly to passenger transport. The airline companies endeavor to identify regular customers in order to improve loyalty by offering them personalized services. This means taking note of their tastes or habits (seating position, meals, papers and magazines preferred, types of films, etc.). One of the limits of one-to-one marketing comes from the assumption that the customer remains constant in this data base whereas he is in fact free to evolve as much in his tastes as his habits. It is not obvious that he wants to discover exactly the same type of service from one trip to another, which is why one-to-one marketing should be applied with care.

NOTES

1. See Chapter 15, Alliance Strategies.
2. See official website www.faa.gov.
3. See Sea Launch in Chapter 15, Alliance Strategies.
4. See official website www.dgac.fr.
5. Source: DGAC. Copyright DGAC-Véronique Paul.
6. See official website www.caa.co.uk.
7. See Chapter 3, Business Marketing Intelligence.
8. See Chapter 4, Market Segmentation and Positioning.
9. Adapted from Kotler, Ph., (1999), *Principles of Marketing*, Prentice-Hall, New Jersey.
10. See Chapter 5, Marketing and Commercial Action Plan.
11. See Chapter 4, Market Segmentation and Positioning.
12. See Chapter 7, Marketing of Services.
13. See Chapter 10, Project Marketing.
14. Source: Alcatel Space.
15. See Chapter 9, Selecting Distribution Channels and Sales Team Management.

REFERENCES

Lawrence, P.K. and Braddon, D.L., (1998), *Strategic Issues in European Aerospace*, Ashgate.
L'Usine Nouvelle, n°2702, 23 September, p 54.

Chapter 2

INDIVIDUAL AND ORGANIZATIONAL PURCHASE

Downstream of the sector, passengers buy air tickets for private or company use. Upstream, companies buy planes, equipment and services. Further upstream, plane or satellite manufacturers-integrators also buy components from their suppliers. Understanding the needs and the processes of upstream or downstream buying is the basis of marketing. This means analyzing the individual psychological, sociocultural and psychosocial factors which enter into the buying decision. The purchase context influences the buyer's behavior. An analysis of buying behavior is also essential in Business to Business, where the purchase is not made for an individual but for a company, involving a variable number of people. To be more commercially efficient, the supplier must take into account the special nature of the industrial purchase, in particular the specific nature of the buying center, the buying process and buying situations. First we will look at the individual purchase and then the Business to Business purchase.

1. *THE INDIVIDUAL PURCHASE*

1.1 *Factors influencing the buying behavior*

Most models present factors influencing buying behavior in a hierarchical and simplified way. In fact, this classification in terms of importance of the factors depends for the most part on the context (Howard and Sheth, 1969). For example in Muslim countries the power of religion is very important and almost certainly dominates other factors such as social class.

In general we can say that there are four types of influence factors (Figure 2.1): sociocultural, psychosocial, personal, and psychological.

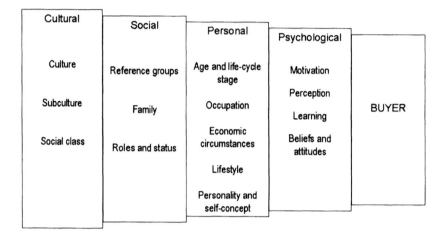

Figure 2-1. Factors influencing behavior [1]

A consumer is first influenced on the cultural and religious level, by the community/communities in which he lives, even before his own characteristics, personality or individual needs come into play. He is also influenced by those immediately around him, usually friends and family. His personal standing at a professional and private level also plays a part in the type of needs he will try to satisfy. Furthermore, psychological factors structure his motivations and perceptions of the different information to which he is exposed. These factors are interdependent although differently weighted. Within a given community, the weighting of the different factors varies from person to person, and for a particular consumer, their importance can also vary from one period to another.

Sociocultural factors

Right from birth, the individual is a part of a community which transmits its values, habits, culture and a way of life. The individual is not a free agent; he exists only by and within the group to which he belongs. In this way he is strongly influenced by the characteristics of his own community:
- the language, more or less widespread which could be the main language of his country or conversely a dialect,
- beliefs, belonging to a religion and to what level this is practiced,
- the type of geographical and climatic environment,
- eating habits, from types of food and meal frequency to methods of preparation and consumption,
- the style of dressing, choice of transport, etc.,
- learning methods and obtaining qualifications.

Added to these influence factors there is the notion of belonging to a more or less homogenous social group, generally defined by type of profession and income level. In India, the castes are an extreme social classification where situations are immutable. On the other hand on the American, European or Asian continents an individual has more chance of changing social class. Within a class, individuals tend to behave more homogeneously. This is why the concept of social class is used in marketing, as a way of dividing up the population into sub-groups showing homogeneity in their needs and purchasing power, in order to narrow down the offer[2].

Psychosocial factors

Beyond belonging to a culture, an individual is part of groups with whom he lives from day to day. This is mainly the family circle, neighbors, friends and people he meets during work or leisure. As for sociocultural factors, the individual is influenced by the different ways of life that he sees around him which can be adopted or rejected. When a new type of product or service appears psychosocial factors play an important role. For example the choice of a new holiday destination will influence the other members of the group. Among the groups which influence the individual, the family usually plays the most important role. A distinction must be made between:
- Immediate family with whom the individual lives from day to day:
  - parents, brothers and sisters when he is a child,
  - partner and children when he is an adult,
- The family relations meaning the ancestral line (grandparents, etc.) and marriages.

THE CHILD AS INFLUENCER

Parents' buying decisions concerning goods or services for the family can be influenced by the children. This can include products or services which directly concern the children (a stay at Disneyland, etc.) but the latter can also influence what the adults buy for themselves. For example school holidays dictating the date when parents can travel somewhere.

The concept of the family depends to a large extent on the cultural environment. Family ties are usually stronger in countries which have a zero or minimal system of social cover age and also outside the large urban areas. Taking into consideration the nature of groups with their shifting interplay of power, marketing aims to target those who exert the most influence.

Personal factors

The particular characteristics of the individual such as age, family situation, profession or personality also influence buying decisions.

- Age is an important criteria influencing customer behavior, regardless of the country. As a general rule, companies must adapt their products and services to evolving demand, especially when this evolution arises from a change in the population structure. This is well illustrated by the appearance of senior marketing.

SENIOR MARKETING

The evolution of the population in developed countries is characterized by a fall in the birth rate and a large increase in life expectancy. Consequently, the over 60's represent a greater and greater part of the population. This age group often has a good income thanks to advantageous pension schemes and represents an increasing part of a country's overall purchasing power.

An analysis of this age group dispels certain home truths: in fact, older people are often favorably disposed towards new consumer products and services. They are less and less conservative, have a stable income and merit a separate marketing approach. On the other hand while people may be living longer, there is no getting around the fact that certain capacities diminish such as visual and auditory acuity or precision of gestures. Therefore senior marketing must allow for certain rational expectations (documents written with a larger typeface, messages which are loud enough, ergonomics of vehicle seats, ease of movement) and certain irrational ones such as receiving special attention, respect for elders, sometimes dealing with somebody who is in the same age group.

Figure 2-2. Sicma Aeroseat: a sophisticated seat answering specific expectations

This 'senior consumers' category is characterized by the dual availability of time and money. Their buying power is paralleled by a large capacity for research and comparison shopping before the actual act of buying. This trend towards " senior marketing " originated in the United States before flourishing in Europe and on the Asian market. At present it is

2. Individual and Organizational Purchase

mainly concerned with airline company marketing departments for travel and tour operators for vacations. Equipment suppliers have also taken steps, e.g. Sicma Aeroseat, Weber Aircraft, seats (individual digital screens, 130/180° seats, armrests with goblets, etc.). During the design phase, this equipment can be integrated to answer the comfort needs of this target population.

- The family life cycle is another important explanatory factor in the 'personal' category. When a young person leaves home and has finished studying, his first job is not normally the moment for long-term investments. When he starts a family, he will be more risk sensitive and will look for financial and geographical stability. In this way, individual ambitions linked to perceived risks will vary according to each person's commitments within the family environment. The birth of a first child normally results in spending being redirected with sometimes the start of savings. Some years later when the children are old enough to leave home, money within the family budget can be re-allotted once again.

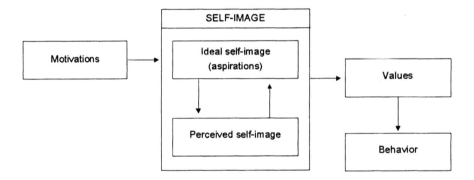

Figure 2-3. Self image

- Social standing and profession, which are closely associated with one's self image, also influence behavior (Figure 2.3). The concept of self-image refers to the way in which an individual sees himself (ideal self image) and how others see him (perceived self image), relative to how much consideration he would like to be accorded (Filser, 1994). The discrepancies noted add on to the initial motivation thus influencing choice criteria and finally buying behavior, whether it be a private purchase or the individual's influence in a organization purchase. The profession of the person has financial effects in the first instance, allowing or limiting the purchase of certain goods, starting with the type of house, the main means of transport used, etc.
- Life styles, which group people according to their opinions, their main interests and their activities, depend to a large extent on those characteristics presented above. A lifestyle classification puts people into homogenous groups[3] which do not take into

account traditional socioeconomic variables, but rather depend on their behavior in terms of preferred products and services, distribution channels and media.

All these factors are widely used by marketing departments: by permitting better market segmentation within the population, they improve the offer made by the company, which can then set up tailor made solutions for each of the main segments[4].

Psychological factors

Buying behavior is the outcome of a sequence going from motivation via perception, attitude and learning culminating in trial or buying (Figure 2.4).

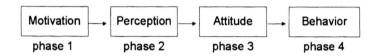

Figure 2-4. The different stages: from motivation to behavior

Motivation: Motivation plays a driving role in the building of desire. These are the deep seated reasons pushing the individual to make his choice. Following on from the work of Sigmund Freud, motivation was studied by Abraham Maslow who proposed classifying needs into 5 categories (Figure 2.5) (Maslow, 1970).

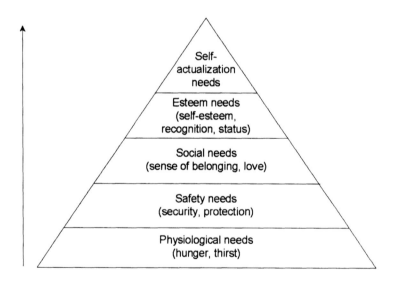

Figure 2-5. Maslow's pyramid of needs

Starting from the theory that needs do not all have the same importance for an individual and that they can be put into a hierarchy, Maslow developed the idea that

2. Individual and Organizational Purchase

it is necessary to have satisfied a first type of need before taking on the next one. Physiological needs correspond to the vital necessity to eat and drink to remain healthy. Safety needs correspond to being able to protect oneself from the elements and from possible danger (animals, criminals, etc.). Once both of these basic needs are satisfied, the individual begins to be interested in the need for belonging and affection (social needs). In this case it is a question of belonging to a community which can be sports oriented, cultural, religious or a certain style of life. Needs related to esteem correspond to a wish to be appreciated by one's close friends and family or within the work environment. These can translate into external signs of way of life or wealth. The fifth category involves the need to accomplish something in terms of personality, by traveling far away or achieving some complex sporting or cultural goal. Whatever the need whether psychological or material, the greater the lack, the stronger the motivation to satisfy it.

NEED CATEGORY	EXAMPLES	PEOPLE
Accomplishment	• Carry out the first flight in a new craft • Accomplish a first flight link up	• Richard Branson (CEO of Virgin) or Bertrand Piccard and Brian Jones for their world circumnavigation attempts in a balloon • Russian cosmonauts some of whom stayed several months in space
Esteem (acknowledgment, status)	• Thanks to one-to-one marketing, it is possible to prove to 1st class passengers that they are recognized, addressing them by name and showing that their habits and preferences have been noted, similarly with their previous comments (food, dress and language codes) • More generally, membership in the " Privilege " clubs according to how many miles traveled	• SkyMiles Silver, Gold Medallion (Delta Airlines) • SkyTeam Elite
Belonging and affection	• Belonging to a nation: the airline's aircraft can be thought of as an extension of national territory where everybody can feel at home • Membership, in a WW2 veterans club or a local gliding club	• US airlines identified by some symbol recalling the national flag
Security	• People in difficulty through floods, fire, embassy evacuation, being winched up to safety by a helicopter	• Firemen, armed forces...
Physiological	• Air drops of food, drinking water, medicines, after fighting or a natural catastrophe	• International Red Cross, MSF, Refugees International, American Red Cross, Care

Table 2-6. Example of a "needs pyramid" which can be satisfied by air transport

Perception: Depending on the various messages received and experiences lived through, the buyer has a different perception of the various services and brands available to him. Following on from the first stage, if the individual is motivated, he will be more likely to increase his level of information to optimize his decision.

Motivated by the purchase of a particular product, he will pay greater attention to the advertising by different brands. It should be noted that the brand concept can modify perception of a product category, even if this does not necessarily seem logical. An individual could have a negative perception of civil aircraft with less than 30 seats but a positive perception of the Crossair company. As a function of his preconceived ideas, an individual will pay more or less attention to the messages put out by the company. For example a Muslim consumer will be much more likely to be motivated by respect for rituals concerning his food and will carefully verify whether there are alcohol or pork in ready-made dishes. This could be an important selection criteria for an airline company. In addition, a consumer often has preconceived ideas about the different competitor brands. When the consumer's brand preference is adversely affected by a disappointing experience, "cognitive dissonance" allows him to adapt the company's message to what he really wants to hear.

Attitude: According to how the company messages are perceived, the potential buyer will gradually build up an attitude concerning the product category and the different suppliers on the market. This attitude will benefit from any previous experience either from a first buy or through the experience of a family member or close friend. An attitude is therefore formed from an overall combination of favorable and unfavorable evaluations whether they be based on objective judgments or emotional reactions. The marketing department needs to know the attitudes of the various customer segments in order on the one hand, to try to correct the unfavorable components and on the other, use those which are favorable. For example a supplier such as EDS can invest in the sponsorship of an event like the Olympics in order to reinforce the technological and performance related image of its systems adapted to the sports context. The attitude of professional users can therefore be backed up by the appropriate choice of communications.

1.2 The buying process

In order to influence a customer segment effectively, marketing must make an effort to understand how the individual customer buying process takes place.

The stages of the buying process: An individual purchase goes through several stages (Figure 2.7).
- *Need recognition:* needs come from stimuli which originate directly with the individual or his environment. This could be the case for example where someone feels the lack of social differentiation, not having had any attention before or during a flight. In addition, company communication can stimulate the need by alerting the individual to the existence of a new service. This will allow the company to identify passengers and get to know some of their wishes.
- *Information search:* as soon as there is a visible need, the individual will start to pay attention to the different sorts of information available within the product

2. Individual and Organizational Purchase

category or services concerned. The search will be more or less active as a function of how involved the individual is in the process. When it is a question of a sophisticated or expensive product such as a computer or a vehicle or a specific brand product which is more difficult to find, e.g. luxury goods, the search could be very active. On the other hand, when it is a question of a more common product, the search will be less intensive. The potential customer generally uses personal sources of information such as family or friends at work as well as business sources (i.e. the companies themselves or their representatives).

Figure 2-7. The different stages of the individual buyer's decision-making process[5]

- *Evaluation of alternatives*: in order to choose, the buyer must first of all identify the attributes which interest him. Here it should be noted that certain attributes will be known or unknown, depending on the characteristics and background of the buyer. The evaluation must weigh up the different attributes i.e. how much importance is accorded to each of them. Researchers have suggested numerous buying models which quantify the evaluation. The "overall consideration" brings the competing brands together in the customer's mind. For the purchase of a PC, let us suppose that the criteria are the price, the memory, the frequency of the microprocessor and the amount of equipment. A mathematical model lets us weigh up each of these four attributes and so assess each of the three competing brands for these same attributes. By weighting the values, an overall mark can be obtained for each brand. If the purchase is based on completely rational reasoning, this is the perfect approach for selecting the brand which corresponds to the customer's expectations.
- *Purchase decision*: an individual buy is never totally rational and therein lies the important limit of the different buy models. In most cases subjective criteria such as brand preference, influence the final choice in spite of any quantitative evaluation. Situational factors at the actual buying site can also modify the buying

decision (promotion, lack of room, sales staff, travel agency atmosphere, etc.). In order to optimize the company's offer and its subsequent communication campaign, the relative importance of the attributes must be precisely determined by the marketing department using qualitative studies[6]. However the finer points of these evaluations are generally lost on the individual buyer when he comes to decide. Two cases are most often encountered:
- The presence of an attribute can be considered as indispensable even only at a minimum level. In this case its absence will be cause for elimination.
- The presence of an attribute can be desired without being essential, and in this case selection is made by comparing the advantages and disadvantages of the different suppliers.

For the customer, the importance of attributes derives mainly from the perceived risk implicit in the purchase (Figure 2.8).

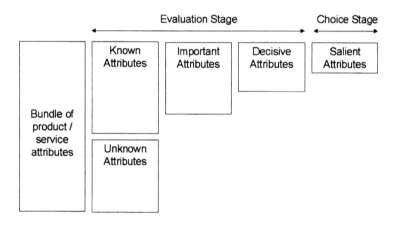

Figure 2-8. Typology of attributes[7]

- *Post-purchase behavior:* it is only after actually buying that the buyer can make a judgment. This will be determined by any eventual difference between what he hoped for and what he actually obtained as regards the different attributes. " Cognitive dissonance " is an individual's capacity to modify his expectations in retrospect in order to reduce this difference, which will in turn allow him to better accept his buying decision. When he is satisfied, he will pass this on to those around him by word of mouth, which remains one of the main communication methods especially in Business to Business marketing. When he is dissatisfied, there are several solutions (Figure 2.9) (Day and Landon, 1977). The negative reaction can be expressed through some simple personal action followed in general by abandonment of the brand and spreading unfavorable word of mouth. The reaction may also include visible actions such as sending a letter of complaint to the supplier, taking legal action and above all creating bad publicity through associations and/or press groups.

2. Individual and Organizational Purchase

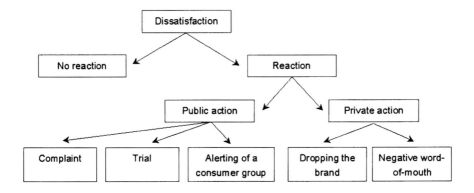

Figure 2-9. Dissatisfaction: the different possible reactions

Roles in the different types of buying: Although they will be less well structured than for an organizational buy, the same main figures that influence the individual buy can be found within a family:
- *The user* who will consume the products or services, for example the child in the case of a theme satellite station such as Disney Channel;
- *The actual buyer* of the product or service. Depending on the culture and type of product, the mother often plays a dominant role here;
- *The decision maker* or the person who decides when there are several possible solutions. When it is a question of general products, the mother or the father plays this role according to their competence in the product categories. If it is a very important purchase such as a house, the decision is generally made by both parents;
- *The influencer* is the person who influences the final decision when this is not the actual decision maker, e.g. children, grand parents, etc.

The analysis of roles in the buying process can highlight the very different types of purchase, from the complex to the routine. While in the former case several people are concerned by the process, in the latter, there is often only one. In a routine buy, the different stages of the process are also reduced to their simplest level which is especially true for frequent purchases of run-of-the-mill services or products.

Figure 2-10. The permanent buy swing on the continuum

The purchase is part of a continuum (Figure 2.10) with:

- on the one hand, the " sociological " buy, based on a desire to display membership of a community (economic decision makers, sports team members, age group, ethnic minority, etc.),
- on the other, the " psychological " buy, based on the need to assert one's personal identity.

The very existence of different product categories and their brands, allows the individual to satisfy a psychological and/or sociological need. The individual is always torn between the desire to be like everyone else or to stand out and emphasize one's originality. For example, concerning air transport, the choice of ticket (first, business, tourist) can be counterbalanced by asserting ones individual identity based on choice of clothes, food, reading matter, etc.

2. THE ORGANIZATIONAL PURCHASE

A purchase will be considered to be Business to Business or organizational from the moment that it is made in the name of a company or organization, regardless of size, from a medium sized company up to a multinational or state company. Professional purchases essentially cover (Malaval, 2001a):
- the purchase of *entering goods*, necessary for making the final product: this can be raw materials (aluminum, titanium, etc.), transformed materials (PTFE Skyflex® joints for portholes, etc.) or components, spare parts, sub-assemblies (cables, braking systems, aircraft windscreens, etc.) ;
- the purchase of *foundation goods*, necessary for the company to function, e.g. office supplies, IT equipment, digitally controlled machine tools and production robots;
- the purchase of *facilitating goods*, from the supply of energy (electricity, water), insurance or banking services to consultations (chartered accountants, auditors, design offices, etc.) via staff restaurants, catering, security services and the transport of goods.

The *buying center* concept encompasses all of the operational and functional managers who take part in the final acquisition decision, and not just the buying department. The interactions between the participants, as well as the weight of each of their social roles, has been highlighted by the work of numerous authors, among them Robinson, Faris and Wind.

2.1 The buying center

Unlike the buying department, the buying center has no formal existence. It groups together all the people who take part in the decision process (Webster and Wind, 1972). Four types of people are usually identified.

2. Individual and Organizational Purchase

The decision makers

This is the person or persons who make the final decision and who can decide in case of disagreement which supplier to choose. In a small or medium sized business, this will normally be the head of the company who will decide, often aided by the person in charge who is seen as trustworthy and competent for the project in question. In larger companies, the "boss" is replaced by the board management committee headed by the chief executive officer or the president. However, in the case of conflict within the buying center, one of the managers will generally have the power to decide one way or another. Information such as this is of great importance to the supplier.

A look at large companies shows that the main way of making decisions is in fact collective, i.e. even a chief executive officer who is strongly involved in technical decisions will prefer to get a collective vote from the board management committee rather than to impose that decision unilaterally. Only after the different propositions have been studied and above all the in-house and external opinions analyzed, will the final decision be made.

The buyers

This is basically the buying department, whose main functions are:
- to collect information concerning the different market solutions available to the company, taking into account in-house know how and financial constraints;
- to find out about the different suppliers capable of adequately answering the company's needs.

This dual function necessitates close, wide-ranging contacts with the market above and beyond the suppliers used and known by the company.

The other main function consists of taking charge of the pre-selection of supplier companies competing for the market: this is the intermediate phase leading to a short list used for the final selection. Having picked out the acceptable and possible solutions, it then remains to choose and convince others of the solution which appears to be the best for the company.

In a growing number of companies the organization of purchases has evolved by putting more emphasis on the final customer, by applying traditional marketing methods to the selection and evaluation of suppliers; this is what is meant by *purchase marketing*[8].

The influencers (prescribers)

The influencers in fact make up the contact group which, from one company to another, varies the most according to the different sectors and the different types of corporate culture. Two types of influencers must be singled out: in-house and external.

The *in-house influencers* are those managers who either favorably or unfavorably sway the other members of the buyer group in relation to this or that supplier's offer. This could be managers from quality assurance, maintenance, marketing, sales or R&D[9].

The *external influencers* also hold sway with the buying center members through the opinions that they voice. Essentially they consist of the customers, the professional experts (engineering consultants, design offices), those in charge of running the airports, the national and international certification organizations, the specialist press journalists (*Flight, Avionics, Journal of Electronic Defense*, etc.). Customers already on the books are also important influencers, even more so when they are a reference in terms of professionalism. This is particularly true for the military market where an order for an arms system (missiles, fighter aircraft, troop carriers, etc.) from an army, lends credibility to the offer and encourages other countries to obtain the same equipment.

The users

The consideration given to users varies markedly according to the size of the company, its culture and geographical location. In general, the amount that customer opinion is taken into account depends above all on their social status. In the aeronautics sector, pilots have a high prestige, quite apart from the fact that they are represented by unions which are often very powerful. Apart from this example, user opinion is increasingly being taken into account. The idea is to listen to those who are going to use the material in question because, as befits someone on the front line, they often make pertinent remarks or criticisms. In addition, taking into account user expectations improves their motivation at work and their loyalty to the company. However, this regard for the user depends also on the national culture. 'Participative' management is particularly well developed in the Scandinavian countries whereas it is almost unknown in southern hemisphere countries such as Latin America. Finally, the weight of user opinion often depends upon the size of the company: special systems for representation generally make users more influential in larger companies.

To conclude, the industrial purchase results from a collective decision taken by a variable number of people, who can be more or less involved in the purchase and who show very little homogeneity between themselves (in terms of education, responsibilities, level of confidence) (Malaval, 1999).

In order to position themselves and negotiate better, it is in the interest of supplier companies to know as much as possible the different in-house contacts within the customer company (exact role, motivations, present position, previous job, education, other personal information). They must also identify the customer company's buying phase and his buying policy (type of negotiation, buying power, dilution of supply risk between the different suppliers, any eventual alliances between contractors, etc.).

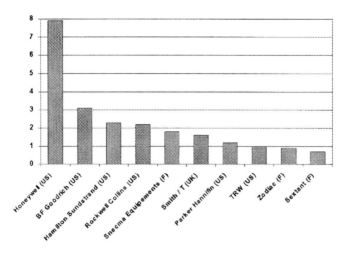

Figure 2-11. The ten leading worldwide aerospace parts manufacturers

2.2 Buying phases

Industrial buying is the result of an iterative process in six main phases[10]:
1. Recognition, or better, anticipation of a need,
2. Definition of the characteristics and quantities necessary,
3. Search and qualification of potential sources,
4. Collection and analysis of propositions,
5. Choice of suppliers and ordering process,
6 Information feedback and performance evaluation.

Certain companies, particularly integrators, select the supplier by putting out to tender[11].

Anticipation and recognition of a need

Anticipation of a need rather than mere recognition allows the company to gain time in terms of the technical details of acquisition. It can orient the in-house decision makers, e.g. the marketing department, towards realistic or really achievable solutions. In the same way, the attentive supplier who can effectively take care of his customer's future problems, benefits from the atmosphere of confidence that he has thus created. The anticipation of needs relies on being aware of all in-house and external information exchanged. This means detecting signs of latent needs which have still not been formalized.

Within technical departments, this could be improving quality by decreasing the amount of waste, or replacing some equipment whose use damages a certain cutting tool, or preparing for the substitution of a particular material whose use will be forbidden under new environmental norms, or protecting oneself against a one-off or long-term shortage which could occur through dependence on a sole supplier (Figure 2.12).

Figure 2-12. The transfer of information from the technical department to the buying department

Within the sales and marketing department, this involves improving customer satisfaction by correcting a product's weak point: integration of more recyclable products, replacement of an ergonomically unsatisfactory control button, a dial which is difficult to read, etc. (Figure 2.13).

Figure 2-13. The transfer of information from the sales and marketing department to the buying department

At the same time, the purchasing department receives information from the suppliers themselves. This could be directly from the supplier's marketing or technical department, proposing a new material or a new process, but it could also be information collected by the supplier from other customers. This will be welcomed by the customer company as long as this information does not come from a direct competitor, because in this case there would be risk of suspicion.

Defining the characteristics and quantities to be bought: the specifications

This means defining expected performance in terms of type of product, materials, tools to take into account and what quantities are necessary for a given period of time. Drawing up the specifications answers this need for precision. Apart from the essential technical specification, the commercial services must be prepared by specifying delivery constraint (times, frequency), as well as the logistic specifications, terms of payment, maintenance and after-sales service conditions. For state-of-the-art products like satellites, the requirements will be drawn up by the technical department (functional engineering, methods section, norms), helped by the users, supervisors, shop foreman, etc. In the case of everyday products, the user department will draw up its own technical specifications. They will be able to supply detailed information about the product and also sometimes about the production process and technical assistance. The more information they have, the more they will be listened to. The more knowledge offered, the more the sales person will be accepted in the problem solving process. His action is an investment for the next phases.

The distinction can be made between suppliers who are " in " or " out " according to whether or not they are already on the company's books. It will be easier for the supplier to offer technical advice if he is already " in " i.e. known within the company. He will be that much better placed to identify possible improvements than a supplier who is "out ".

Searching for and qualifying potential sources

This phase consists of drawing up a list of potential suppliers. This must include not only the "in" suppliers currently on the company's books, but also those, "out", with whom there have already been exchanges in the past as well as those who have so far never supplied the company.

The search for these suppliers should not be restricted to a given technology: a competitive solution can be obtained through different technological channels. In the plastics molding industry for example, most supplies are concerned with either thermoforming or injection molding. Both techniques can be competitive depending on the qualities (flexibility, resistance) and quantities needed for the materials. Consequently the search for suppliers must cover both types. One very popular method for drawing up a list of possible suppliers consists of noting the most dynamic companies present, as reflected by their communications strategy, in the trade press. In addition, the impression they make as exhibitors in shows and the factory visits they propose are further reasons to include them on the list of competitors or not.

Certain companies are even more dynamic: here, rather than waiting for proposals from the suppliers, they make the first step. This method presupposes a different state of mind. Even when the buying structure is a large company or government administration, it should not become over confident or adopt the "easiest solution" rather, it should develop a purchase marketing approach. The latter consists of applying methods developed downstream for customers to suppliers (Bénaroya, 1997). The ultimate aim is to make the suppliers want to work with the customer company even under technically or financially limiting conditions.

Collecting and analyzing propositions

The collection of information must be objective and scrupulous, not based upon preconceived ideas and trying not to take previous experience into account. A grading table can be made and applied to the different propositions, scoring each criteria, and these values should be weighted as a function of the essential character of each of the different characteristics.

For anything concerning foundation goods, negotiations can last several months. There are frequent exchanges of proposals and counter proposals between the company and its suppliers. Quite apart from the rational arguments, the supplier who produces a made to measure solution is at a great advantage.

It is in this phase also that the respective advantages of in-house production and external supply will be compared: *Make or Buy ?* Each time the company can produce more economically than the lowest buying-in price, the in-house solution will be retained. It will also be favored in the case where the company wishes to keep control of what they consider to be strategic know how.

"MAKE OR BUY ?", THE EXAMPLE OF THE SUPER TRANSPORTER: FROM SUPER GUPPY TO A300-600ST [12]

During the seventies, Airbus favored the " buy " route. What they did was to transport fuselage parts from the Hamburg (Germany) production site to Toulouse (France), using the McDonnell Douglas Super Guppy (Boeing B337 Stratocruiser) which had been developed to carry American Titan rockets. About fifteen years later, Airbus adopted the "make" route by developing the Airbus A300-600 Super Transporter, nicknamed the Beluga, to transport fuselage parts in a more practical and efficient way.

Figure 2-14. The Airbus Super Transporter

Today, this plane is actually a new offer from Airbus for transporting very large parts or objects (Airbus Transport International): no more costly dismantling of power lines to let a convoy through, or avoidance of a river or sea route where there are low bridges. The Airbus Super Transporter flies over difficulties like these. Generally considered to be the plane with the largest freight hold on the market, the AST is the best adapted for the transport of oversize elements. An example of its extraordinarily large transport capacity was when it flew different, enormous, full petrol storage tanks between two densely populated areas of Western

Europe. The same feat on the road would have taken several weeks compared with the AST's one hour flight time. The cargo thus arrived faster and more reliably at its destination.

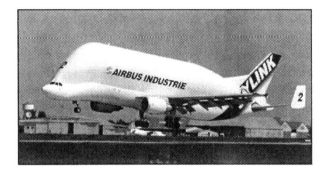

Figure 2-15. The Airbus Super Transporter initially destined for the transport of fuselage parts between Hamburg and Toulouse

Choosing suppliers and the ordering process

Choosing the ordering process depends on in-house constraints, notably technical ones. For example *just in time* production requires a special organization which the supplier must adhere to. The choice also depends on the propositions and possibilities of various suppliers as a function of their own logistical organization:
- proximity of supplier's production site ;
- stock distribution platform.

It should be noted that the buying procedure is only terminated once the product has been delivered and received by the company, checked by the user department and declared to be ready for use. A change in ordering method is obviously possible, but it will involve a re-negotiation of the agreement between the company and its supplier.

Information feedback and performance evaluation

A complete buying process necessitates a checking phase. It is only by feedback from the working environment that the buying center and especially the buying department, can evaluate whether they have made the right choice. The two main areas are marketing-sales and the production department. The latter is the first in line to warn of anomalies or differences in the quality ordered and that actually received.

Apart from this quality follow-up, the production department can feed back information concerning ease of setting up or a bad choice in terms of protection during transport, etc. However, the essential feedback will be from the sales department. Do customers notice a difference ? Are they more satisfied ? Respect of the requirements brief will be judged from a technical and a commercial perspective:

speed of delivery, regularity, flexibility, repair service capacity, etc. In order to facilitate information feedback, the buying department must do everything to catalyze the process, notably by explaining that a supplier's position can never be taken for granted and that changes are always possible depending on reactions.

From the supplier's point of view, the ideal would be to meet the customer company as soon as possible. If the potential supplier is involved right from the outset, it is possible for him to influence the contents of the specifications, or even buying procedure by the grouping together or the separation of the equipment/supplies ordered. Beyond the phases, there is another characteristic which renders business to business marketing even more complex: the importance of the respective representatives varies according to the type of purchase. For an identical buy, the fact of buying it for the first time or the nth time modifies the amount of influence of each of the members of the purchase center.

2.3 Different situations

Three main types of buy can be distinguished (Robinson and *al.*, 1967):
- the straight rebuy,
- the modified rebuy,
- the new task.

Straight rebuy

The straight rebuy pattern is the most often encountered and corresponds to an almost unchanging need. In this context, the company's evaluation criteria are well known as are the appointed suppliers. Commercial relations are stabilized thus favoring the already established partners. To reinforce his position, the accredited supplier can go as far as to offer automatic restocking in order to make it even more difficult for a potential competitor. In the same way, the trend towards *just-in-time* production management tends to reinforce loyalty. In most cases, for this buying pattern the decision maker remains the buyer or the head of the buying department, when it concerns "everyday goods". When it is a case of relatively sophisticated equipment, users exert the greatest influence ahead of the buyers. They are offering the company the benefit of their experience.

Modified rebuy

This situation involves the desire to modify the response to an existing need. This is true for example of a product in the mature phase, for which the company would like an improvement in quality and/or reduced costs. This situation can also hide an underlying dissatisfaction with the present suppliers who, if they had only recognized the company's expectations soon enough could have proposed a modification or at the very least a direction in which to conduct improvement research.

2. Individual and Organizational Purchase

Here, the company will try to find out more information about other possible procedures and other possible suppliers. This is an opportunity for a supplier who is not already on the company's books to attempt to be the best regarding the particular improvement. It could also happen that this very same supplier is at the origin of the questions concerning the previous deal... When changing a supplier becomes a possibility, the number of people to contact within the company increases: the buyer retains an important role, but the Production, Technical, Engineering and Quality Departments for example are concerned by the choice. They must verify that the proposed improvement does not mean a failing on another criteria.

New task

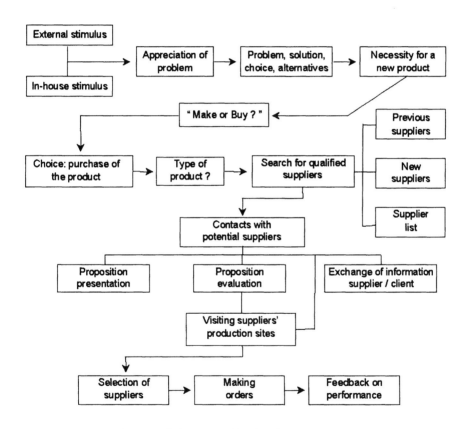

Figure 2-16. The new task buying decision procedure

This involves the most number of people within the buying center. There are numerous interactions between the marketing department, the technical department management (Method, Engineering, Quality), and the Production and Buying Department. Here, the risk is at a maximum, which is why the opinions of the influencers and advisors predominate. The buying department's opinion no longer takes priority over that of the technical management. This is the most open type of buying situation for the non-appointed suppliers. In order to obtain new solutions, information is crucial and members of the buying center should spend a lot of time collecting and selecting it. As can be seen in Figure 2.16 it is here that the *"Make or Buy?"* alternative becomes apparent: either in-house production of the necessary material or else buying in from an outside source (Reeder and *al.*, 1991).

If the project is new and innovative, it could be useful to keep complete control. By making a particular spare part or component, the company will acquire new know how which will allow it to get to know the limits and demands of the type of material used. In addition, choosing in-house production could be the most discreet solution. According to how successful the product is a few years later, the company can still choose the buying-in solution. And this new task situation also corresponds to a new product launch phase and in particular to those for which price is less important than quality.

2.4 The behavior of professional buyers

In the wake of work by Levitt, many researchers have focused on the behavior of industrial buyers. Robinson and Faris's model (Figure 2.17) also called the Buygrid model, uses both the three buying situations (Buyclass) and the different phases of the buying process (Buyphases).

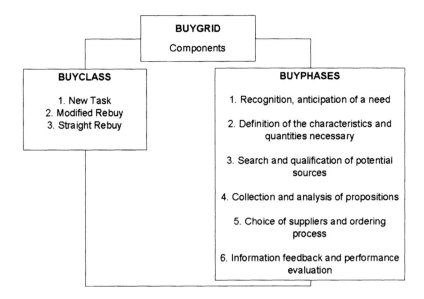

Figure 2-17. The (simplified) Buygrid model

2. Individual and Organizational Purchase

This model stimulated a sequential analysis of the salesperson's responsibilities. Before meeting his customer, the salesperson must ask himself:
- In what buying phase is his customer?
- What type of purchase does the project represent for his customer, and consequently, who are the people really concerned and what is their respective influence?

Other models have been developed which integrate:
- individual characteristics of the decision makers (training, past experience, degree of satisfaction with previous purchases…),
- factors associated with the nature of the particular company (sector, size, etc.),
- factors associated with the product (response time to need, level of perceived risk…),
- the level of mutual long-term involvement for both customers and suppliers.

In the industrial context, the customer is active, but the supplier also participates in defining the products or services offered: given the latter's increasing complexity, a supplier can contribute to elaborating the customer's requirements ; by doing this, he will be better able to stand up to the competition. This is especially true in the case of bid determined markets[13].

While researchers have sought to create models for the industrial purchase, it is also important to consider the non-Cartesian criteria which contribute to implicit buyer expectations. In fact, the industrial buying process and purchase have long been qualified as rational compared to the more emotionally driven behavior of individual consumers. This concept stems from the economic rationale underlying buying in the industrial context, in particular regarding the search for the lowest cost. However, this attitude varies considerably depending on the situation.

The explicit criteria generally considered include price, specifications conformity, delivery times, payment, repair/replacement capacity, and diverse quantifiable commitments. In addition to these rational criteria, the customer's implicit expectations must be identified (Table 2.18) so as to optimize the negotiation underway (Malaval, 2001b).

Rational criteria	Non-rational criteria
Price level Delivery times Specifications conformity (Norms) Quality Precision of estimates and summary notes at the end of project After-sales service Proximity of distributor or supplier sales department User safety Availability of managers	Prestige and reputation of supplier How long the relationship goes back Membership in an ethnic, linguistic, religious or political community Family relationship Investment role of the supplier in the country Educational background of people involved (universities, engineering schools, business schools…)

Figure 2-18. Main rational and non-rational criteria

These criteria can include:
- the reputation or prestige of the potential supplier. The very concept of "reputation" can vary considerably from one company to another, and the supplier must understand what this means in the particular context: working in the same geographic area, in the same sector of activity, for a particular government administration or company ;
- the size of the company: people have prejudices about working with a big company, or on the contrary, a small one. Whatever the size of the customer company, buyers can be reassured or doubtful about the idea of working with a very large supplier, just as they can be concerned about the lifetime of a small company, or expect better service.
- the nationality or geographic location of the supplier: this can be both an irrational criteria (regional preference) and rational (mental and physical proximity which can allow more efficient after-sales service) ;
- conscious or unconsciously desired similarities with the supplier's staff in terms of ethnicity, culture, religion, or educational background, among others.

From one country to another, whatever the development level, these non-rational factors come into play.

The concept of perceived risk plays a significant and ambiguous role in analyzing buyer behavior (Figure 2.19). In fact, there are risks perceived by the:
- company, in terms of the reliability of supplies, protection from exchange rate fluctuations,
- buyer himself, in terms of resistance to his final choice from a particular department, being accused of going for the easiest solution, suspicion of partiality, etc.

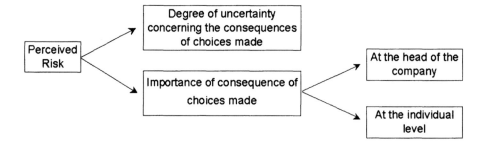

Figure 2-19. The main components of perceived risk in the industrial buying situation

According to Choffray's analysis of perceived risk (Choffray, 1979; Salle and Sylvestre, 1992), the buying department determines a hierarchical classification of risks and procedures corresponding to the estimated risk level. For example, when there is a high degree of uncertainty and the choice has important consequences, the perceived risk is highest. In this case, the buying department signals this to the company and the buying procedure will necessarily go through the different phases described earlier.

2. Individual and Organizational Purchase

Figure 2-20. A Lockheed-Martin campaign based on the reduction of perceived risk

2.5 Bidding

In the context of public markets, the authorities establish regulations so as to optimize selection of the best possible supplier; thus they ensure publication of the bid and impartiality between the candidates, while trying to minimize the risk of corruption. There are three types of public buying:
- *Negotiated contract or mutual agreement:* below a certain amount, the manager chooses the most competent company candidate as he sees fit, without going through any formalities. This selection can be made with or without competition. This procedure, which is the simplest, is the most common in terms of orders or work projects. It offers greater flexibility without top-heavy administration, but it does not allow for a posteriori control.

- *Formal procedure by allocation (bidding):* this is based on the principle of the lowest bidder. The administration establishes specifications which are distributed to candidate companies. Beyond a pre-defined time limit, the company offering the most cost-effective solution is the winner. This procedure includes three phases. A first selection in a public session based on administrative criteria, a second non-public phase which involves eliminating candidates on the basis of technical criteria. The third phase in a public session which includes the reading of financial propositions, whereby the least expensive one is selected. To encourage competition, this procedure is accompanied by mandatory advertising in specified media (official press and/or trade press). This procedure is less and less common. While this method effectively reduces the chances of corruption, it can incite suppliers to save money on production costs, which can result in a lower quality product or service in spite of specifications.
- *Formal procedure for invitation to tender:* this procedure is becoming increasingly common in most countries. It has much in common with the procedure described above, in particular as regards advertising. However, it corresponds to the logic of the best tender: after comparing proposals submitted by different companies, it is not the lowest cost solution that is chosen but rather the best one in terms of the supplier's commitments. Another difference lies in the fact that selection sessions are not public.

By definition, public markets depend on each country's or each free exchange zone's particular regulations. Decision makers are often torn between the desire to favor local players (who have an important economic role, but also a social or political one in the community) and the desire to optimize economic competition to obtain the most advantageous conditions.

In the framework of public markets, two types of players dominate:
- politicians, who are generally elected officials. They need short-term results before the next elections. They can be sensitive to qualitative arguments such as a project's potential detrimental effects for inhabitants, a risk of pollution, etc.
-civil servants and specialists whose role is to ensure the continuity of projects in light of eventual political changes. This second population is mainly sensitive to rational and quantifiable arguments.

It should be noted, that in many countries, private companies have borrowed from these procedures by launching similar consultations with the aim of rationalizing their strategic purchases.

The role of "commercial engineers" consists mainly in informing political leaders and experts about the different techniques and options available for the project in question. However, this must be preceded by building contacts and networking (Cova and Salle, 1999). The more the engineer gets involved upstream, the more access he will have to information. He will then better understand the explicit and implicit expectations of the main decision makers. On this basis, the company employing him will be able to develop a more attractive proposal, as it is closer to the needs of the decision makers.

2.6 E-procurement and the development of the marketplaces

From communication to sales

Presence on the international market ensured that aeronautics and space companies rapidly became interested in the opportunities offered by Internet for research and broadcasting information (external and in-house communication)[14]. Although a few parts manufacturers had dabbled in e-business, it was mainly the airline companies who set up real trading sites in the aerospace sector. This commercial use of Internet centered on the sale of tickets on-line via a "one to many" outlet, and became an extra distribution channel[15].

E-procurement or managing purchases with the Internet

For the aeronautics sector, Internet rapidly became a tool which facilitated the supply process. When the latter uses the Internet it is called e-procurement, in other words the on-line management of the company purchases. It consists of detecting simply, rapidly and efficiently potential suppliers, evaluating what they have to offer and then selecting the best adapted and competitive solutions. It uses a data base and a management system (ERP).

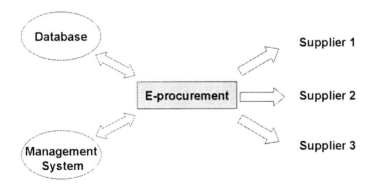

Figure 2-21. The principle of e-procurement

The main advantages for the company are:
- a reduction in costs,
- control of end-user (employee) spending and redirection of orders to approved suppliers,
- arguments for price negotiations with suppliers,

- an acceleration of the buying process (choice of supplier, negotiations, real time order validation),
- less risk of errors (fewer intermediaries).

Thanks to the suppression of the administrative tasks, the purchasing department can concentrate more on negotiation and partnership with the suppliers.

For the suppliers, e-procurement simplifies customer relations, increases the order volume and allows independent on-line management of the products or services offered. Although e-procurement allows an improvement in the efficiency and rapidity of transactions by giving greater flexibility, this new supply mode does not seem to have modified buying behavior. Professional buyers favor known suppliers who have an established brand, and those with whom the company has already worked.

Marketplaces

There has been a large increase in the number of websites in the aeronautics sector, designed specially to carry out commercial transactions between a large number of purchasers and a large number of sellers. Called " marketplaces ", these sites bring together on a virtual market place, several people with common interests linked to their activity ("many to many"):

- parts manufacturers (Original Equipment Manufacturer) and system manufacturers,
- aircraft builders,
- parts and component suppliers,
- the operators (airline companies, business aircraft, etc.),
- aeronautical maintenance and service centers,
- distributors.

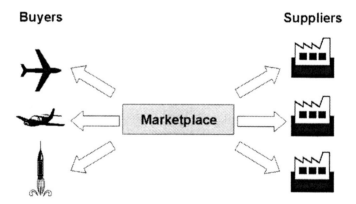

Figure 2-22. The principle of the marketplace

The general principle is simple: the buyers register on the site and put forward their bids (specification, quantity, quality, price, etc.). The site then searches for qualified suppliers, carries out any necessary follow up and puts competitors on line. Suppliers are contacted as soon as there is a bid which concerns them. Generally speaking, the bigger sites also offer other services such as supplying information about the activity sector, on job opportunities...

After the companies, the constructors and the parts manufacturers, it is the airports' turn to go on the Internet. A case in point is the airport management concern ADP (Aéroports de Paris) in association with, its British (BAA, which manages London Heathrow, Gatwick, Stansted, Edinburgh, Glasgow and Aberdeen), Danish (Copenhagen airport), American (Dallas-Fort Worth, Houston, Pittsburgh and Indianapolis airports, and Australian (Melbourne airport) homologues who have developed a common site called World Airports. The latter contains information about services available and a marketplace through which the airports can put forward their bids and group up their purchases in order to obtain better tariffs from their suppliers. The project partners represent an annual traffic of 430 million travelers (i.e. 25% of the world market). Every year they spend several billion Euros on the development and functioning of their infrastructures.

3. CASE STUDY: THE AIRCRAFT CONSTRUCTOR'S APPROACH TO THE AIRLINE'S BUYING CENTER

One of the main characteristics of the industrial purchase is that the amount of business that a company does with a particular supplier depends on the activity level of his customers situated downstream. This interdependence of players within a given sector has two important consequences:
- the first is unfavorable: a drop in a given sector's activity will affect all of the suppliers down the line. Thus, when the airline transport market slows down, aircraft constructors have fewer orders for new planes, which in turn affects parts manufacturers and suppliers downstream.
- the second is positive: there are several different levels of marketing influence possible. The direct customer can be targeted, but also his customer's customers, and so on. In other words, to better control demand, the company can widen its marketing range to successive customers situated downstream.

This multi-leveled strategy can be illustrated using the example of an aircraft constructor. A simplified diagram of the aeronautics industry highlights the major players in the sector.

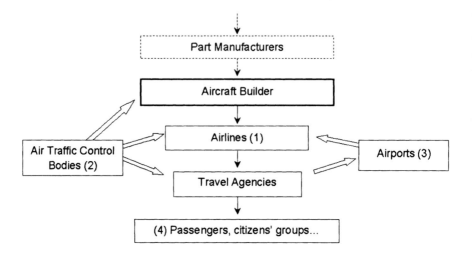

Figure 2-23. Simplified example of a supply chain in the aeronautics sector

3.1   First action level (1): the customer airline

Of course, the first step is to target the direct customer, the airline. This means identifying the buying center, but which managers need to be contacted? What types of sales arguments should be developed? Each of the managers involved in the buying decision needs to be contacted, carefully taking into consideration their specific needs and key motivations. This calls for a genuine influence strategy designed to create a positive, receptive attitude within the customer structure.

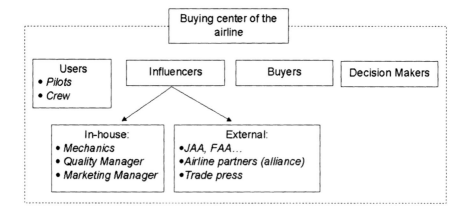

Figure 2-24. The buying center of an airline

2. Individual and Organizational Purchase

Influencers

• The marketing director

The aircraft constructor can provide country-by-country market studies on passenger and freight traffic. This requires using macro-economic information (changes in imports and exports, evolution in the number of telephone calls and their average duration...) to establish clear and simple indicators, which can help the marketing manager to choose destinations and associated services. The aircraft constructor, who has shown how important the customer company is, will gradually build trust. The results of the first study will raise the issue of modifying the network or the frequency of flights.

Other studies, for example on consumer behavior (business or tourist travelers, nationality, etc.) can further strengthen this process. The constructor assists the airline's marketing department by providing it with reasons for preference, satisfaction, or dissatisfaction of the main customer segments. Thanks to this information, the airline's marketing department can modify its network, or focus on greater on-board comfort (more seat room, individual digital screens, better customized meal service, etc.).

• The operating manager

Above all, the operating manager is concerned with operational costs. However, before approaching this subject, the constructor must check that it is feasible to operate the new equipment that he is proposing (maintenance, training, know-how).

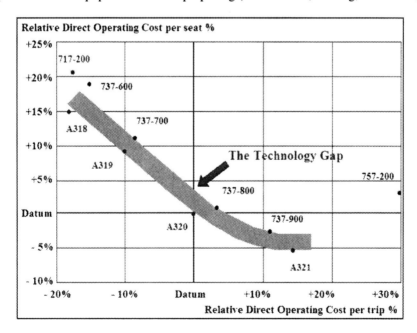

Figure 2-25. Examples of cost studies of the A320 family presented to the operational directors[16]

Compatibility with equipment, ground support, and the main airports used must also be verified: compatibility with the different fuel filling systems, loading and unloading of baggage, connection of maintenance systems in real time (spectrometry of oils…), cleaning services. The operating director is primarily concerned with productivity. Right from the start, the constructor must understand how his customer reasons, and what are the main costs analysis that he uses. Besides flight range and passenger and freight capacity, is the average length of stopover time a criteria used by the airline customer? As the relationship develops, the sales representative must understand what are the unsatisfied needs of his customer and his weak points in comparison to other airlines.

After identifying them, another appointment should be made to present comparative information from other players. The difficulty lies in preserving the confidentiality of the information gathered from other airlines to establish costs analysis and other elements of comparison. If not there is the risk of losing the confidence of the prospective customer airline. One solution is to produce average weighted figures from a selected sample of 4 or 5 competing airlines. No information is revealed about any one of the companies in particular, rather the average performance extracted from the sample, should allow the customer to improve a specific ratio. As in project marketing[17], the process is to start from a latent demand and reveal dissatisfaction. This will act as a lever to build a more efficient solution together.

- The maintenance manager

In the same way, the maintenance department must be addressed by a maintenance specialist. Analysis grids can be used to break down budgets according to the price of spare parts, the cost of storage or transport, cost of labor, and the cost of worker training. Maintenance-oriented arguments need to be developed every time that the constructor proposes an advantageous solution for the company (logistics, simplification of tasks…).

- A similar procedure must be followed with:
- the quality department (i.e. presentation of quality measures perceived by passengers),
- the logistics department (i.e. arguments concerning reliable stocks nearby, or rapid delivery times),
- training[18] should also be the focus of special attention, demonstrating its importance in the supplier's action plan.

- The sales manager of airline could be interested in the sales results of competing airlines. By the same token, only significant performance figures should be provided to show a potential gain. At first, the marketing department of the airline might not appreciate this input which could seem like meddling. But because it simplifies its own job, that is to convince in-house sales manager, this involvement is generally appreciated when it is carried out with care. The supplier can help the sales manager to increase his influence on the marketing department.

2. Individual and Organizational Purchase 55

While the marketing department will favorably influence the buying department or the general management, depending on the nature of the main decision maker.

At this level of analysis, it is clear that the influence strategy must integrate a varied number of players who can " sell from the inside ". This genuine lobby for the supplier will likely be in conflict with other managers who in turn are defending the position of a competitor. All of the managers involved in the buying process are part of the " buying center ", which is not limited to buyers but rather includes decision makers, in-house and external influencers and users. The outside influencers include the trade press, design offices, shareholders, elected officials, and civil servants. The influencers can play a role through their recommendations or opinions, the choices envisaged by the decision makers, the buyers or the users. In certain cases, aircraft constructors do not hesitate to address the shareholders of airlines through their main representatives. The financial reports with simulated results and a re-formatted fleet were presented so as to accelerate the decision-making. As far as users are concerned, besides the technicians, of course the pilots are the first concerned by the choice of the aircraft constructor.

- The pilots

These users are receptive to testimonies from other well known pilots and are particularly attentive to arguments concerning the comfort of piloting, the ergonomics of the control panel, the user-friendliness of digital screens relative to electromechanical monitor... Furthermore, the " commonality " of different models in a range make it easier to change from one aircraft to another by reducing the time required for training. Efficiency, duration and place of training are also additional factors that should be supplied to pilots. The " commonality " also makes it easier to acquire new skills and to accelerate the pilots' career development. The entire influence strategy used at this first level aims to increase the respective power of each of the customer's managers encountered by providing them with information and arguments that will strengthen their own positions, or in other words, through *empowerment*. By transmitting expertise, the supplier empowers his contacts, with the aim of making them " ambassadors through interest ".

3.2 Second action level (2): air traffic regulatory bodies

For the aircraft constructor, the second essential level of action consists in approaching the regulatory bodies to obtain the necessary certification:
- for each apparatus and model,
- for each destination and geographic situation.

This means carrying out test flights in accordance with these organizations' specifications. It is essential to plan for this step by maintaining a constant relationship with the different organizations. The aim of the constructors is to favor their technical solutions and arguments in terms of greater safety (linked to the installation of their equipment), thereby setting a new standard which will become

56   *Aerospace Marketing Management*

the norm and will be an additional constraint for future versions of aircraft proposed by the competition.

3.3 Third action level (3): airports

In particular, the airport authorities must be approached in case of an important technical change. A new range of aircraft requires validating ground resistance to different runways and often the building of access roads (bridges, roads).

The essential managers to see are those in charge of operations, investment and new projects, maintenance, quality, communication, and human resources.

3.4 Fourth action level (4): passengers and citizen groups

Figure 2-26. Airbus press ad to reinforce the TV campaign[19]

2. Individual and Organizational Purchase

Aircraft constructors have another influence channel, which until now has been little developed: the end user or the passenger. Aircraft makers have not focused enough on trying to influence the final choices of passengers, whether regarding the purchase of personal or business tickets. With this in mind, Airbus has developed a corporate campaign for selected national American TV channels. The aim is to show target populations that the A340 first and business class has better cabin comfort. In both cases, the aim is to create positive feelings among passengers with the final objective of generating preference.

Aircraft constructors mainly invest in corporate communication, in particular to establish their "citizenship" or contribution to the community. Thus in Europe Boeing explains in an ad in a daily economic newspaper that it is creating jobs in particular in the Massif Central (France) and in Bavaria (Germany). This is illustrated with employee testimonies from subcontractor companies. At the same time, Airbus communicates in the US about the amount of equipment purchased from American companies starting with General Electric engines (in partnership with Snecma), Pratt & Whitney... Furthermore, Airbus reaches out to citizens' groups such as associations of people who live in airport zones, thereby demonstrating the company's commitment to environmental protection, in particular by controlling noise levels.

Figure 2-27. The Airbus campaign in the US

Figure 2-28. Environmental Protection: Airbus commitments to the environment

In this way, the supplier seeks to create an "influence network", starting with citizens' groups who influence the airport management, who in turn influence the airline management. The airline's purchase of the aircraft result from the effectiveness of the influence strategy on the buying department and the entire buying center.

4. CASE STUDY: AIRING'S APPROACH TO THE BUYING CENTER OF THE SULTAN OF M

Contrary to what most people think, highly powerful political leaders cannot generally make complex buying decisions such as the purchase of a plane, by themselves. In many ways approaching a special private customer like the Sultan of M is very similar to dealing with an airline; however, understanding the particular cultural context is even more important.

At the Airing marketing department, people recall with amusement the failure of a 1979 negotiation with one of the Persian Gulf Emirates. At this time, commercial engineers trained in an extremely rational approach used traditional arguments to present the advantages of the Airing proposal:
- fuel savings compared to the aircraft already used by the Emirate,
- the possibility of replacing three aircraft with two Airing planes capable of

carrying approximately the same number of passengers as well as the same cargo.

Upon reflection, they realized that the two arguments were not only irrelevant but actually negative for Airing since:
- fuel is an unlimited resource for the Emirate, a major producer. In addition, it is not acceptable to evaluate the royal family's consumption.
- the fact of addressing pilots, by suggesting the suppression of a plane was immediately interpreted as a threat, for themselves but also for the technicians and everyone's extended family.

The Airing marketing department had become too used to its traditional customer airlines and had not known how to adapt its approach to the specific needs of the Sultan. The Bobus proposal was selected.

Some twenty years later, Airing marketing had learned how to cater to the special requirements of private customers with enormous buying power. This is true for the sale of an Airing to the Sultan of M. For each request from the technical department or one of the religious leaders, the Airing marketing department responds positively, taking into consideration the fact that the most important thing is to outdo competitors, without focusing on costs: if the customer is satisfied, the bill will be paid. Besides developing traditional contacts such as with the aviation director, the security manager, pilots, and the maintenance manager, Airing used a tailor-made approach to reach other very important players.

- Right from the start of the negotiation, Airing understood that the throne had already been clearly defined. In particular a very heavy chandelier had already been chosen, made from hundreds of precious stones. The first question asked by the head of the technical department was "Can the chandelier be mobile during take off and landing?" The answer was negative, the chandelier had to remain perpendicular to the ceiling at all times. Given this factor, the engineers had to design a new structure for the upper part of the plane capable of resisting the weight and integrating the required fixations.
 During the first test flight, the technical department manager noticed the noise caused by the chandelier's stones, which would be disturbing for prayer in particular during take off. It was decided to remake the chandelier, mounting the stones one by one.
- The space selected by the Sultan for his room had initially been planned for the air conditioning system. This was a "non-negotiable" point, thus the material had to be placed in the back of the aircraft, creating an imbalance in the distribution of the plane's weight. Airing marketing thus asked the technicians to add weight to the front of the aircraft (!) so as to reestablish the planned on center of gravity. This was shocking for the technical department who were usually asked just the opposite. The extra weight was obtained by replacing materials for the bathroom. The Corian® from Du Pont Nemours for the bath was "advantageously" replaced by solid marble, gold faucets, etc.
- Long discussions with the religious leader raised certain questions: "What would

happen if his Royal Highness died during the flight?" After consulting the sacred texts a solution was found so that the head of the Sultan would always be oriented towards Mecca. The bed was installed on a gyroscopic platform which using a magnetic device maintaining the bed in the desired direction.
- Falcon hunting was a favorite pastime of the Sultan. That is why the person in charge of the Royal birds was not to be neglected. Falcons are sensitive animals difficult to transport. A part of the cargo area was set up for the falcons integrating protective measures in case of temperature differences, vibrations and pressure changes. The Airing system answered the specific needs of the customer: that the falcons be in good health and operational right after landing.

This exotic but real example, emphasizes two essential points:
- there is no one all powerful person, single decision maker. There are multiple people involved in the final decision with varying degrees of influence;
- in fact, this example falls under the perspective of project marketing. Airing marketing department understood the importance of taking each of the customer organization's needs seriously, no matter how unexpected. In each case, the head of marketing empowered the person he was dealing with by providing pertinent information. When the head of Airing did not have enough know-how, he to turned to outside experts. For example, for a better knowledge of falcon behavior, university theses as well as expert advisers were consulted.

5. *PURCHASE MARKETING*

Purchase marketing, which was first developed in the seventies and applied in the late eighties, improves company performance thanks to more effective buying departments. This means going beyond the traditionally static supply role (waiting for sales visits by suppliers, getting the lowest price, etc.) and considering the buying function not from the administrative angle as a cost management center (operating budget, buyer salaries, computer and office equipment costs of the buying department, etc.) but as a profit center. Focusing outwards, like the marketing department that intervenes downstream, purchasing marketing is responsible for keeping abreast of the best the market has to offer in terms of products and methods. It aims to improve company performance by seeking out the most efficient suppliers, products and services. Without an active purchasing strategy, the company stands to miss out on other suppliers too busy working with competitors, who perhaps can offer higher performance solutions (delivery priority, favored status, etc.).

Better quality more adapted materials, components, or ingredients are very likely to improve the performance of the finished product. Equipment and new production technology can significantly lower the cost price of products manufactured by the company. Through its relationships with current and potential suppliers, the buying department must be aware of competitors' technological advances, but also developments in other sectors, that could be adapted to the company's needs. Thus the buying department has an extremely dynamic role, helping to improve quality and performance and creating value for the company. An example of this is Alcatel Space, which in rethinking the design of its satellite platforms, has applied this process to its internal organization. Guided by the new concepts of the new ISO 9001 norm, the company was able to establish a basis for rationalizing the organization, which quickly produced rational and concrete supply approaches.

5.1 The different conceptions of purchase marketing

There are different conceptions of purchase marketing (upstream marketing, reverse marketing, inverse marketing...) (Fenneteau, 1992) of which the two main types are :
- "Reverse marketing" or applying marketing techniques " backwards or opposite " to the traditional usage. Directed towards customers, downstream marketing has proven effective, hence the idea of applying it in reverse by focusing on suppliers. This involves carrying out market studies on suppliers, similar to those that are traditionally used with customers. In addition, communication campaigns are carried out at supplier conventions or reverse shows: the company presents itself to current and potential suppliers, specifying its main expectations and needs for a particular project.
- "Mirror marketing" which consists in putting the company in the position of the supplier. This second conception is closer to the marketing perspective, aiming this time to appear attractive as a customer. The aim is to sell the company's needs, making the supplier want to work together.

In reality, these two conceptions overlap, as can be seen by the action methods used, developed further on.

5.2 Purchase marketing objectives

Like traditional marketing, purchase marketing has several specific objectives :
- Supplying the necessary equipment both in terms of quantity and quality;
- Keeping track of the different upstream markets and finding out about suppliers to determine which are the most competitive and the most innovative for which products;
- Knowing how to present the company efficiently and attractively to potential suppliers, clearly explaining the company's different areas of activity;
- Specifically presenting the company's needs, in particular those that are not satisfied, formulating the expectations of the main contacts involved;
- Making suppliers want to work with the company, providing arguments on the company's buying potential, the commercial benefits to the supplier of working

with the company (prestige, leader status in the sector) and on potential shared technical concerns or values (quality, technological innovation…). The idea being to stimulate competition between different suppliers, so as to develop a balanced supplier portfolio, similar to a desirable customer portfolio;
- Making it easier for suppliers to draw up proposals: instead of holding back on technical data, provide them with maximum information, so that their proposals can be as effective as possible as far as production capacity and quantities are concerned as well as from a quality perspective: what are the necessary options ? how to personalize ?
- Providing the company with a competitive advantage, by finding a more efficient supplier, who is encouraged to make a product that is better adapted to the needs of the company and /or incited to make an effort on price to work with the company ;
- Being proactive: thanks to the exchange of information between the in-house departments and suppliers, the buying department can and must propose modifications: for example, to the production department concerning a particular material or process which would improve production performance, or to the marketing department concerning a particular presentation or aspect of the products offered which would give a higher margin or added value in comparison to the competition.

Although purchase marketing also aims to better control the balance of power with suppliers, above all it is based on marketing rationale, which tries to favor cooperative rather than conflicting practices. This outlook is more common in industrial companies, and in particular in the high tech sector, than in sectors producing everyday consumer goods.

5.3   Means of action

The buyer acts as an interface between suppliers and in-house departments (Figure 2.29). He represents the company as a customer of the suppliers and as such, he is in charge of " selling " the needs of the company to suppliers. He also represents outside suppliers vis-à-vis in-house departments, handling the " sale " of supplier proposals to different departments: production, marketing, etc.

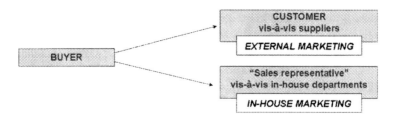

Figure 2-29. The buyer's double position

2. Individual and Organizational Purchase

The buyer has different marketing means available in relation to these two targets.

Purchase marketing and the outside target

Besides the traditional methods for gathering information on suppliers (trade press, publishing of bids, regular attendance at the main professional shows), new tools have developed such as:

Putting needs on-line: the main advantage of the Internet for the company lies in the capacity to present specific needs and on-going projects to suppliers all over the world. This can lower buying costs by favoring competition and making it easier to compare suppliers' proposals in light of specifications. Nevertheless, this tool is more suited to general purchases than to technologically sophisticated or delicate parts (i.e. space sensors, the side panels of satellites, etc.). In this case, the company is more concerned with the reliability of suppliers rather than low prices. In fact, many secret service departments have gone into Internet spying in a search for valuable information (in particular specific and sophisticated purchases). For security reasons, companies only reveal non-sensitive information when putting their needs on-line.

Organizing supplier conventions: the company organizes a day during which it presents its general policy, main orientations and objectives, in-house structure, culture, core values, as well as its commercial and financial profits according to activity. Current suppliers are invited as well as potential suppliers with whom the company is considering working. The organized reception of participants will make it easier to get to know them to work together later.

Participation in Reverse shows: in these shows, as the name suggests, the customers or contractors present their company and their technical requirements. The nature of the show attracts specialized potential suppliers, more concerned by the activity sector. At these shows, suppliers can contact different contractors and vice versa.

After having gathered information and met with potential suppliers, it is necessary to regroup data by segmenting suppliers in terms of the sought after advantages. They can be *classified* according to technical expertise, type of technology used, production capacity, delivery time, geographical area, etc. Each supplier should be considered as unique, as if he were a customer. Each has his advantages and inconveniences, thus the one that will be the most efficient in terms of the company's specific expectations should be selected. The best, most effective suppliers need to be listed, so as to be able to offer the best products. A very efficient *segmentation method* is to analyze the suppliers' marketing strategy. Thus the buying department can compare the advantages highlighted by the different suppliers and then verify their validity with the technical departments.

Purchasing managers trained in *sales techniques* are more efficient (Gauchet, 1996) and better able to react methodically to supplier's arguments, which will be easier to analyze. In addition to better understanding the representatives involved, buyers will be more able to sell the company's needs, and to incite the supplier to propose a solution that will genuinely satisfy requirements.

With this in mind, it is important to mention how suppliers are welcomed. In contrast to the bad habits that are often seen, the company must pay attention to the reception of suppliers by:
- planning appointments no later than three months ahead of time, even for a first meeting,
- by respecting the appointments made,
- by providing a comfortable room where the supplier can sit and wait, equipped with up-to-date documents on the company and possibly with a telephone, fax and Internet access.

Ways in which the company could adapt: the most effective negotiation strategy is the win-win situation, where each party aims to win, while making sure that the other, supplier or customer, is also satisfied with the outcome. This strategy starts with the assumption that every company is more successful when it is surrounded by successful customers and suppliers. This is why a buyer can offer certain advantages to a supplier with whom he wants to work in spite of disagreement about price. Among those arguments that can be made by the buying department during the negotiation, are decreasing the number of listed suppliers so as to increase the business for each (Fenneteau, 1992), providing the supplier with technical assistance (when the latter is smaller), a commitment to working with this supplier for a designated period of time (capacity to plan on the purchase volume over time), help in diffusing an innovative product or process, especially if the customer company is itself a leader in the sector.

Purchase marketing and the in-house target

Essentially, this involves favoring the communication of technical information between the different departments within the company. The buying department is one of the main players with that of production but, most of the time, the intensity of the information exchange depends on the attitude of the marketing department.

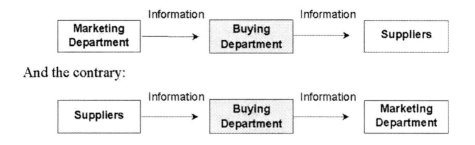

Figure 2-30. The essential role of information in purchase marketing

2. Individual and Organizational Purchase

It is important that the marketing and R&D departments inform the purchasing department about on-going projects as early as possible. In fact, in this case, the latter will be able to anticipate the information gathering and supplier pre-selection phase. Furthermore, he will be able to orient the suppliers selected so that they will be able to fine-tune their proposals and options so as to best satisfy the company's requirements.

By the same token, the buying department must communicate the information on suppliers as early as possible. In fact, the information collected will help the company to make progress. It is thus essential for the buying department to go outside of the company, visiting shows and factories so as to discover new product and process applications.

To facilitate information transfer, two rules are advisable:
- for the marketing department, the buying managers should be invited to participate in project meetings right from the start ;
- for the buying department, factory visits should be arranged with suppliers and marketing managers should be invited to attend trade shows.

Useful information can also indirectly come from competing companies in the same sector, but even more often from companies outside of the sector. If the marketing department is informed early enough, it could influence elaboration of the products offered. In general, it is important that requests from the marketing department should be satisfied with rapid solutions and the shortest possible achievement times.

Diagram 2.31 summarizes the main information exchanges between purchases, studies, production, and after-sales service at Aerospatiale-Matra (EADS): these exchanges allow the company to better position itself relative to competitors.

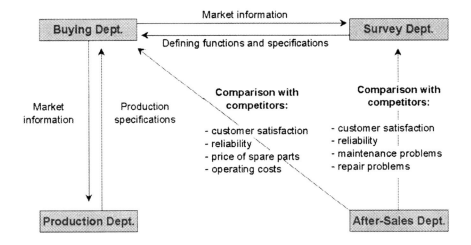

Figure 2-31. Exchange of information between the different departments[20]

PURCHASE MARKETING AT EADS

EADS, a world leader in aeronautics and space (Airbus programs, Eurocopter, Ariane...), early on developed a buying policy based on the principles of purchase marketing (in particular in the former French Aerospatiale-Matra). This involved anticipating the end of the sector's growth cycle (end of the eighties) and the resulting price wars, intensified international competition and pressure on costs and purchase. The president of the group considers that the "changes in markets and the competition incited us to set up a more responsible organization from an economic point of view, more customer oriented". The development of the competitiveness of products leads to being more demanding (Figure 2.32).

The performance and competitiveness of companies in fact depends 50% if not more on their suppliers. The reduction of cycles and stocks, the optimization of financial flux, quality improvement, cost control, are all factors of progress which involve suppliers or partners of the company, and in turn their most important contacts, the buying department.

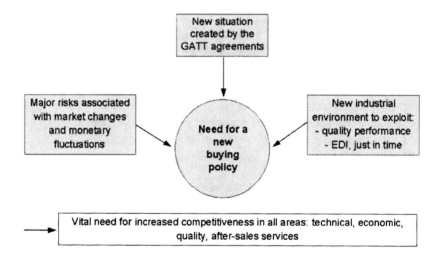

Figure 2-32. Reorganization in its context[21]

The buying policy was accompanied by a reorganization of the buying departments according to activity and project. Responsibility for price estimates (in-house and external costs), supply, sub-contracting purchase were of course maintained and assigned to the different structures. For example, one unit was in charge of buying everything for the structure of the plane, in particular the purchase of aluminum, of titanium, but also mechanical and electrical components. The aim of this new organization was to better take into consideration customer expectations in terms of competitiveness, reactivity, and flexibility. This meant getting involved in the customer's project as far upstream as possible (a

process that is increasingly crucial), while being ever-more aware of cost issues. The priority objectives were established:
- aim for minimal global costs by in-house optimization of supply,
- build durable relationships with selected high-performing suppliers,
- guarantee the competitiveness of purchases through better knowledge of the markets and competition.

EADS set up pluridisciplinary teams (design office, quality, methods...) under the responsibility of a buying manager. A genuine project leader, the latter coordinates and orients all of the exchanges and decisions involving suppliers, with a direct operational link to those who design the aircraft, prepare or pilot its production. This process-oriented approach inevitably leads to rationalizing the organization, simplifying interfaces, for the greater benefit of the supplier who is much more clear about the coherence of the information that he receives.

Obtaining lower costs starts with optimizing parts right from the design stage by integrating the suppliers' technical constraints. It means reducing the number of standards on the parts: in other words, "recuperating" parts and components from one aircraft to another, even from one range to another, so as to decrease the number of stocks. Reducing the number of suppliers means that to be more demanding, EADS must still develop the average volume of orders with its suppliers. The most dynamic among them will thus acquire different know-how. In-house optimization of supply involves ensuring that the techniques used for data exchange facilitate information transfer between the contractor and his supplier, thereby approaching just-in-time management. As far as potential synergies with partners, other sources of reciprocal satisfaction need to be established with selected suppliers.

The building of long-lasting relationships with high performance suppliers calls for dynamic qualification: EADS encourages its suppliers to invest in qualification. Regular evaluation will serve as a means of selecting future markets. Sharing the market risks means a reciprocal commitment, so as to share the burden of costs when there are strong market fluctuations or cancelled orders.

For long-term contracts, the contractor commits to a long enough period of time to enable his supplier to amortize invested material. In terms of competitiveness clauses, the supplier commits to investing on a regular basis, so as to be able to supply the best solutions. Transparence and the respect of commitments, improve the level of mutual confidence between the contractor and his suppliers.

Guaranteeing the competitiveness of purchases above all requires cooperation between the buying department and the design department. The marketing department and a better knowledge of competitors plays a role close to that of cooperation with the design department: purchase marketing and the different types of technological scanning make it possible to situate the choices of the company in relation to competitors.

The purchase marketing process was initiated to help improve purchasing performance. For EADS, purchase marketing is based on both internal and external customer needs. This requires setting up a dynamic system of market studies covering on the one hand, customer needs – the airlines – in terms of functions, and on the other, supplier proposals in terms of know-how. The process also consists in trying to exploit innovative ideas which sometimes come from other sectors.

- **The objectives of purchase marketing at EADS**

To observe as thoroughly as possible the technologies used, methods chosen, services provided, which will give the least expensive and most reliable solutions. The more pertinent the choices, the easier it will be to sell the solutions to airlines. This is why the objectives of purchase marketing at EADS are to:

- precisely define the buying act as a function of the market resources,
- study improvements to obtain minimum global costs over the complete life cycle of products sold (standardization, modernization, etc.), the purchasing price being one of the elements of global cost,
- use knowledge of the market to negotiate progress plans with suppliers or to introduce new suppliers,
- prepare for product evolutions,
- help in decision-making: make or buy, choice of technologies, products, suppliers...
- to continually be aware of what the competition is doing.

This involves collecting, structuring, storing and communicating data.

The buying department must ask itself, "in relation to what, with whom, to what purpose, and in what way will the purchasing act be carried out?". While this may seem trivial, in fact these questions are complex, for example when the issue is a Flight Management System for a fighter plane that will be used for 25 years, or solar panels for a telecommunications satellite which is supposed to have a 15 year lifetime without any possibility of repairing or replacing it. Thus the type of purchase must be very precisely defined.

And contrary to preconceived ideas, the most important step for purchase marketing takes place above all in the purchasing company. In fact, while there is always one supplier to meet a particular demand, it is vital that the purchasing company very early on eliminates the most constraining choices in terms of technology. This will reduce the number of potential suppliers to a handful. It is best as much as possible to opt for standard, less expensive, and more accessible solutions. It should be remembered that the purchasing price is often fixed 80% right from the design stage (technical specifications). The aim is to avoid going over budget due to modifications. John Lehee, an Airbus manager said, "Ask yourself, when you suggest a modification, what will it bring to the passenger in seat 41", a symbolic way of saying that customer satisfaction does not necessarily lie in technical perfectionism.

However, technical considerations often win out over economic ones owing to the extremely severe technological constraints. One can think of the launching conditions of a satellite. This explains the extreme importance of going to the source to find solutions that will generate gains, and looking to standardize them. As delivery deadlines are increasingly tight, it is often too late at the launching of a program to look for innovative solutions, which by their very nature are riskier and more time consuming.

Consequently it is important to look with the supplier for possible ways of improving the product and then freeze them and write the characteristics into the product flow chart. This is a radical change in perspective favoring the bottom-up approach, constructing a complete

2. Individual and Organizational Purchase 69

system as opposed to the traditional top-down approach which breaks down the whole of the upper level and then defines a cascade of interface and performance specifications, down to the basic model.

The act of buying is not only concerned with a product but with the technical performance and know how on which related products or a complete system depends. A failure in orbit can rapidly ruin a solidly established reputation and cause a long term decrease in market share. Similarly, the systematic replacement of a part or piece of equipment on a whole fleet of operating aircraft can entail an outlay of several tens or hundreds of millions of dollars. The act of buying is very closely associated with the concept of risk: technological risk (meeting performance and reliability objectives), industrial risk (human resources and means of production), financial risk directly or indirectly linked to supplier performance. The purchasing department plays an essential part in controlling this risk by an in-depth evaluation of the real capacities of the companies upstream (financial solidity, product/market strategy, shareholder structure, etc.) and also by the rigor of the contractual agreement on signature and the strict control of these engagements right up to the end.

Once the limits of face to face negotiations have been reached (adjustment of margins and provisions), there remains a whole gamut of cost reductions that the supplier can not make by himself. This is why EADS made the GREENLOOP system which is an Intranet connecting up hundreds of suppliers to the production center and needs assessment system, often going to the point of linking individual production management systems with those of EADS. The company has in this way made considerable savings which have not encroached on the profit margins of the GREENLOOP partners, but on the contrary have improved them!

Companies are not robots which react in a mechanical way to various requests. The quality of relations between individuals whether at board level or engineers in daily contact during the developmental stages or manufacture, often play a very important role. A deeply conflicting situation on the edge of breakdown or court action can sometimes be unraveled by bringing in new people with no preconceived ideas who, because they want to get things moving, start with a sound base and then reconstruct a badly weakened relationship often dating back many years. The buying manager must know how to mediate in order to get an agreement. The idea of purchase marketing is at the heart of this sort of process which aims to replace balance of power, threats or retaliation by a search for in depth understanding of reciprocal expectations, an unfettered analysis of the respective difficulties and drawbacks, with the objective of finding a balance.

Company industrial partnerships are without doubt the most successful form of the buying act, but also the most complex and delicate to set up and manage. They obey some simple rules:
- the partner companies must have complementary strategies and skills,
- it must be a long term relationship,
- the overall profit from the partnership must exceed that which could have been obtained individually,
- under no circumstances must the partners lose sight individually of competitiveness requirements.

The buying department is the company's fighting force on the downstream markets, but it must know who to defend and to what aim. Buying is effective when we can see clearly what (what product, performance, life expectancy ,etc.) and for whom. The Airbus system was constructed over 30 years on an admittedly solid and complex industrial system, but one in which each participant had their role. The Germans from DASA would not think of manufacturing the wings produced by their British BAe partners. Similarly it would not cross Aerospatiale-Matra's mind to make engines.

A company has become mature once it has made a clear choice between "Make or Buy". Knowing how to control what has been delegated to others because they will do it just as well, or better and thus freeing human and financial resources for other activities closer to the company's main skills. This strategic notion of Make or Buy is essential for the construction of an efficient purchasing policy. Purchase marketing is there to feed it with essential information concerning supplier market resources. Conversely, once the main lines have been drawn, purchase marketing can offer coherent buying strategies.

Market surveillance

Market surveillance is an important component of purchase marketing. EADS makes the distinction between on-going general market surveillance and dynamic surveillance, targeted by a precise need[22]. Basically this entails studying all the company's customers in order to define the types of needs and, in the event, the means to be put into place. It is thus a question of collecting information in order to know the market and the competition better. Subsequent analysis of this information makes for a more effective purchase marketing action plan.

The diagram 2-33. shows the three main axes given to EADS purchase marketing.

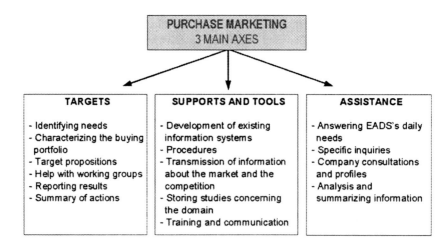

Figure 2-33. The main axes of purchase marketing[23]

EADS's purchase marketing field of investigation, and thus its market surveillance system is worldwide. It is a question of being informed about the opportunities on offer from the different world suppliers and taking the initiative. In addition, market surveillance can sometimes concern the other activity sectors, e.g. automotive.

EADS offers an interesting example of the application of different market surveillance systems: technological, competitive, commercial and environmental[24].

An electrical sub-system equipment example

In the electrical sub-system equipment domain and especially for connectors, the example of benchmarking applied to an earth module can be cited. This module is designed to connect aeronautical equipment. By taking solutions found in the automobile sector, the design of this module has been reviewed and improved. This optimization has been made possible by linking the skills from the design office and the purchase department.

As can be seen from the diagram in 2.34, the first result of this optimization has been the considerable reduction in the number of parts used in the module.

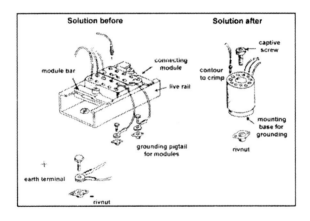

Figure 2-34. Descriptive and simplified diagram of an earth module[25]

But in addition to reducing the number of parts, the product performance is improved. The module offers more connection possibilities while at the same time taking up less space and being lighter. It gives a reduction in fitting time and easier access and maintenance.

In this way this new solution allows an economic improvement, particularly in terms of initial price and cost of setting up. It has meant getting a product with improved performance which, developed for the ATR program, has been extended to include the Airbus program where it is original equipment starting with the A320.

NOTES

1. Adapted from Kotler, Ph., Armstrong, G., Saunders, J and Wong, V., (1997), *Principles of Marketing*, 2nd European Edition, Prentice-Hall.
2. See Chapter 4, Market Segmentation and Positioning.
3. See Chapter 4, Market Segmentation and Positioning.
4. See Chapter 4, Market Segmentation and Positioning.
5. Adapted from Filser, M., (1994), *op. cit.*
6. See Chapter 3, Business Marketing Intelligence
7. Adapted from Filser, M., (1994), *op. cit.*
8. See further, 5. Purchase Marketing.
9. See 3. Case Study: Approaching the Airline's Buying Center.
10. See in particular Robinson, P. *and al.*, (1967), *op. cit.* and Sheth, J.N., (1973), A Model of Industrial Buyer Behavior, *Journal of Marketing*, 37, p 50-56, October.
11. See further, 2.5. Bidding.
12. Source: Airbus
13. See 2.5. Bidding.
14. See Chapter 12, Selecting Media.
15. See Chapter 9, Selecting Distribution Channels and Sales Team Management.
16. Adapted from Airbus documentation
17. See Chapter 10, Project Marketing.
18. See Chapter 14, Building Loyalty: Maintenance, Training and Compensations.
19. Extract from *Flight International*, (2000), 31 October – 6 November, p 8.
20. Source: Purchase & Logistics Dept. EADS.
21. Source: Purchase & Logistics Dept. EADS.
22. See Chapter 3, Business Marketing Intelligence.
23. Source: Purchase & Logistics Dept. EADS.
24. See Chapter 3, Business Marketing Intelligence.
25. Source: Purchase & Logistics Dept. EADS.

REFERENCES

Bénaroya, Ch., (1997), Étude comparée de l'efficacité perçue des outils du marketing achat, *Mémoire de DEA*, Université Toulouse 1.

Choffray, J.-M., (1979), Perception of risk in the industrial purchase, *Revue Française de Gestion*, 22, p 24-30, September ; Salle, R. and Sylvestre, H., (1992), *Vendre à l'industrie*, Paris, Éditions Liaison.

Cova, B. and Salle, R., (1999), *Le marketing d'affaires*, Paris, Dunod.

Day, R.L and Landon, E.L., Toward a Theory of Consumer Complaining Behavior, in Woodside, A.G., Sheth, J.N. and Bennett, P.D., (1977), *Consumer and Industrial Buying Behavior*, New York, Elsevier.

Fenneteau, H. (1992), Les caractéristiques de l'acte d'achat et la logique du marketing amont, *Recherche et Applications en Marketing*, Paris, PUF, vol. n°7, n°3.

Filser, M., (1994), *Consumer Behaviour*, Paris, Précis Dalloz.

Gauchet, Y., (1996), *Achat industriel et marketing*, Paris, Publi-Union.

Howard, J. A. and Sheth, J. N., (1969), *The Theory of Buyer Behaviour*, New York, Wiley.

Malaval, Ph., (1999), *L'Essentiel du marketing business to business*, Paris, Publi-Union.

Malaval, Ph., (2001a), *Marketing business to business*, Paris, 2nd Ed., Pearson Education.

Malaval, Ph., (2001b), *Strategy and Management of Industrial Brands*, Boston, Kluwer Academic Publishers.

Maslow, A.H., (1970), *Motivation and Personality*, 2nd Ed., Harper & Row.

Reeder, R.R., Brierty, E.G. and Reeder, B.H., (1991), *Industrial Marketing*, Englewood Cliffs, N.J., Prentice-Hall.

Robinson, P., Faris, C.W. and Wind, Y., (1967), *Industrial Buying and Creative Marketing*, Allyn and Bacon.

Webster, F.E. and Wind, Y., (1972), *Organizational Buying Behavior*, Englewood Cliffs, N.J., Prentice-Hall.

Chapter 3

BUSINESS MARKETING INTELLIGENCE

Marketing studies are used to measure satisfaction, evaluate potential, discover new openings, verify the segmentation method adopted, reinforce sales arguments, etc. For all of these, useful information needs to be collected and the aim of market surveillance is to enrich the information system (IS). In the light of the strategic importance of marketing surveillance and studies, it is essential to set out the objectives and methods as well as the main techniques needed to acquire, protect and exploit pertinent information. However it is not enough to just have a department for market studies. The main thing is to set up an action plan to best exploit the results.

1. THE INFORMATION SYSTEM

An organization's information system comprises the sum of human, material and method-based means concerned with dealing with the different forms of information encountered. The company's main objective is to gain the required knowledge about another organization in order to make decision making and subsequent implementation more effective. It must therefore gather, process and reconstruct the information (Malaval, 2001), in other words:
- produce information,
- improve the functional coherence of the organization,
- and above all help in decision making.

First, the type and source of information required must be determined and then this must be collected, communicated internally and cross-checked. Finally, before being exploited, the data must be given a rank order and updated (Figure 3.1). Thus the search for and treatment of quality information must be directed towards gaining predefined results, such as cutting costs, attacking a new market segment or launching an innovative, new product.

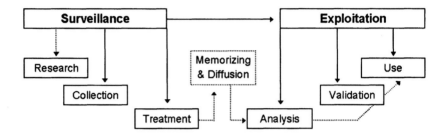

Figure 3-1. The information management process[1]

Information only really exists when it is communicated and exploited, but it often passes through informal channels. Certain information is strategic and possession is often synonymous with risk anticipation, reduction of uncertainty, competitiveness and power. This is the reason why companies try to maximize both the quality and quantity of their information system. To this end they use two different but complementary approaches: surveillance and studies. In fact there are few actual marketing study departments within companies of the aeronautics and space sector. There are two main reasons for this:

• The presence of large technical studies departments which mainly take care of technical questions and to a lesser extent economic ones;

• A reduced customer base allowing commercial staff to directly carry out studies on customers.

However, in the case of a new project requiring a one-off study, this is often farmed out to an external firm.

DASSAULT AVIATION: INCREASING FALCON CUSTOMER SATISFACTION BY USING AN EFFICIENT INFORMATION SYSTEM[2]

The Dassault Aviation group, well known for its defense division, is also present in the business aircraft sector where its Falcon division has 45% of the market competing with manufacturers such as the Canadian, Bombardier or the American, Gulfstream. Dassault Aviation has delivered 1500 Falcon type airplanes to almost 800 customers worldwide (almost 70 countries) with the majority in the United States (62%). Falcon customers are very demanding: among the main members of the buying center, apart from airline passengers (business men, personalities, etc.), there are financial directors, technical and sales teams,

3. Business Marketing Intelligence

heads of operations, etc. It is essential to set up a personalized system with appropriate contacts between the sales team and the different members of the buying center. This means an offer which is as personalized as possible and takes into account customer imperatives. Thus hand in hand with the sales force there is a customer service department and another in charge of the "customization" of the product.

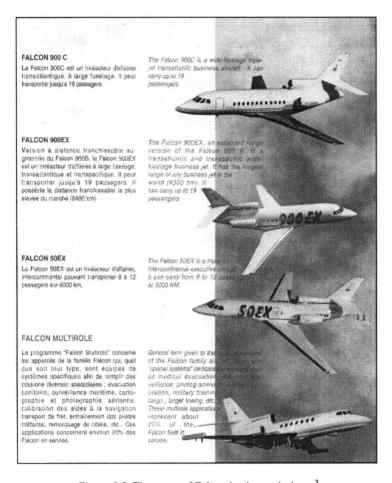

Figure 3-2. The range of Falcon business airplanes[3]

Dassault, which intends to maintain its position as leader, is faced with competitors who have greater customer satisfaction for service quality. Consequently, a Customer Relationship Management unit (CRM) has been set up, with the aim of increasing customer satisfaction by establishing a more efficient information system. The group has a technically reliable product but cannot give entire satisfaction to its customers unless this is accompanied by total service quality (spare parts, instructions, adapted documents, qualified repair personnel, etc.). The Falcon customer service director puts it another way: "customers don't forget interrupted flights, regardless of the reasons". Therefore there must be an overall increase in the efficiency and reactivity of the units in customer contact (repairs, training, documentary services, maintenance centers, crew, etc.) who, when there is a problem, have a tendency to pass the buck. The establishment of a worldwide information service has provided an answer to one of the main difficulties, which is free circulation of disparate information within the company (files, reports of visits, letters, complaint faxes, etc.) coming from in-house (colleagues, salespersons, etc.) and external (customers) sources. The information can thus be retransmitted internally, dealt with and then redirected to the customers. In particular, this allows a more accurate view of the main problems affecting each customer and the creation of more pertinent solutions.

2. MARKET SURVEILLANCE: ACTIVE LISTENING

The specific aim of market surveillance as an information system tool, is to gather information and reinject it into the strategic context of the company with a view to favoring decision making. It consists of observing and analyzing the scientific, technical, technological, legal and regulatory environment as well as the present and future economic impact, in order to determine the threats and opportunities for development. It gathers information for the following main areas:
- The products, to the extent that the information leads to the use of new materials and the creation of new ranges;
- The procedures, meaning the setting up or re-setting up of industrial automation;
- The commercial organization, where the information can allow new distribution channels to be established;
- The management, e.g. implementing the "just in time" method.

Depending on the importance given by the company to market surveillance, the latter may be thought of as a simple, passive and normal activity, a useful and participatory activity or a strategic activity in its own right (intelligence or active market surveillance), always assuming that specific means and objectives have been defined. Because of the diversity of observable domains, four types of industrial scanning have been developed, each with the aim of analyzing and diffusing information: competitor surveillance, commercial surveillance, environmental surveillance and technological surveillance.

2.1 The different types of surveillance

It is important to remember that surveillance is also a key component of purchase marketing[4], in terms of detecting potential new suppliers, new materials, etc.

3. Business Marketing Intelligence

Competitor surveillance: This focuses on the newcomers on a given market, the strengths and weaknesses of research and development, manufacture, products, costs, prices, distribution, communication, organization, financial capacity intensity of competition: number and size of competitors, market growth, products, costs and prices, production capacity, distribution (networks, cover, etc.), supply (ordering systems, etc.), sales force (size, organization, motivation, training, results, etc.), patents and research investment, financial capacity, entry and exit barriers, etc.

Commercial surveillance: This is concerned with the suppliers and customers present on the market. It analyzes growth rates, customer needs, products or services offered; it monitors the evolution in costs of materials, the introduction onto the market of new skills, the development of relations between customers and suppliers and customer solvency. In industries working predominantly with bids for tender and on projects[5], commercial scanning is also observation of "project opportunities". Here, it is a case of detecting commercial opportunities a long way upstream in order to anticipate customer needs and create a favorable environment for when the actual bidding process is launched.

Environmental surveillance: This concerns factors other than those directly linked to the company's particular speciality: economic, political, sociocultural and geopolitical, regulations, legal and legislative aspects, mostly in terms of evolution and trends, etc.

Technological surveillance: This targets technological information coming from fundamental and applied research. It involves searching for details concerning products, services, substitution processes and technologies linked to the company's particular area.

Design and construction of...	Perimeter product
Launchers	Satellite and capsule launch
Satellites	Telecommunications, data transmission, earth observation
	Military
Civil airplanes	Medium long haul (more than 170 seats)
	Short haul (90 to 170 seats)
	Regional transport (50 to 90 seats)
	Business jets (less than 50 places)
Military airplanes	Fighters
Helicopters	Civil – all categories
	Military – all categories

Table 3-3. Main sectors to be analyzed in the aeronautics and space sector[6]

The necessity to innovate forces the company to stay abreast of any information to do with its activities and those related to them (Table 3.3): to be able to innovate means knowing what the company does itself and what others are doing.

Surveillance allows technological breakthroughs to be surmounted and market evolutions, anticipated: strategic decisions can therefore be taken in order to ensure the company's future competitiveness. In addition, it is pointless to spend money on costly research if innovative solutions have already been found (detection of ideas, processes, know how, patent applications, cooperation agreements, etc.), or proven to be a dead end in the experience of certain competitors.

2.2 Setting up surveillance

Surveillance implies observing a large number of parameters. For it to be effective, the framework within which information is sought, relative to the company strategy must be very precisely defined. This active surveillance requires the organization of coordinated information collection and its piecing together, in order to obtain something reliable and useful.

Figure 3-4. Surveillance, a process leading to action

The first phase consists of defining the information needs otherwise the essential data may be lost in the crowd. Information needs are usually defined according to three main stages:
  • Defining the objectives sought: this is carried out as a function of the factors critical for success, the company strategy, its short and medium term objectives, its strengths and weaknesses, opportunities and threats and by analyzing the activity sectors, etc.
  • Drawing up a research plan consists of organizing and allocating objectives around a few major directions: necessary questions, people interested by the information, response precision required, information expire date, etc.

3. Business Marketing Intelligence

• Defining indicators: is the selection of a group of observable facts which could constitute information, if only partial, on the axis under consideration.

> • Adapt the company to its environment
> • Innovate
> • Increase knowledge
> • Favor cooperation with other partner
> • Use information as a weapon
> • Manage interpersonal relations
> • Protect the company and information
> • Increase profitability

Table 3-5. Surveillance's main objectives

After this first essential phase comes the definition of budgets and resources available (materials, personnel, etc.). The selection of the most appropriate and effective research tools is carried out according to the objectives and means already defined. The chosen sources and information must first be evaluated then sorted, put into a hierarchy, cross referenced and classified. The processing will in the main consist of:
• Ordering the references,
• Statistical analysis, e.g. of patents,
• Putting the data onto a spreadsheet in order to be able to set up a data base.

This treatment of information phase is essential. Information research is like a jigsaw puzzle: the main information is often difficult to obtain but it can be pieced together by cross referencing scraps picked up here and there. Even if the jigsaw is confidential, the pieces are usually not. For example analyzing the price, the distribution network and the communication policy of competitors, the company can deduce their strategy.

Figure 3-6. General diagram showing information management

The information once processed, structured, analyzed, summarized and validated must be passed on. Good information is that which arrives at the right moment, in the right ears which can then exploit it. The information collected and processed will permanently enrich the company's information system and, in addition, will allow corrective or complementary action to be taken on decisions pending. High quality information is not enough, it must be used.

The use of the information could allow an increase in the number of projects and patents filed. Above all it favors an exchange of information in particular between the marketing department, and research-development and production, decreasing the time needed to set up new products while at the same time reinforcing company cohesion.

For effective use, a scanning system must be backed up by an organization which can classify and store relevant information. This is why Aerospatiale-Matra have developed the Docland Documentation Center.

DOCLAND: AEROSPATIALE-MATRA'S INFORMATION AND DOCUMENTATION CENTER (EADS) [7]

- A traditional library, documentary capital in the field with more than 30 000 documents (books, congress minutes and extracts, complete collections of magazines and articles, reports, etc.) to answer the immediate, basic reference needs.
- A management service of more than 2000 subscriptions and systematic detailed study of about ten main aeronautical titles.
- To go further, a documentation service run by a team of specialists who make more than 1400 information searches per year mainly on external data bases, and who put together and circulate more than 100 profiles to some 300 subscribers.

3. INFORMATION SOURCES

In-house information

A large amount of information is available within the company itself. Internal audits, studies and measurements of the different company functions, the detection of malfunctioning, the identification of formal and informal human relation networks, are all operations that the company can run internally in order to collect and circulate information. An information service which is efficient must be just that from an internal point of view: surveillance, before setting its sights on the exterior, must be applied to the company itself. With a market study for example, it is in the company's interest to exploit work carried out by the different departments (customer files from sales, activity summaries, financial department data base, information collected by the production department, R&D information on products and technologies, etc.).

Desk research

Secondary sources are existing documents (publication, data banks, professional organizations, etc.). Looking for secondary sources is less costly than primary sources. It

3. Business Marketing Intelligence

is essential to identify these sources and evaluate their validity and relevance[8] because this affects the value of the information collected (Figure 3.7).

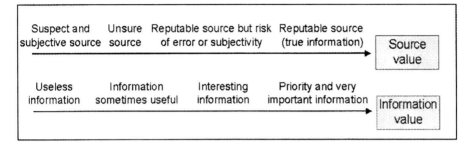

Figure 3-7. Scale of values for sources and information

The information source can be evaluated as a function of:
- richness: a confidential letter is richer than a magazine;
- performance over time: some information comes in real time, another is obsolete; some has only a fleeting value, another is more long term;
- reliability: an exhibition stand is not obviously representative...;
- capacity to provide information: a source is more or less independent, competent...;
- discretion: a source which can supply information without alerting competitors is highly valuable.

Field research

Other information can be found hidden within the company. These are primary sources made up of all the information collected in the field directly from contacts including users, influencers, buyers, distributors, etc. Ad hoc and permanent studies like panels also come under this primary source heading.

3.1 Main information sources

Open sources

Open sources make up about 70% of all information, and cover for example:
- The press; scientific and technical magazines, and specialist magazines: Air & Cosmos, Avionics Magazine, Aviation Week & Space Technologies, Business & Commercial Aviation, Defense Daily, Defense News, Flight International, Airline Business, Jane's Defence Weekly, Revue Aerospatiale, Air & Space Europe, Space News, Defense Briefing, Aviation Security International, Air Cargo World, Air Cargo News. A cheap source, accessible to everyone, it is by definition non confidential. The information appears after a certain time lapse;

- Data banks: they provide information on all subjects spanning a wide geographical range and time period. There are several thousand in the United States (Dun & Bradstreet, CompuServe, Dialog, Lexis-Nexis, Investex, etc.),in Europe and Asia, mainly scientific and technical, but also legal, sales and marketing, economic and social. Connection is quite cheap but their use is expensive. Updating is often long, and accessing is complex and should be handled by specialists;
- Patents: filing and publication give precious information on innovations available and the strategies of competitors. However the information contained is 70 or 80% technical and thus difficult to analyze without the help of a specialist;
- Organizations: Statistical Abstract, Industrial Outlook, Business Statistics, Census of Manufacturers, Standard & Poor's Industry Surveys; The ICAO (International Civil Aviation Organization, United Nations agency) in particular publishes estimations for total civil airline traffic (expressed as number of passenger-kilometers) for regular airline flights: 2791 billion passenger-kilometers (1999). 2956 billion passenger-kilometers (2000), almost 3118 billion passenger-kilometers (2001 forecast);
- The associations: Bundesverband der Deutschen Luft- und Raumfahrtindustrie (www.bdli.de); Aerospace Industries Association of Canada (www.aiac.ca); Aerospace Industries Association (www.aia-aerospace.org); European Association of Aerospace Industries (www.aecma.org); The Society of British Aerospace Companies (www.sbac.co.uk); Groupement des Industries Françaises Aéronautiques et Spatiales (www.gifas.fr); Eurospace (www.eurospace.org); Airplanes Electronics Association (www.aea.net); International Air Cargo Association (www.tiaca.org); Pilot Association (www.pilote.org);
- Newsletters circulated by post or via the Internet such as Orbital Report News Agency (www.orbireport.com); AeroWorldNet (www.aeroworldnet.com); USA space (www.france-science.org/usa-espace); Cargo Facts (www.cargofacts.com).

There are also sources like:
- Market study bureau producing multi-customer or tailor-made studies: Forecast International / DMS, Frost & Sullivan, Airfax / Merlin Associates (www.airtrading.com);
- Books: these have strong added value because they carry structured information for long-term use;
- The different media and supports: articles, TV reports, public relations films, congresses, trade shows[9], exhibitions;
- Directories such as World Aviation Directory; AWST: Aerospace Source Book; Jane's ABC Directory;
- Thousands of Internet sites; Airforce Technology (www.airforce-technology.com); Aerospace Online (www.aerospaceonline.com); Space Business (www.spacebusiness.com); Air Transport Intelligence (www.rati.com);

3. Business Marketing Intelligence

- Search engines: X-CD Business Technologies (www.x-cd.com); Launchspace Directory (www.launchspace.com); Space Business Archives (www.spacearchive.org);
- Commercial tribunals are also potential sources.

Most of these sources are usually well informed, rich and easily accessible. And because of this, the information supplied has less value than that gathered from closed sources.

Closed sources

Accounting for about 20% of information, closed sources are also called "informal", they are legally accessible and are thought of as closed because they are more difficult formalize. They include:
- Visits, meetings, missions and study trips;
- Relationships with customers;
- Company networks;
- Competitors; sales materials, exhibitions and shows, open days, product information, customer/supplier collaboration, in-house newspapers;
- Suppliers and buyers;
- Colloquia and congresses: providing specialized information, they are an opportunity to pick up market trends, develop contacts and exchange information;
- Research theses, prospective job candidates;
- Sales staff, after sales service, etc.

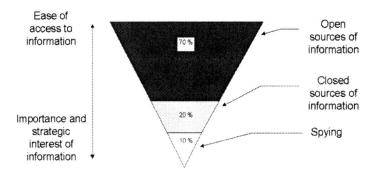

Figure 3-8. Value of sources and information

In general the more formalized the information, the quicker it becomes dated and the less interest it has. The value of sources and information increases with their reliability, confidentiality and freshness.

Other information methods

Among the methods used by companies to collect information, benchmarking, which comes from the surveyor's word for a fixed point, is very important[10]. This consists of close and careful surveillance of one particular participant within the environment: perhaps a supplier, a customer, a potential new competitor, etc. (Figure 3.9).

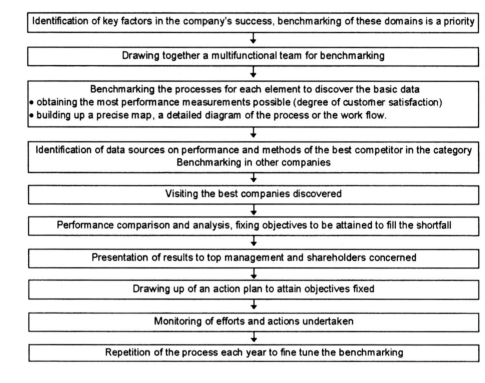

Figure 3-9. The process of benchmarking

The analysis is carried out department by department. It is possible to compare companies from various industrial sectors, as well as their products, services, processes and organization. The analysis of the product and its performance (benchmark) as well as its dismantling (reverse engineering), provide a clearer understanding of the technology used, production methods and the structure of the corresponding costs. In this way, the company has operational, already tested "excellence models" towards which it can aim. The information collected contributes to the internal functioning of the company, thanks to the application and adaptation of these successful models.

Information is expensive, which is why certain companies try to group together to obtain it. Benchmarking is often carried out within a group of about ten companies who accept to pool their information in order to get systematic, ratio by ratio comparison. Those who get a significantly higher score accept to demonstrate how it was achieved. Another way is to sub-contract to a corporate investigator, who are like intelligence strategy private detectives or counselors. The latter rely on a bulging address book and networks of highly specialized contacts to obtain " gray " or " semi open " information. This may be a discrete long-term process perhaps including infiltration, or very rapid, more akin to a commando operation.

For the most part, information can be obtained legally but there is a fine dividing line: to be sitting by chance in a restaurant next to a couple of competitors is legal, but to ask to be placed next to them is spying. About 10% of information is estimated to be collected illegally, using numerous methods derived from state intelligence agencies (hacking, breaking in, taps, indiscreet photos, false visiting cards, false identities, garbage sorting at the headquarters, going through suitcases in hotel bedrooms, etc.). Spying however, does not stop at the technological level, but also covers the sales, financial and strategic activities of the company.

3.2 Information protection

Whether they have been victims of spying or are merely conscious of the potential dangers, more and more companies are concerned by information management. Considering the enormous strategic importance of certain information, many companies have set up information structures and protective measures. Protection is a necessity unless one wants to see competitors beating the launch date with an identical product. Even if complete protection is illusory, the company can still safeguard itself against intelligence gathering:

• Prevention of pirating using for example the services of a security company or a sophisticated alarm network;

• Internal security with the personnel signing confidentiality of information and competition clauses;

• Misinformation or communication of false or misleading information and its circulation to the exterior (filing of patents designed to lead competitors down the wrong path, " doctored " data given in press interviews, etc.);

• Legally, the company is allowed to safeguard itself against counterfeits of brands, products, processes, drawings, models, logos, software, etc. Filing a patent as well as registering with a public notary allows the company to justifiably take legal action against any such miscreants once identified. Taking out a patent on an invention enables indeed to exploit it exclusively and legally (legal protection against counterfeiting). However, these legal guarantees cannot prevent wholesale plundering, but they do allow it to be identified and tracked down and above all they authenticate the original inventors.

Although it is necessary to protect strategic information, it is also important not to refuse all broadcasting of information to the exterior. Information is enriched and lives when it is exchanged. For example, the granting of licenses and the transfer of technology from one company to another allows the latter to innovate, gaining access to new markets with less risk. When a company grants a manufacturing license it passes on its techniques, but increases its profitability thanks to the wider use of its technology. New production units are made with minimal investment and risk, and export of goods from the parent company to the license holder's territory are developed.

Similarly, the joint venture[11], is a common company created by partners with the aim of combining skills and exploiting information together. The alliance or the partnership lets companies cooperate but the links are only as strong as the participation agreements. They keep a certain independence while at the same time pooling their experience and enriching the technological hard core.

MARKET STUDIES MADE BY BOEING AND AIRBUS

Airplane manufacturers like Boeing with its Current Market Outlook (CMO) as early as 1970, or Airbus more recently with its Global Market Forecast (GMF), carry out in-depth studies in order to anticipate the market demand, and so evaluate what types of airplanes will be necessary and for which destinations. These studies predict the activity growth rate for civil airplane transport for the next 20 years, information which interests not only the airlines but also the suppliers (parts manufacturers, engine manufacturers) and the service companies in the sector. Each constructor offers an analysis based on its market vision and own values. Therefore the two studies cannot really be compared, with each offering particular results based on specific criteria. Airplane segmentation for example, can be made according to the number of seats or the all up weight. Similarly, the studies are presented by large geographical zones or more specific regions, etc. However, because they are dealing with the same market, they are extremely useful for the people involved in the sector to know the main trends for the future. The overall value of the market is globally convergent for both, around $1400 billion over 20 years. On the other hand the share out of different airplanes varies: Boeing predicts a strong growing regional airplane market while Airbus sees growth in the wide body jet segment. Obviously the manufacturers then develop their offer as a function of their predictions: Boeing aiming more towards regional airplanes and Airbus adding a wide body jet to its product portfolio.

- **The Airbus Global Market Forecast 2000-2019**

Airbus estimations are made in terms of number of passengers and passenger/cargo transport fleets. All the information gathered and calculations from numerous economic, statistical and technical elements, are put together in a study of the global market, published and available on the Airbus website, and called the " Global Market Forecast ". This is not a top-down type macroeconomic analysis but a precise and detailed microeconomic study. The following have been covered for the period 2000-2019, (plus an intermediate 2009 date):

- world airline transport demand, passenger/kilometer development for passenger traffic, ton/kilometer development for cargo traffic;
- demand in terms of capacity, expected number of passengers and volume of freight;
- number of airplanes in service, the evolution in terms of use per airplane type;
- market share in number of seats per country and large zone;

3. Business Marketing Intelligence

- type of cargo transport expected (specialization, etc.);
- saturation level of the air traffic control systems;
- aircraft capacity in terms of passengers and freight;
- forecast for the present fleet (airplanes retired, replaced, converted).

The study covers 228 airline companies and their 49 subsidiaries, representing 98% of the world fleet as well as 187 cargo transporters. Nevertheless, the data from the Community of Independent States (CIS: Russia, Ukraine, etc.) is not included due to its unreliability. The report looks at the growth rate in passenger transport, increase in productivity and the increase in the flight frequency/size of airplane ratio, over a total of 10,013 flight sectors connecting 1,896 airports in 82 distinct international markets. The study shows (Figure 3.10) that 4,601 airplanes should be retired from the market between 2000 and 2019, while 15,364 should be purchased. Apart from this, 3,174 civil airplanes will be maintained in service using retrofits and 2,389 others will be converted into cargo carriers.

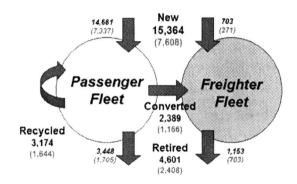

Figure 3-10. An estimate of the evolution of the world airplane fleet from 2000 to 2019

The market estimations in value by size of airplane show (Figure 3.11) that the 300 to 400 seat category of airplanes at $319.3 million is just as important as the 100 to 175 seat category ($327.4 million).

Among the results, Airbus gives a demand forecast for general air traffic (5% annual growth) and then a main market forecast in terms of revenue per passenger/kilometers (RPK). The first ten markets represent a total of 68% of the revenue for 2018: United States (16.8%) Europe-United States (13.3%), Asia-United States (8.8%), Inter-Europe (7.6%), Europe-Asia (6.5%), China (3.9%), Europe (2.9%), Africa-Europe (2.4%), China-Asia (2.3%), Inter-Asia (2.2%). The highest growth rates are those for the Chinese market (8.4%), Europe-Asia (>6.5%), Asia-United States (6%) and Pacific-Asia (6%), etc.

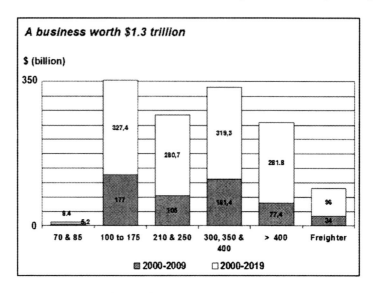

Figure 3-11. Estimation of airplane sales by size

As the American domestic market is fully mature, its rate of growth (2.6%) will be less than the Chinese domestic market (8.1%). The growth of the different markets will reduce the relative weighting of the American market from its present 26% to about 17% in 2018 (Figure 3.12). Air traffic will still be concentrated, with the 9 largest markets generating two thirds of the RPK.

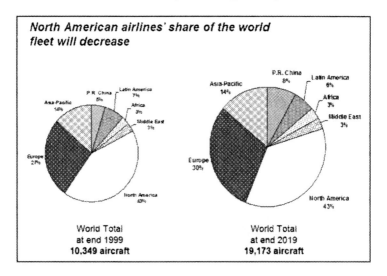

Figure 3-12. Fleet evolution

Another forecast made in the GMF concerns the growth in airline company seating capacity, and this varies by region. The result will be an average growth in the number of seats per plane and in the number of airplanes. The renewal of airplane fleets is then analyzed. Companies prefer, mainly for economic and marketing reasons, to replace the passenger airplanes in their fleet before they reach the end of their economic life. The old airplanes are then often converted for cargo transport or continue taking passengers with another airline company. Airline airplanes are replaced on average after 24 years, but disparities exist according to the zone studied. The youthfulness of the renewed fleets in Europe and Asia-Pacific present business opportunities for manufacturers and suppliers of equipment and services. Cargo planes have a life of 37 years for single aisle jets and 35 years for wide bodies. From the data processed, Airbus is able to make a quantitative estimate of the number and type of airplanes which will be bought over the period under consideration and thus predict the market evolution (Figure 3.13).

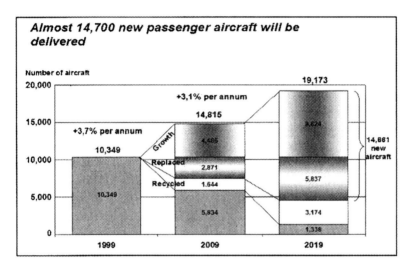

Figure 3-13. Analysis of delivery of new airplanes

This market is estimated at around 14,768 airplanes, or $1.3 billion over 20 years, split up as shown below (Table 3.14).

Airplanes type (Number of seats)	Number of airplanes	%	Value ($ billions)	%
70, 85 & 100	1913	13	43	4
125, 150 & 175	6477	44	303	26
210 & 250	2830	19	237	20
300, 350 & 400	2340	16	340	29
> 400	1208	8	263	22

Table 3-14. World market estimation in terms of numbers of airplanes and turnover (Airbus)

Table 3.14 shows the pre-eminence of single aisle airplanes (from 125 to 175 seats) for unit sales. On the other hand, the value of sales shows that 4 segments are almost equivalent due to the high price of the large airplanes.

• **The Current Market Outlook from Boeing**

Since the 1970's, Boeing has undertaken the publication of an evaluation of the civil airplane market based on a large number of sources of information (AEA, Association of European Airlines, Airclaims, DOT Form 41, DRI-McGraw Hill, Jet Information Services, OAG, IATA, ICAO, AAPA, Association of Asia Pacific Airlines, WEFA and Boeing sources). If at the outset it was concerned with future airplane deliveries, the Current Market Outlook study has progressively taken into account the market evolution (economic modifications, new players, new markets). Thus, in its latest edition, the CMO includes in its market approach, all the service providers used to improve the efficiency of the world's airline fleet. The first editions of the CMO presented market growth within a regulated aeronautical context. Now, most flights take place in a liberal and global environment: abolition of numerous national barriers, liberalization of whole economic sectors (telecommunications, energy, postal services, transport, etc.), world alliances between airline companies and privatization of airports and air traffic control, increased possibility for airlines to choose their destinations, fix their tariffs and determine their service offer.

These changes have led the airlines to become more efficient, more convincing and to increase customer loyalty, while at the same time reducing costs. They therefore look for suppliers who are capable of bringing them global solutions (integrated systems, processes and infrastructures) to help them improve the working efficiency of their fleet. Airline working expenses include all those activities aimed at attracting customers and ensuring the transport of passengers and freight to their destination. Within these activities, there is a group of support services which are indispensable for the maintenance and growth of the fleet. The product support market and service provision (around $87 billion) has a greater value than the civil jets market (around $70 billion). Its growth should continue at the same rhythm as that of the world airline company fleets, or of their operation, depending on the market segment being considered.

Boeing's CMO analyses the "Demand for Aviation Services". Forecasts are made which allow the actual market to be compared with the situation in 20 years time in terms of heavy maintenance, company services, engine repairs, equipment repairs, airplane modifications, repairs of engine or airplane parts or elements, crew training, restructuring of airplanes or airport/road services. Boeing makes an effort to study the infrastructures market, airports, airline companies and air traffic control; covering large geographical zones, this analysis covers the final demand, whether passenger or cargo.

Airplane type (number of seats)	Number of airplanes	%	Value ($ billions)	%
Regional jets	3347	15	90	6
Single aisle	12720	57	600	40
Widebody	4909	22	615	41
B747 size	1339	6	195	13
Total	22315	100	1500	100

Table 3-15. World market estimation in terms of numbers of airplanes and turnover (Boeing)

3. Business Marketing Intelligence

It takes into account traffic and market trends (reduced price offers, increased frequencies, types of airplanes used, etc.). In this way Boeing can offer airlines a section of the analysis on "Demand for Air Travel" and/or "Demand for Airplanes". Here the market is estimated at $1,500 billion for the next 20 years, or 22,315 airplanes with a total active fleet in 2019 of 31,755 (Table 3.15).

The renewal rate is obtained as a function of the age of the airplanes in the world fleet and by type (cargo, long haul, medium haul, single aisle, etc.). In this way, 4,230 airplanes will be withdrawn and replaced by equivalent sized ones but also by smaller and larger ones. For example, an airline could replace its B727-200 (156 seats) with 757-200 (201 seats). Almost a quarter of the $1500 billion civil airplane market concerns replacement of old airplanes, with the remaining three quarters allotted to adapting the offer to passenger and cargo market growth. Almost 22,315 airplanes will be bought and 9,440 of the present park will still be in service (Figure 3.16).

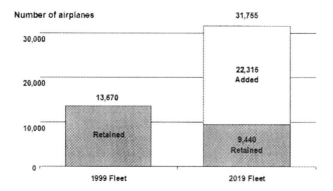

Figure 3-16. Estimate of the number of new airplanes sold in the next 20 years

Boeing's study also, increasingly incorporates the regional airplane (less than 70, seats) segment, which is developing strongly on the American market. These airplanes will become the company's strategic partners by ensuring non-stop flights, permitting greater geographical coverage from the hubs of the main airlines, taking traffic outside rush hours, replacing the large airplanes on low density destinations and replacing propeller driven planes. According to Boeing, the wide body airplane market for the main destinations over the next 20 years, will be limited to 1,010 (of which two fifths will be 747-400 sized and a quarter will be for freight). In total, only about 330 airplanes with greater than 500 seats will be bought over the next 20 years. The proportion of wide bodied airplanes will decrease from 7 to 6% unlike their intermediate cousins which will go from 19 to 22%. The main development will concern the single aisle airplanes which will fall from the present 67 to 57% of the market, and small commercial airplanes whose share will increase from 7 to more than 15%. Variations are to be expected as a function of the geographical zone under consideration (Figure 3.17).

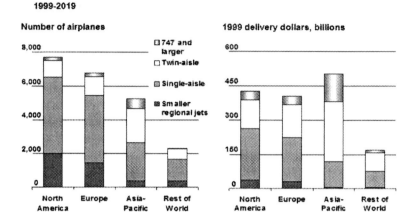

Figure 3-17. Clear differences in demand

The volume and type of demand varies from one region to another: in North America - the most well developed market - and in Europe, there will be a strong demand to renew the fleet. In America, 85% of the estimated 7,835 airplanes will be regional or single aisle type. This value will exceed 80% of the 6,800 airplanes predicted for Europe. The Asia-Pacific airline companies will develop their fleets using mainly wide body jets (the number of planes delivered being three times greater than for North America over the same period).

• The two studies made by Boeing and Airbus give essential information as to the demand and its evolution, whether concerning the passenger market, cargo, airplanes or services.

	GMF	CMO
Interest	• Airline companies • Part manufacturers • Engine builders • Service provides	• Airline companies • Part manufacturers • Engine builders • Service provides • Airport infrastructures, air traffic control
Demand	• Passengers • Cargo • Airplanes (A/C > 70 seats)	• Services • Travel • Cargo • Airplanes (including regional airplanes < 70 seats)
Period	20 years (1999-2018)	20 years (1999-2019)
Market estimation	$ 1300 billion	$ 1500 billion (25% renewal, 75% for market growth)
Airplanes bought	14,768	22,315
Airplanes retired and replaced	4,436	4,230
Present airplanes still in use	5,557 (retrofit and converted for cargo)	9,440

Figure 3-18. Some comparative data from the two CMO and GMF studies

4. THE MAIN TYPES OF STUDIES

Marketing studies have several objectives and in particular, with the long term in mind, that of communicating and selling (communication plan). In the shorter term, the object is to gather pertinent and measurable information which can help the decision making process (product plan). Indeed, studies are not an end in themselves, they only make sense if they give rise to an action plan. The role of studies is to enrich the company's information system (market knowledge and trends, customer expectations and needs, competitors and their strategies, etc.). The information collected is then structured to simplify analysis (customer behavior, exact nature of relations between different members of a supply chain, strengths and weaknesses of the company and its competitors, strengths and weaknesses of actual offer, etc.). The results obtained show which actions should be prioritized, whether for customers or products. These decisions will give rise, among others, to technological and industrial policy choices.

The distinction is made between quantitative and qualitative studies. To define them the simplest way is to start from a general diagram showing the objectives of different types of study.

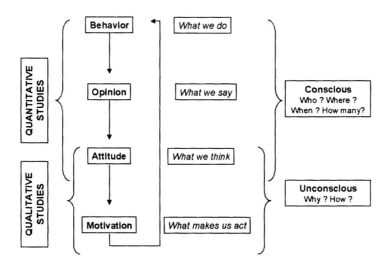

Figure 3-19. The different stages: from motivation to behavior

This diagram shows the boundaries between the two studies. The qualitative studies must be made before the quantitative ones since they improve their effectiveness. The first stage to study is that of motivation: what are the real, deep-seated reasons driving the buyer to make his choice. This is the object of the qualitative studies.

4.1 Qualitative studies

In particular, qualitative studies analyze the causes (why?), in other words the motivation behind customer buying behavior. The answers to these questions are generally not clearly understood by those replying; in fact, professionals questioned do not necessarily know their real, more or less conscious motivations. This is why qualitative studies need special methods of information collection such as the non-directive interview. This can go deeper, analyzing motives for satisfaction or dissatisfaction, the origins of preferences, etc., and give more discussion on eventual projects. It can take four different forms:

The face to face interview: An arranged meeting, the face to face interview is the most appreciated technique because it reinforces the status of the interviewee. Visual aids can be used and the investigator can interact directly and give any necessary explanations. Under these conditions, the interview could last up to an hour. The downside is the cost, all the more as things are utterly dependent on the interviewee's timetable: even with an appointment, the person may have another more pressing engagement.

The telephone interview: This is also a non-directive interview which requires, even more than for a telephone survey, an appointment and some in-house contact. However, contrary to popular belief, the telephone interview can give good results from the moment that the person answering is concerned by the problems raised. It is perhaps preferable to the face to face interview because it is quicker: it is difficult to spend more than 20 to 25 minutes on the phone. There is also an obvious cost advantage, because there are no traveling expenses.

The group interview focus group: This is a face to face interview organized with a small group of 5 to 10 professionals, discussing a given topic. This type of interview can be better organized, with one or two leaders, including the head of the company in question, a video recording and possibly an agreeable setting for the meeting. The main advantage is the interactive nature of the discussion which will depend to a large extent on the composition of the group, made up of a variable number of buyers, influencers and distributors.

This type of interview is expensive because travel and possibly reception expenses must be reimbursed. The professionals are generally not paid and so it is a good idea to select one particularly well known figure from the sector who will entice others to come, if only to meet him. The focus group is especially well adapted for studies about a subject of importance to those present, e.g. the speed of circulation of some new technology or material.

On the Internet, there is the on-line focus group and it must be said that exchanges in newsgroups and forums offer potentially useful material for company marketing directors. The free expression and spontaneous reactions can be lexically analyzed (keywords, most common associations, vocabulary used). The advantage of on-line groups is the reduced costs (no participant traveling expenses, no data transcription costs) but the participants are paid. They are especially useful for testing new service or product ideas.

The Delphi Method: This consists of setting up a group of professional experts as in the previous case, however the interviews remain independent: there is no group meeting. Each person is consulted initially, and a summary is sent to each of the participants who are thus able to position themselves and their opinions relative to the general view. Repetition of this will favor a convergence of opinion sometimes arriving at an agreement. The advantage is that each person can step back from his own opinions and reduce the risk of over impulsive reactions. This method is well adapted to studying new technologies and markets. The length of the process, up to 2 or 3 months, has to be accepted, and the group members must be carefully selected.

It should be noted that the Delphi-leader method has developed on the Internet, where experts are replaced by participants who are the opinion leaders within the category of products concerned by the study.

A recent development shows that studies which summarize and include both quantitative and qualitative aspects are more frequently used. From a sufficiently large number of interviews, i.e. about a hundred, qualitative information can be gleaned. Above and beyond the quantitative lessons to be learned about a particular subject, a word analysis measuring the percentage of those answering having used a specific argument or having made a particular judgment on a product or company can also be employed.

4.2   The quantitative studies

Quantitative studies concern the three other stages and in particular that of behavior. The main questions are:
- What: what category of products is concerned? what suppliers and in what quantities ?
- Who: what profile and how many ?
- When: seasonal cycles, long term trends, etc.
- Where: what types of channels, what geographical zones ?

The persons questioned generally know the answers to these questions which concern the whole of their activity sector. The only difficulty is " selling " them the study, by showing them what they can gain from it, so that they participate freely.

And it is because of this that easily processed, directive questionnaires can be used for these three stages. The competitive price of this method allows it to be applied on a large scale. About 200 questionnaires are used for the professional market and closer to 2,000 for a general public study. When the quantitative study is based on a directive questionnaire, it is wise to carry out a qualitative pre-study. This allows verification of the question order and vocabulary choice to make sure that the whole document can be understood by respondents.

THE INTERNATIONAL TOURISM STUDY: A KEY ANALYSIS FOR AIRLINE DESTINATION STRATEGY

Analysis of the world tourism market reveals a certain number of fundamental trends. It turns out that the main tourist destinations are:

- Europe which takes more than half the world's tourists,
- the American continent (Canada, United States, Mexico and the Caribbean) accounting for 20%,
- the Pacific and Far Eastern Asia which has 15% of the market.

Looking closer, the most numerous tourists are Europeans, and in particular the Germans, British, French and Italians. Their main destinations are for the most part inter-European, even if there is an increasing trend towards the American continent and Asia.

North American tourists tend to travel mainly within the American continent (~3/4), in Europe (~15%) and in Asia (5%).

The Pacific-Asian tourists favor their own region (~3/4) and travel equally as much to the American continent as to Europe (~10%). Nonetheless, those from Japan go to Asia (44%) and North America (1/3).

In this way it is possible to distinguish between different types of homogenous tourists according to their type of behavior in terms of preferred destination.

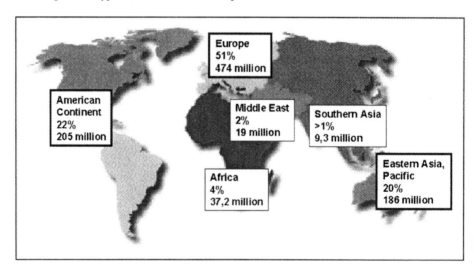

Figure 3-20. Tourist load regional forecast for 2010 (expressed in raw numbers and percentages going to the regions)

3. Business Marketing Intelligence

To: From:	Europe	American Continent	Eastern Asia Pacific	Southern Asia	Middle East	Africa
Europe	91,3	14	12	41,5	19	34,1
American Continent	5,4	78,6	7,4	9,2	2,8	2,9
Eastern Asia Pacific	2,3	6,8	78,1	13,7	18,4	1,5
Southern Asia	0,1	0,2	1,8	28,1	3,1	0,3
Middle East	0,4	0,2	0,3	5	50	5,1
Africa	0,5	0,2	0,4	2,5	6,7	56
All	*100*	*100*	*100*	*100*	*100*	*100*

Table 3-21. Tourist distribution according to origin and destination (in %)[12]

Directive documents for quantitative studies

Studies using the postal system or Internet: At the outset these have certain drawbacks. First of all they have a very variable rate of return, from 2 to 50% according to target and skill at making people reply: explanatory mail, phone call, promise of survey summary dispatch, even a participation present as in professional catalogues. Apart from this, it is uncertain whether the questionnaire has been filled in by the target person, and in addition, this type of study must use the directive mode and there can be no making up for errors if the person misinterprets the question. The questionnaire must therefore be tested beforehand. On the other hand these studies (postal system or Internet) do not impose a specific appointment time on the person and also the cost is independent of the geographical range of companies targeted. As it is the least expensive type of study, a greater number of replies can be aimed for.

On certain flights, airline companies sometimes distribute a questionnaire to passengers in an effort to better understand their expectations and habits. The questions are simple, clear and can be easily answered by ticking the boxes and using the smiley ratings (Figure 3.22).

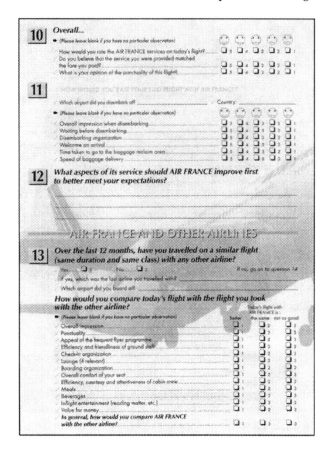

Figure 3-22. Example of questionnaire given to passengers

Internet: Internet allows rapid and cheap studies to be carried out on customers with an e-mail address (Aragon *and al.*, 2000):
- telephone assistance users,
- Intranet and website satisfaction studies,
- external market studies (pre-tests),
- post -training satisfaction studies, etc.

Internet studies must respect a certain number of values specific to the media (language, practices, ethics, etc.) (Nicovitch and Cornwell, 1999). Different types of studies are used: sending e-mails containing questions (simple to set up, manual sorting), sending computer self-administrated interviews as attachments, sending an e-mail indicating the address of an Html form available on a web server, etc. Internet is a powerful collection and processing tool added to which there is the power of the network, potentially capable of reaching hundreds of millions of geographically dispersed responders. The latter can receive text, pictures and sound simultaneously,

3. Business Marketing Intelligence 101

and the attitudes and behavior of the internauts can be recorded almost instantaneously (Galan and Vernette, 2000).

Directive study by telephone: A variant of the preceding one, the telephone study has the advantage of being rapid. The target person must agree to a telephone appointment at his convenience, and after the maximum 15 to 20 minute interview, processing and use can be immediate. The cost relative to face to face interviews, only varies to a small extent with the location of the companies and is thus less expensive than traveling to the actual site.

Nevertheless, the telephone study is not always welcomed and it is better to have a contact in the company who can recommend someone. Finally, it is still not generally possible to use visual aids to show documents or photos or perhaps samples, except by using a webcam.

4.3 Permanent and ad hoc studies

An ad hoc study is a specific one exclusively ordered by a customer and so cannot be sold as such or adapted for another client. As the study agency needs to recoup all its fixed charges, this type of order is very expensive.

On the contrary, permanent studies are characterized by regular procedures which allow for frequent updating of scores. By comparing results, the company can follow the evolution of the main measures: an analysis of static numbers leads to a dynamic analysis.

Most of the qualitative studies are of the ad hoc type, whereas most of the permanent studies are quantitative.

Among the permanent studies are data panels coming from consumer goods sector. A panel is a measuring tool based on a representative sample of customers. It is confidential because the participation of a company in the sampling must not be known to the customers. These techniques are therefore meaningless for companies which only work with a few customers and prospects, because here they need to make an exhaustive study of all potential and actual customers. Whether the company is in the aeronautical or space sector they will use panels as soon as they need to address the general public. Cases in point are airline companies studying passengers or satellite integrators studying end customers who use telecommunication services. Conversely, a parts manufacturer deals in general with about twenty potential customer companies and so the market study will be an exhaustive, customer by customer one.

There are three types of panels:
- User panels,
- Audience panels,
- Distributor panels.

The first two can be used directly in the aeronautics and space sector.

The user panel

The user panel is a repetitive study on a permanent and representative sample of consumers. The aim is to find out for a particular category of products, the quantities consumed in volume and value and their brands. This allows an analysis to be made of product or service purchases by brand, place of buying, over time, and consequently to determine the market share and its evolution. It also shows up brand loyalty and transfers which can thus be measured, as can the effects of an advertising or promotional campaign. The development of Internet has favored the use of these panels with the internaut population.

The quantities purchased (QP) are expressed for a 100 households and can be broken down into quantities purchased per purchasing household and number of purchasers (NP):

$$QP = \frac{QP}{NP} \times NP$$

For actual value measurements, the sum spent (SS) allows an analogous analysis to be carried out using the prices of the goods. The ration " SS / NP " replaces " QP / NP in the above equation and corresponds to the sum spent by the purchasing household.

Example: a subscription to a cluster of satellite TV channels

$$1500 \text{ €} = 50 \text{ €} \times 30$$

⇒ 30%: number of households taking out a subscription to at least one satellite chain during the study period. This is the penetration level also known as the trial level when it concerns new services /products. By extrapolation, 70% are not consumers of this category of service.

⇒ 50 € : mean purchase for purchasing household.

Use: With this information, the marketing manager can decide which variable is worth concentrating on, generally with the object of increasing the amount spent: either he can try to increase the penetration level or the amounts spent by the purchasing households.

For instance, an airline company is studying manager customers (Northwest Airlines in the United States). The penetration level for all companies taken together corresponds to the number of individuals who have taken the plane during the study month (54% in the example). On average the amount spent by each passenger comes to 1000 € .

3. Business Marketing Intelligence

Airlines	NP	SS / NP	Total SS	SS Market share in %
All companies	54	1000	54,000	100%
A	32	840	27,000	50%
B	15	1100	16,200	30%
C	8	1400	10,800	20%

Table 3-23. Comparison of the amount spent and the number of customers per company

For company A, the penetration level is already very high (32%). However, the amount per passenger stays low (840 €): it is this second variable which should therefore be acted on:
- Is the product judged to be sufficiently good ?
- If yes, how can customers be kept loyal to company A ?

For company C, the results are the opposite: there are few buyers (8%) spending a lot (1400 €), which can be seen as satisfaction with company C. After having verified that price is not the main restriction on purchasing, C's penetration level (8%) must be raised somehow, since the " all companies " level is at 54%. A promotional technique[13] could be effectively used to attract new passengers to this company (reduced rate subscription for the first year, free hotels, etc.).

Thanks to its cyclical nature, often monthly or bi-monthly, a panel of users represents a useful barometer for the marketing department. As with all other studies, it is only meaningful if it feeds an action plan. In particular, this means knowing whether the company should above all try to increase existing customer loyalty (company A in the example) or whether above all it needs to recruit new customers (as in the example for company C).

The audience panel

The audience panel looks at all advertising campaigns in trade press, billboards, TV[14]. It doesn't take into account communication campaigns in tradeshows, nor company documents or sponsorship. So with these restrictions, it gives information about the investments made on a category of products and services and their evolution, as well as the breakdown between the main players. The share of voice is the company's advertising expenditure proportional to the total advertising spending of the sector.

$$\text{Share of voice (A)} = \frac{\text{Advertising expenditure (A)}}{\text{Total advertising expenditure of all competitors}} = 18\%$$

Company A thus represents 18% of the spending on advertising for this product category. This figure must be compared with the market share obtained. The comparison between the market share and the share of voice over several years, allows the effectiveness of the mix adopted by the company to be analyzed. As a general rule, groups such as Nielsen, Information Resources Inc (IRI), TN-Secodip, GfK offer their customers this sort of competition monitoring.

The distributor panel

This is based on a representative sample of distributors. It is useful each time that a company uses external distributors to circulate its products. The example of tour operators for airline companies or equally, spare parts and small equipment wholesalers can be cited. This type of panel measures the quantities sold per category and per company and so gives a good idea of sales volume, market share and price levels. Two types of calculation are necessary to use a distributor panel:

• Sales calculations using mainly the market share in volume and value,

• Calculations on the presence of the product at points of sale, therefore measuring the effectiveness of the sales team.

The "numerical distribution" (ND) corresponds to the percentage of sales outlet carrying product A.

$$\% \text{ ND (A)} = \frac{\text{Number of the sales outlets carrying product (A)}}{\text{Total number total of sales outlets}}$$

This calculation has the drawback of only considering the number of sales outlets carrying the product irrespective of actual sales. For this reason, "potential sales" is used, in other words, the potential sales value of the distributor where the product is carried. This calculation is made using the product category turnover for all brands together.

$$\% \text{ PS (A)} = \frac{\text{Turnover of the general product category of the distributors carrying product}}{\text{Turnover of the general product category of all the distributors}}$$

For example, if the ND (A) =50% and the PS (A) = 80%, this means that product A is carried in 50% of distributors and that these represent 80% of the turnover for the category in question. In this example, the distributors have therefore been well selected by the sales team and so the distribution measurements initially give a rating for the effectiveness of the sales staff.

"Potential sales" is more relevant since it takes into account the sales potential of the distributors where the product is carried. This value is a useful measure for judging when to launch an advertising campaign. In other words, it is pointless to advertise until the product is sufficiently carried. In general, companies invest in advertising when there is a presence level of >50%.

It should be noted that PS corresponds to the probability for an end user of finding the product when he visits a distributor.

Demand: By dividing the sales by the "potential sales", the following equation is obtained (Figure 3.24).

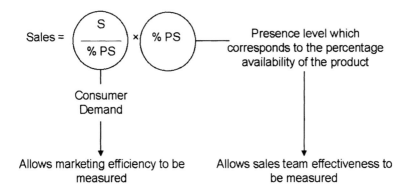

Figure 3-24. Calculating demand

Calculation of the demand is very realistic. Indeed, if Sales =2000 passenger-tickets and PS =50%, then S/PS= 4000 passenger-tickets. This is the theoretical level of sales for 100% presence of the product at the sales outlets. If it is improbable to imagine a rapid development in "potential sales" from 50 to 100%, all that is required is to take the sales levels corresponding to 1% of the PS, i.e. 4000 / 100 = 40 tickets. This figure also enables sales for the following year to be better forecasted as a function of the development of in-shop presence: this is the responsibility of the sales team (Table 3.25).

	A	B
S	4000	2000
PS	10	50
S/PS	400	40

Table 3-25. Comparison in the sales performance of companies A and B

The first line shows that company A is leader with double the sales of company B. The second line then shows that company B has 5 times the number of listing outlets as A while on the other hand, the demand S / PS leads to the conclusion that with equal presence, company A sells 10 times more than B. This can be due to:
- A's product/service which better suits to the local customers,
- A's price level,

- A's communication.

Company A enjoys a marketing mix which is 10 times more effective than B's, thanks to its marketing strategy. On the other hand, company A has a big problem concerning its distribution strategy. However, A's case is preferable to B's because it is easier to change the presence level than the consumer demand, whose development can only be done in the medium term.

Another measure of the demand is given by calculating the holder market share, which takes into account total sales within the category. The absolute quantity of sales on the top line is divided by the total for all brands together, giving the market share:

$$\frac{S/M}{PS} = \frac{\text{Market Share}}{PS} = \text{Holder market share}$$

This market share corresponds to the share of the distributors where the product is listed. For example, two companies A and B have the same market share but a different distributor value (PS) (Table 3.26).

	A	B
Market share	12	12
PS	50	100
Market share / PS	24	12

Table 3-26. Comparison of market share and holder market share for brands A and B

In this case, company A is better off:
- it is more efficient, selling the same quantity while being 2 times less present,
- it has room for maneuver because it can raise its presence level,
- it can use its holder market share score (twice as high), to obtain listed status with other distributors.

4.4 The other objectives for studies

In addition to being a key source of information, studies are a means to attract customers, establish contacts, strengthen sales argument.

Using studies to attract customers

A company which asks a customer or potential customer to participate in a study demonstrates a real desire to work together, especially if it is a highly specialized study, clearly targeting genuine unsatisfied expectations. In the case of a small customer, often the very act of asking him to be part of the study confers on him an importance to which he is unaccustomed. Thanks to this type of study, the company can better show its commitment to customers and prospects. At the same time, it can confirm via the study the image that its own managers have gained or otherwise with the customer.

Contact studies

Carrying out a study which includes a prospect or a company with whom there have been no previous business dealings can give rise to new contacts. But above all it can be the opportunity to demonstrate work methods and discuss potential proposals. This function is particularly important in sectors where activity is organized into operations and functions according to the Project Marketing model, as for example in the Buildings and Public Works sectors when constructing and installing airports. After a joint construction site, several years can go by without the customer needing the company. The study can produce a certain continuity in relations and re-establish contact, in between successive work sites.

Studies to strengthen sales arguments

According to the type of study, the company will be able to make direct use of it as a sales argument, or indirect use as a support factor for its sales presentation. With this in mind, the studies carried out are usually oriented in three ways.

Technical based studies: This is a comparative follow up of the technical characteristics and above all performance of the company in relation to its competitors. The strong points highlighted are generally used as sales arguments and incorporated into the technical presentation documents for the products and materials.

Sales based studies: This type of study is destined to explain to the customer how companies which have already purchased the product or service are faring, and who are the users. The essential point here is to know how the product is used, possible applications and above all the reasons for customer satisfaction or dissatisfaction with the different materials (as much those of the competitors as those of the company).

Financial based studies: Careful observation of a large number of customers allows the ratios of costs and use, analyzed by type of situation, to be calculated. Companies as different as Boeing or Pratt&Whitney regularly use these performance arguments, quantified, to allow their customers to position themselves and to have the most possible information for choosing the correct material.

```
┌─────────────────────────────────────────────────────────────────────┐
│ 1.1    Operating costs - containerised version                      │
│                                                                     │
│ Aircraft type ...........           Fleet size .........            │
│ Annual utilisation:     ......... flt cycles,   ......... km sector length,   ......... flt hours. │
│ Average pax load.........          ULD ratio for baggage : freight .........  : .........         A│
│                                                                     │
│ Additional fuel consumption due to extra MWE/OWE (See note 5)      │
│                                                                     │
│      ..... IATA cont. G at ....... kg tare weight   =              │
│      ..... IATA cont.H at ..... kg tare weight      =              │
│      ..... Pallets at ................. kg tare weight  =          │
│      Cargo systems at ........... kg                =   ........... ( 64kg per A320, 92kg per A321 increase in MWE)│
│      Total increase in MWE/OWE      =      kg.                     │
│                                                                     │
│ With fuel burn at .......... kg per block hour per kg of installed weight therefore:   (See note 11)│
│  ........ kg of installed weight  x .......... kg of fuel  x .......... flight hours per year  x ........cents per kg of fuel│
│      therefore annual cost of additional fuel consumption per aircraft  = $│
│             Annual cost of additional fuel consumption per aircraft │
│             using ULD ratio in line A (See notes 1 & 6)    = $      B│
│                                                                     │
│ Manpower for loading/unloading baggage and freight                  │
│                                                                     │
│ ..... men x ....... hours x ............flights per year x $.......... per man hour = $│
│ ..... men x ....... hours x ............flights per year x $.......... per man hour = $│
│ ..... men x ....... hours x ............flights per year x $.......... per man hour = $│
│             Therefore annual manpower requirement per aircraft      │
│             using ULD ratio in line A (See notes 1 & 6)   = $       C│
│                                                                     │
│ Maintenance costs for ULDs and GSE   (See note 9)                  │
│ Assume ....... % of the investment per year for the fleet (line 1)  = $│
│     Maintenance costs of cargo loading system                       │
│ Assume $5 per aircraft flight-hour                        = $       │
│             Therefore annual maintenance costs per aircraft  = $    D│
│                                                                     │
│ Ground handling contracts                                           │
│             ............. turnrounds at $ ............ each  = $    │
│             ............. turnrounds at $ ............ each  = $    │
│             ............. turnrounds at $ ............ each  = $    │
│             Annual ground handling costs per aircraft               │
│             using ULD ratio in line A (See notes 1 & 6)   = $       E│
│                                                                     │
│ Additional ULD controllers                                          │
│                                                                     │
│ ..... Men at $.................. per year including overheads  = $  │
│             Therefore annual cost of controllers per aircraft       │
│             using ULD ratio in line A (See notes 1 & 6)   = $       F│
│                                                                     │
│ Total annual operating costs per aircraft (B+C+D+E+F)   = $         G│
└─────────────────────────────────────────────────────────────────────┘
```

Figure 3-27. Example of a quantitative study grid for the running costs of cargo airplanes

"Tailor made" studies

In order to carry out such studies, the company must know the real figures for the customer company over a sufficiently long period of time to be valid. This presupposes a high level of confidence at least as far as confidentiality is concerned.

In the machine tool sector for example, a manufacturer could ask a customer or prospect to supply him with production statistics for the last twelve months. Since the company has normally got sophisticated software for production analysis, they can calculate essential performance parameters such as:
- the hourly output, the output per employee,
- the percentage time stopped (for maintenance, because of breakdowns, etc.).

Using this data the company offers a more objective analysis: without mentioning names, it allows the customer to see where he is against the average for

3. Business Marketing Intelligence

a list of his competitors. Using this tailor made study, the company can fine tune its proposal and demonstrate to the customer why he needs such a machine with its particular potential and essential options rather than another superfluous one, etc. This comes down to defining the best product offer.

Airbus carries out wide ranging market studies in order to determine future developments and so prepare an offer which corresponds to market requirements. This involves in-depth market analysis and evaluation of the main future trends. This means studying the airplane user market, i.e. the passenger and cargo airline companies. This market depends on numerous economic, political and technological factors. The studies made by Airbus of the global air market, allow useful data to be available for different airline companies (prospects or customers).

FROM THE AIRBUS A3XX PROJECT TO THE A380: THE PRIMORDIAL ROLE OF STUDIES

Within the framework of the A3XX project[15], Airbus have carried out qualitative and quantitative studies in order to:
- Better understand the future needs and expectations of passengers,
- Publicize the merits of the two types of airplanes: single and double aisle,
- Evaluate the acceptability of an upper deck on the A3XX.

Figure 3-28. The Airbus A380 project: the A3XX

The methodology used was the following:
- Period: April to October 1998,
- Place: 8 cities in 3 continents: Tokyo, HongKong, Singapore, Paris, London, Frankfurt, etc.
- Study sample: 1200 travelers accustomed to long haul flights,
- Respondent: 150 in each city, half economy class passengers, half business class,
- Method: presentation of 3 different upper deck versions for the A3XX and one for the B747.

Representing about fifteen rows of seats, these mock ups have been made in kit form for dismantling and transport.

Figure 3-29. Mock up of what the interior will look like

Passengers were asked about their travel experiences. The study questions covered motivation, behavior and opinion;
- " What does a passenger feel like sitting in this seat ? "
- " What things are influencing his impressions ? "

To avoid any bias, the people questioned did not know the name of the company which required this information. Both types of study, quantitative and qualitative were used.

- **The quantitative studies**

The respondents filled in the questionnaire alone. The latter contained several " who, how, what, where " type questions. Using symbols to make answering easier, yes/no and multiple choice questions were used frequently: knowledge of type of airplane, knowledge of the airline company, reactions to the cabin design, the on-board restaurant service, the space available and the personal space, preferences concerning a national transport company, importance of armrests, perception of the airplane manufacturer, etc. It is interesting to look at the reasons leading a passenger to prefer a particular airline company.

Expectations	*Europe*	*United States*	*Asia*
Pleasant crew	42	32	40
Service quality	37	37	8
Pleasant service	25	18	27
Meal tray quality	25	26	29
Comfort	19	19	6
Safety	11	5	15

Figure 3-30. The reasons for preferring an airline company

The above scores show the importance of the crew's role in customer satisfaction. They also demonstrate by their satisfaction levels, that safety is taken for granted (low scores: " obviously expected " and " essential ") and also service quality and comfort in Asia.

- **The qualitative studies**

Each of the respondents was also questioned personally on topics ranging from airplane characteristics to those of users. For example, managers prefer pre-reservation of seats, wide body airplanes and wider, more comfortable seats. This type of qualitative study has played

an important role in the design of the cabin and arrangement of the seats. In qualitative studies, the opinion and nationality of each passenger is noted. Each question is doubled up in order to know:
- the behavior of the passenger,
- the expectations of the passenger.

In addition, as part of its consultations with the airline companies concerning the design of the A380, Airbus has constructed a life-size model (without the wings) in order to test *in situ* the different interior fittings and discover any problems of space and confinement. This A3XX, much better than a virtual model, is 60 meters long, 7.10 meters wide and 8.50 meters high, allowing customers, marketing managers, designers and hostesses to avail themselves of a " real " cabin and test new ideas (bars, restaurants, gymnasium, nursery, office, infirmary, relaxation room, etc.).

• **Passenger behavior: segmentation of the airline's customers**

Studies carried out with future passengers give a better understanding of their behavior and expectations. This allows the airline companies to better define their services for future passengers on board the airplane. Below is an example of a typology obtained from these studies.

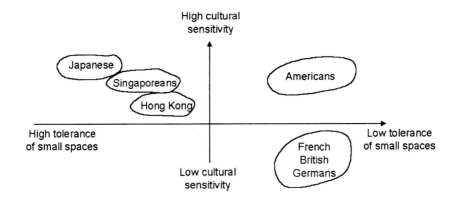

Figure 3-31. Passenger typology made using the criteria of cultural sensitivity and tolerance of confined spaces

Analyzing this diagram and taking into account the spread of individuals, it is possible to interpret the axes as follows:
- Vertically, passenger's cultural sensitivity which shows itself as a stronger or weaker preference for an airline company of his own nationality,
- Horizontally, as a greater or lesser tolerance to confined spaces.

Some conclusions can be drawn. The appropriation of personal space (intimacy) is in fact more important than the volume of space available. At the end of the study, Airbus identified a certain number of expectations common to manager class passengers who for example attached a great deal of importance to the presence of a table or place designed specially for laptops where they could work and send and receive e-mails. It should be noted that this sort of equipment exists already in first class on the most recent airplanes, e.g. Airbus A340 and the Boeing B777. Similarly, these passengers would also like an automatic distributor/minibar, relaxing/massaging seats and special areas for children.

The diagram produced therefore allows segmentation of the international customers studied as a function of these two axes. It transpires that Americans prefer US companies and are intolerant of lack of space on board. The Japanese also prefer their national companies and appear to be more tolerant of confined spaces. As for the Europeans, they could more easily opt for a foreign company but cannot tolerate lack of space on board. This information can be used to advise airline companies on choice of interior fittings. Americans want self-service areas and a place where they can talk to other passengers. By comparison, the Japanese want above all to avoid physical contact with their neighbors and hence the proposal will be for wider spaced seats in first class and large armrests.

Figure 3-32. Representation of the A3XX aircraft

NOTES

1. Source: Malaval, Ph., (1999), *L'Essentiel du marketing business to business*, Paris, Publi-Union.
2. Source: "Dassault ou l'envol de la satisfaction client", www.planeteclient.com, 23/10/2000.
3. Source: sales documentation, Dassault
4. See Chapter 2, The individual and organizational purchase
5. See in Chapter 1, 3-2 Different types of marketing
6. Source: adapted from Gemini Consulting.
7. Source: Purchase & Logistics Department, Aerospatiale-Matra.
8. See The Global Market Forecast 2000-2019 of Airbus and the Current Market Outlook of Boeing
9. Chapter 12, Selecting media, 1. Trade shows.
10. See in Chapter 2, 5. Purchase marketing, An electrical sub-system equipment example
11. See Chapter 15, Alliance strategies.
12. Source: *World Tourism Organization.*
13. See Chapter 11, Communication policy.
14. See Chapter 12, Selecting media.
15. Thanks to enough MOU, the A3XX project has been officially launched the 19th December 2000, under this new name A380.

REFERENCES

Aragon, Y., Bertrand, S., Cabanel, M. and Le Grand, H., (2000), Méthode d'enquêtes par Internet: leçons de quelques expériences, *Décisions Marketing*, 19, 29-37, January-April.

Galan, J.-Ph. and Vernette, E., (2000), Vers une quatrième génération: les études de marché on-line, *Décisions Marketing*, 19, 39-52, January-April.

Malaval, Ph., (2001), *Marketing business to business*, Paris, Pearson Education France.

Nicovitch, S. and Cornwell, T.B., (1999), An Internet Culture ?: Implications for Marketing, *Journal of Interactive Marketing*, 12, 4, 22-33.

Chapter 4

MARKET SEGMENTATION AND POSITIONING

The two fundamental marketing tools are segmentation and positioning. Basically, segmentation is splitting up the market into homogeneous sub assemblies in order to be able to choose an appropriate strategy for each one (targeting). Then, it is a question of making the clearest distinction possible between oneself and competitors in the mind of the customer. This is the object of positioning.

1. SEGMENTING A MARKET

Segmenting a market consists of splitting it up into homogeneous sub assemblies. Homogeneity can be defined mathematically by saying that any two elements from the same segment must be as alike as possible, whereas those from two different segments must be as different as possible. Measuring homogeneity, i.e. the differences or similarities, consists of looking at the behavior of customers relative to the company's products or services. There are several segmentation methods (by type of customer, type of product, etc.) according to whether the company deals with the general public or professionals. What is important is to know the objectives sought (improving the relevance of the offer, the communication, the approach, etc.).

1.1 The objectives of segmentation

Using successful segmentation means that the company can offer something which will perform well for the customer segment chosen. Conversely, an analysis without segmentation will only give too general product or service proposals that are ineffective. There are therefore several objectives to segmentation (Malaval, 2001).

Improving customer knowledge: Firstly, it is easier to carry out an effective study after segmentation. The relation between studies and segmentation is interactive : studies allow the segmentation to be sharpened, or even to adopt new criteria for the initial split up. Secondly, relevant segmentation allows studies to be developed which correspond better to customer expectations. In addition, by grouping together in a relevant way actual and potential customers with common characteristics, segmentation allows for a better understanding and better prediction of customer reactions to different marketing campaigns.

Making the product offer more relevant: By breaking up the market into homogeneous segments it is possible to set up a range of products which better correspond to the expectations of the customers from each segment. Their motives for satisfaction or dissatisfaction concerning previous products, competitors or otherwise are easier to distinguish. Right from the design of the product, segmentation allows faster development of the different varieties. Each manager can work from a customer-based viewpoint to fine-tune the finished product.

Allows better targeted communication: Segmentation also leads to an improvement in the effectiveness of advertising strategy. The fact of having defined the different customer segments means that the company knows the main motivations of its customers better. Consequently it is easier to choose which argument to put forward for which segment. In addition, it has better knowledge of the sources of information used by its customers whether general public (daily newspapers or magazine, radio, TV, etc.) or professional (professional press, exhibitions, technical documents, etc.). In this way, segmentation allows choice of the most effective supports for the communication objectives.

Allows a more effective sales approach: Defining homogeneous customer segments can lead to better selection of distribution circuits and thus specialization of a part of the sales force. Whether it is a question of selling tickets to the general public or equipment to airline companies, segmentation gives a more effective sales approach.

Contributes to the pricing policy: In markets where price emerges as the determining factor, it can be used as a criteria for segmentation. This could lead the company to propose different offers as a function of the price[1].

Overall therefore, segmentation turns out to be a highly important tool within the Marketing Plan, with repercussions on all the variables cited. For the Sales Actions Plan, it highlights which priorities to adopt for each segment[2].

THE SEGMENTATION OF THE SATELLITE MARKET

The space sector is characterized by high technical and industrial complexity which necessitates international cooperation between the various players. Several activity sectors are concerned by satellite solutions but the expectations of the customers vary from one main area to another. Because of this, a precise analysis of the satellite market must be made by launch

4. Market Segmentation and Positioning

companies as well as satellite manufacturer-integrators, in order to anticipate the evolution in the demand. To this end, segmentation is an absolutely essential tool allowing, for example, the identification of two macrosegments (Table 4.1):

- One stable segment grouping together telecommunications, narrow band multimedia applications, meteorology, data collection and localization, and navigation. In general, these activities are in the mature phase of their life cycle with known customers, planned needs for the future, econometric models used, etc.

An emerging sector including earth observation, scientific information and digital radio. These activities, in the launch or pre-launch phase are characterized by the uncertainty implicit in their experimental nature, by the difficulty of making forecasts and by the conflict which exists with other solutions.

Stabilized segment		Emerging segment	
Sectors	Criteria	Sectors	Criteria
Telecommunications: exchanges between a distant transmitter and receiver, fixed or mobile, on the earth or in space	Communication quality, bi-or-unidirectional, real time or time lapse, capacity, regional or global cover	Earth observation: transmission from space of data concerning the earth and its environment	Quality of service, download time, resolution, integration within the geographical information system, service continuity
Narrow band multimedia: with digitalization of the electrical signal numerous functions can be carried out by the same support	Direct TV, data transmission, telecommunications	Science: supplying the scientific community with information in addition to that obtained from other terrestrial sources	Diverse areas, experimental, technological possibilities
Meteorology: gathering and transmission to earth of information necessary for forecasting	Meteorology, climatology	Digital radio: " no boundaries " radio broadcasts	Fixed and mobile users, new channels
Data collection/localization: collection of information concerning the nature and situation of a terrestrial object	Environmental surveillance, search and rescue, fleet management		
Navigation: directing a mobile element to a given destination by calculating its position and trajectory	US system, Russian solution, international cooperation, strategic dependence		

Table 4-1. Segmentation of the space market: characteristic of the sub-segments[3]

1.2 The main segmentation methods

Segmentation by type of customer and type of product

Companies are increasingly abandoning segmentation by products in favor of segmentation by customer. The product oriented approach is translated by a state of mind turned more towards the company rather than the customers (Table 4.2). The tendency observed is towards a customer oriented approach (Beane and Enis, 1989).

Segmentation by type of:	Products	Customers
Advantages	• Easier to set up, often starting from an existing base	• Customer needs must be analyzed • Leads to the set up of segments by advantages sought
Disadvantages	• Risks being too oriented towards the company know-how rather than customer needs: risk of conservatism • Risks proposing an offer out of touch with customer needs and late in terms of competitors	• Risks dispersing design and production efforts • Risks getting in the way of a productivity approach • Frustration when the company's offer does not correspond to the segments thus defined

Table 4-2. Comparison of methods of segmentation by products and by customer

It must be said that in certain cases the two approaches, customer or product oriented, come to the same result. A case in point is when there is a small number of suppliers in a market and they are all rigorous in their segmentation. In air transport for example, 15% of passengers state that the availability of vegetarian meals on-board is a decisive criteria in their choice of airline. This same figure (15%) comes up when an analysis of on-board meals is made as a function of their technical characteristics. Regardless of method therefore, the size of the segments is very close.

Company	Market	Environment
Network Fleet (size composition, average age of planes, orders) Locality (main base) Type of structure (private/public) Domains of activity (sub-contracting of certain activities) Financial results Role, strategy, objectives (leader, challenger, niche) Alliances Culture, history (links with partners) Operational constraints	Constraints from regulations Size Growth perspectives Sectors (freight, passengers/leisure, passengers/business, etc.) Conceivable evolution Competition	Changes linked with transport (High Speed Train, etc.) New players New customer markets New airplanes (possible effects on the results) Regulations (ETOPS, noise, traffic rights, etc.) Airport saturation Saturation of airspace Political changes

Table 4-3. Examples of segmentation criteria in the aeronautics market

Top-down segmentation

This method consists of taking the total market volume and breaking it up again and again as a function of the criteria chosen. This will give a certain number of segments, i.e. 12 on Figure 4.4. From here it is then a question of calculating the volume, the turnover and the potential profit for each segment. Then the company selects certain segments as a function of its know-how and sales priorities.

This method has the advantage of being sure from a quantitative point of view, since it is derived from a real situation. The calculation can just as easily be made using the number of customers as the number of products purchased. Descending segmentation is particularly suited to products with a large distribution.

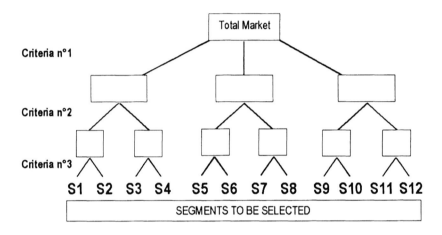

Figure 4-4. Market segmentation using the top-down method

AN EXAMPLE OF TOP-DOWN SEGMENTATION APPLIED TO CATERING

Because the airline companies are concerned about the quality of their service to passengers, they regularly give out in-flight questionnaires in order to measure the level of satisfaction and find out the main areas where improvements are expected[4].

Among the questions asked, apart from passenger profile (age, sex, socio-professional group, etc.), a certain number concern in particular the on-board catering (especially on long haul flights). Passengers can thus express their food preferences freely and the data allows in particular the identification of the main expectations, and the construction of a typical passenger profile as a function of eating habits.

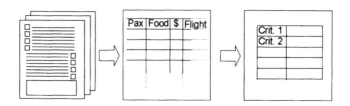

Figure 4-5. From data collection to highlighting of a segmentation criteria

This information, once collected and ordered allows definition of the main segmentation criteria and then evaluation of the number of passengers per segment.

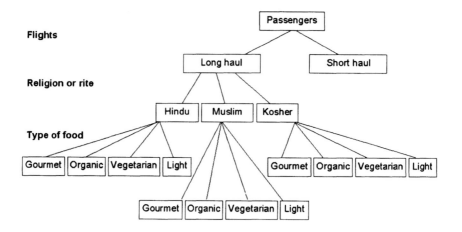

Figure 4-6. Example of top-down segmentation applied to on-board catering in civil airlines

For meals, the "on-board catering market" can be successively segmented according to the following criteria:
- The type of flight: does the passenger expect an in-flight meal ?
- Religion or rites: dishes must respect methods of preparation according to religious requirements (Muslim, Hindu, Jewish, etc.)
- Type of food: what type of food is preferred: traditional for gourmets (reflecting the country of departure or destination), or more special (organic, vegetarian, light) ?

The airline company and/or the on-board catering company then select the segments (targeting) as a function of:
- the potential activity for each segment,
- the marketing and sales objectives and resources of the airline company.

In this way the segment of passengers wanting a traditional, gastronomic and gourmet meal is that with the strongest potential (particularly compared with light or vegetarian traditional gastronomic meals). Another segment seems to be particularly attractive for long

4. Market Segmentation and Positioning

haul flights (e.g. between North America and North Africa) and that is the traditional Muslim meals one.

For passengers, several segmentation criteria are used in particular (Table 4.7) e.g. reason for journey, passenger's country of origin, etc.

Aim of journey
Business trip:
   - company
   - independent
   - conference, stimulation/motivation
Leisure trip:
   - holidays
   - visiting friends
   - visiting parents
Personal trips
	Length of journey
	   - connections
	   - direct
	
	Level of use
	
	Cultural and national origin
	
	Lifestyle

Table 4-7. The main segmentation criteria for passengers

This gives more precise profiles of the main types of passenger, business or leisure (Table 4.8). It is thus possible to produce a more adapted offer thanks to the characteristics identified.

Business passenger	Leisure passenger
PROFILE - Low relative sensitivity to price - Socio-demographic characteristics (age, sex, social background) - Large number of trips, experienced passenger - Strong expectations and demands - Interesting profile for the companies (in terms of profitability) but less than leisure passengers (more numerous)	PROFILE - Highly sensitive to price - Highly seasonal demand - Very different socio-demographic characteristics from the " business " segment - Expectations in terms of flight frequency, times and accessibility of seats, relatively undemanding - Interesting profile for the companies (in terms of volume): possibility of redirecting passengers into uncongested periods
OFFER - High frequency, small airplanes, ease of access (luggage) - Good standard of comfort (space, services) - An adapted timetable for business destinations all the year - Ease of payment and reservation, simplified tickets (e-tickets) - Coordinated timetable for specific destinations (code sharing) -Yearly services with peak daily and hourly traffic	OFFER - Low prices - Wide-bodied planes (low frequency) - Ease of access, lower (luggage) - Times and dates less important - regional airports or low cost solutions - Average standard of comfort - Non-coordinated times for direct flights - Important peak use - highly seasonal service - Main point is value for money

Table 4-8. The offer to be made in order to give a better service for business and leisure passengers

Bottom-up segmentation: from individuals to segments

This method, also known as typology, consists of grouping together customers or products as a function of relevant criteria. In this way they can be shown on a graph. There are two possible cases.

The company actually knows the criteria which are relevant to use for splitting up: For the freight market, this could be, for example, the operating range of the airplane and its freight capacity. This is a frequent case in business to business where relevant criteria are easier to determine (Figure 4.9).

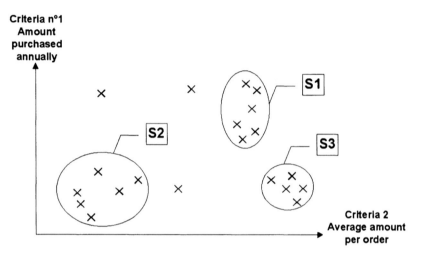

Figure 4-9. Market segmentation by the bottom-up method

Figure 4.9 shows an example of market segmentation using the ascending method. Each customer is positioned on the graph as a function of his performance on each of two criteria:
- criteria n°1: amount purchased annually in the category under consideration,
- criteria n°2: average amount per order.

A segment is obtained by grouping together customers close to each other on the graph. Three segments can be identified, S1, S2 and S3:

- Segment S1, a high priority area corresponding to customers with a high annual potential who make large orders, inciting the supplier company to offer a specific mix ;

- Segment S2 groups together customers with a low annual potential making smaller and more frequent orders. The price will be higher here, with a more elaborate service to fulfill their expectations;

- Segment S3 groups together customers who only buy a relatively small but expensive amount annually. This means reduced costs and a moderate service.

4. Market Segmentation and Positioning

The three customers situated outside the segments must be analyzed separately if they are important to the company, otherwise they will be offered the mix of the nearest segment.

- When customers are scattered on the graph with no clusters possible, this means that the criteria used are not relevant. Here, the whole analysis must be started again using other criteria and this could imply carrying out a qualitative study to find variables which are more "segmentable".
- When we are dealing with a market made up of a restricted number of professional players, traditional segmentation is not used. Each customer is taken to be a separate segment in its own right, needing a separate and specific approach. The airline market for Airbus and Boeing, the two main aeronautical manufacturers, is a case in point. Representing a total of 271companies (excluding C.E.I.) it can be analyzed and targeted individually.

The segmentation of civil aircraft can be carried out using the Boeing and Airbus product portfolios established as a function of passenger capacity and operating range (Figure 4.10). This gives the following segments:
- very large carriers,
- large, long haul carriers,
- wide body, medium size carriers,
- single aisle, medium size carriers.

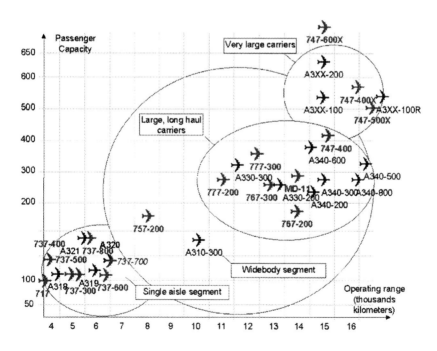

Figure 4-10. Segmentation of the civil airplane market as a function of the passenger capacity and operating range

The company does not know the relevant criteria for splitting up: Using a customer oriented approach, the company must study this population in order to define the segments. Having identified them in terms of age and socio-professional category, it is a case of questioning customers as to their attitudes and opinions and their behavior concerning the media and distribution circuits for the products and services offered. It is then possible to set up a customer typology using this information. The "identity cards" for each group of customers can be very detailed according to the number of criteria covered and the number of replies received.

BOTTOM-UP SEGMENTATION: THE EXAMPLE OF THE BUSINESS JET MARKET[5]

The business jet market is characterized by a varied offer of airplanes from about twelve main manufacturers: Envoy 3 (Fairchild Dornier), Learjet (Bombardier), Hawker (Raytheon), Falcon (Dassault), Citation (Cessna), Gulfstream, Galaxy, etc. Although they are sometimes thought of as a company privilege or an unjustified luxury, the business jet is nowadays associated with a dynamic and efficient company image. It is true that unlike commercial jets, these business jets are very often on the ground which represents an extremely high immobilization cost (fixed and variable charges and depreciation) approaching a million dollars annually. However the business jet must above all be thought of as a flexible, rapid and efficient working tool, which allows management teams to travel to their own or their customer's different company sites around the world. The largest market is to be found in the United States with more than 4,000 airplanes, followed by Europe with 1,600 (of which 500 are in France, 400 in Germany and 300 in the United Kingdom). Over the last five years, more than 2,000 new business airplanes have been delivered (including turbo-props).

Generally a market groups together customers with similar needs (even if these clients are not themselves homogeneous), who differ in the advantages that they are looking for, the specific functions that they need or the amount that they are ready to pay. Thus on the business jet market the main customers are:

- private personalities ;
- company directors, CEO ;
- management teams.

In order to understand the business jet market it is necessary to segment it to be able to better target one or several categories of customers identified, with the best possible adapted offer. But in addition this segmentation governs the capacity of the company to prepare its future products/services. Thus recently in the United States new players have appeared on the business airplane market with the concept of fractional ownership. They offer their customers all the advantages of having a plane, for only part of the cost. They are aiming at customers who do not really need to own a plane permanently because they will only use it a small amount, but who nonetheless need one for their business (flexibility, productivity). The main fractional ownership companies are Netjet (250 airplanes), Dubbed Private Fleet (140 airplanes), Flexjet (70 airplanes), TravelAir (50 airplanes) or Flight Options and JetCo.

4. Market Segmentation and Positioning

A geographical segmentation of customers can be made according to whether they come from the North American continent, Europe, the Middle East, Asia-Pacific or South America. Each of these regions is distinguished by characteristics in terms of market growth, company purchasing potential, tax regulations, monetary value, etc. Certain segmentations have been tried using the number and type of engines on the business airplanes but they did not work. At present, more informal criteria are used (ease with which one can stand up in-flight, ease of movement around the plane) to highlight different possible offers. In general, market segmentation for planes is according to capacity (in size and distance) or price and these criteria in fact reflect the expectations of customers in terms of type of airplane. In this way, the business jet market can be split up into five segments (described below) and can be represented in the following way (Figure 4.11).

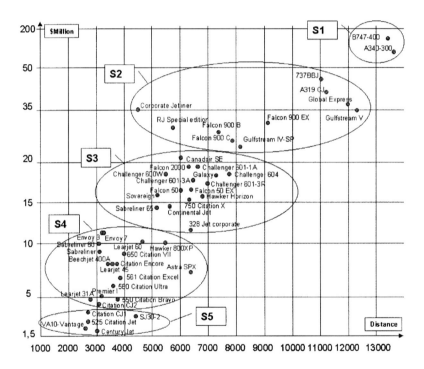

Figure 4-11. Segmentation of the business jet market

- Segment 1: With a price of more than $150 million, the business jet is destined for VIP's (Table 4.12). The exact price is often difficult to obtain when the plane has specific or even secret equipment as is the case with Air Force One for the President of the United States. For presidential airplanes, national manufacturers if they exist are preferred (B747 in the United States, A340 or Concorde in France, etc.). This is a very narrow market segment, with heads of state mainly opting for the hire of a plane rather than its purchase.

Manufacturer	Plane	Price $ Million	Passengers	Distance km	Engines	Speed km/h
Boeing	B747-400	167,5-187	416-524	13570	4	910
Airbus	A340-300	150	295	13750	4	

Table 4-12. Segment 1: Airplanes worth more than $150 million

- Segment 2: With a price of between $23-45million, the "large cabin bizjet" is luxuriously equipped, large (less than 100 seats) and capable of flying long distances (7,000 to 12,000 km). Examples are the Boeing Business jet (BBJ) and the Airbus Corporate Jet (ACJ) (Table 4.13).

Manufacturer	Plane	Price $ Million	Passengers	Distance km	Engines	Speed km/h
Airbus	A319 Corporate Jet	40	10-40	11500	2	870
Boeing	737 Boeing B. Jet	45	12-60	11100	2	870
Bombardier	Global Express	37,5	8-19	12040	2	924
	Corporate Jetliner	35	3-30	4518	2	837
	RJ special edition	28	14-19	5780	2	837
Dassault	Falcon 900B	26,5	12-19	7401	2	870
	Falcon 900C	25	12-19	7908	2	870
	Falcon 900 EX	29,4	12-19	9050	2	870
Gulfstream Aerospace	Gulfstream IV-SP	24,6	3-19	8019	2	870
	Gulfstream V	35	3-19	12582	2	870

Table 4-13. Segment 2: Airplanes worth between $23– 45 million

Designed for governments or private companies. The interior can be fitted out as offices, meeting rooms, individual work space, private suites with bathroom facilities, sports training facilities and communication areas. These airplanes offer a sufficiently large amount of space to house an entire working team.

- Segment 3: Costing from $11–20 million, this is the "super-mid-size cabin bizjet" (Table 4.14). With a seating capacity of 13 to 19 seats it is designed for intercontinental flights (5,000 to 7,000 km) e.g. non-stop transatlantic, Boston - London, or non stop coast to coast in the United States. Examples are Bombardier's Canadair SE and the Challenger 600W, the Falcon 50 and 50 EX from Dassault, the Cessna Sovereign or the Hawker Horizon from Raytheon. These airplanes allow business leaders to visit their different sites and to negotiate with their customers on transcontinental flights. This is the segment with the highest growth potential.

4. Market Segmentation and Positioning 127

Manufacturer	Plane	Price $ Million	Passengers	Distance km	Engines	Speed km/h
Bombardier	Canadair SE	22	3-19	6074	2	837
	Challenger 600W	18,6	2-19	5562	2	837
	Challenger 601-1A	19,5	2-19	6661	2	870
	Challenger 601-3A	17	2-19	6549	2	870
	Challenger 601-3R	16,8	2-19	6959	2	870
	Challenger 604	18	2-19	7750	2	870
	Continental Jet	14,25	8	5700	2	870
Cessna	750 Citation X	15,6	8-12	6351	2	979
	Sovereign	16	8-12	5221	2	822
Dassault	Falcon 50	17	9-19	6058	3	870
	Falcon 50 EX	17,25	9-19	6645	3	870
	Falcon 2000	19,5	8-19	6396	2	870
Fairchild Dornier	328 J. Corp Shuttle	11,9	2-30	6396	2	727
Israel Aircraft	Galaxy	18	2-19	7310	2	892
Raytheon	Hawker Horizon	15,8	2-13	6819	2	892
Sabreliner	Sabreliner 65	14	12	6819	2	846

Table 4-14. Segment 3: Airplanes costing between $11 – 23 million

- Segment 4: Costing from $4-11million; this is the " mid-size cabin bizjet " (Table 4.15). It has a capacity of 6-12 seats for an operating range of 3,000 to 5,000 km. Examples are the Learjet 60 and 45 from Bombardier, the Astra SPX from Israel Aircraft International, or the Sabreliner 60 from Sabreliner. They are designed for private passengers, government or company teams wishing to do business within a given region (Europe or the United States). The main advantage of these airplanes is their great flexibility and reduced operating costs.

Manufacturer	Plane	Price $ Million	Passengers	Distance km	Engines	Speed km/h
Bombardier	Learjet 60	10,5	2-10	4758	2	859
	Learjet 45	8	2-9	3673	2	827
	Learjet 31A	4,6	2-7	2869	2	870
Cessna	550 Citation Bravo	4,6	7-11	3702	2	745
	560 Citation Ultra	6	7-8	3630	2	743
	561 Citation Excel	6,8	7-11	3852	2	796
	650 Citation VII	8,7	7-13	4037	2	882
	Citation CJ2	4,4	6-7	3110	2	740
	Citation Encore	8	7-10	3704	2	796
Fairchild Dornier	Envoy 3	11	2-12	3226	2	755
	Envoy 7	11	2-12	3295	2	718
Israel Aircraft	Astra SPX	11	2-9	6476	2	892
Raytheon	Premier I	4,9	2-6	3216	2	881
	Hawker 800 XP	10,5	2-15	5475	2	815
	Beechjet 400 A	8	2-9	3450	2	827
Sabreliner	Sabreliner 60	10	10	3147	2	835
	Sabreliner	9	10	3147	2	835

Figure 4-15. Segment 4: Airplanes costing between $4 – 11 million

• Segment 5: Costing from $1,5-4 million; this is the "lightweight bizjet". It has a capacity of 2-6 seats for an operating range of 2,000 to 3,000 km (Table 4.16). Examples are the Citation CJI from Cessna, the Sino Swearingen SJ30-2 or the Vision Aire VA10-Vantage. These type of airplanes, designed for short distances are the most economical to run.

Manufacturer	Plane	Price $ Million	Passengers	Distance km	Engines	Speed km/h
Cessna	525 Citation Jet	3,3	4-5	2750	2	718
	Citation CJ1	4	5-6	2730	2	703
Sino Swearingen	SJ30-2	3,7	2-6	4600	2	850
Century Aerospace	Century Jet	1,9	4-5	3000	1	664
Vision Aire	VA10-Vantage	2	4-5	2600	1	653

Table 4-16. Segment 5: Airplanes costing between $ 1,5 – 4 million

1.3 Other segmentation methods used in B to B

Wind and Cardozo segmentation

A method for segmenting industrial markets in two steps was proposed by Wind and Cardozo ; their model is based on the use of macro and micro-variables (Saporta, 1987). The first phase which is called "macro-segmentation" is considered to be the easiest and most cost effective method for grouping industrial customers in homogeneous segments. It is based on the use of general variables such as activity sector (food, mining industry, etc.), company size (number of employees, machines, turnover, etc.) and company geographic location (number of sites, nationality, head office, agencies, etc.). Generally, these variables are easily observable, in particular using secondary sources. At this stage, it is not necessary to carry out individual interviews with customers. Unfortunately, this macro-segmentation is generally not sufficient. In fact, the different types of company organization and buying behavior are not explained in any depth using such general criteria.

The second phase, called "micro-segmentation", consists in re-dividing each macro-segment into more limited groups as a function of the characteristics of buyers. Thus micro-segmentation requires better knowledge of the customer, beginning with his needs and the type of relationship he has with suppliers. Essentially, the following variables of micro-segmentation are analyzed:
 - the degree of use (new users, non-users, already users, customers, potential customers, etc.),
  - the degree of customer expertise (how far back does the know-how go),
  - the type of purchase (new task, straight rebuy, modified rebuy),
  - the type of buying procedure (level of centralization of decisions),

- the level of influence of people and departments in the customer companies,
- desired advantages (relative importance of customer expectations),
- personal characteristics (training, experience, etc.).

Micro-segmentation requires in-depth information about customers: thus the sales representative can adapt his product/service to the particular micro-segment selected. However, the method is inconvenient in that it uses primary sources[6]: the sales rep must gather data during personal interviews with their many contacts, which makes the process long and costly.

An example of segmentation on the cargo airline market can be presented taking a sample of six cargo airlines working in particular in Asia, Africa and South America. In this case, the variables chosen were the customer's sector of activity for the macro-segmentation and buying process for the micro-segmentation. The percentages correspond to the relative weight of each segment in total potential turnover (Table 4.17).

In %	Electronics	Textile	Fruits & vegetables	Other	Total
Completely centralized buying	3	4	11*	0	18
Buying centralized by establishment	18*	15*	7	2	42
Buying entirely decentralized by department	17*	10*	3	10*	40
Total	38	29	21	12	100

* Target market

Figure 4-17. Segmentation of the cargo market of a sample of six airlines

- Once the macro-variable has been chosen, the activity sector makes it possible to classify customers into four areas: electronics, food (fruit and vegetables), textiles, and others.
- The micro-variable of the buying procedure includes three levels:
  - purchasing of the transport service is completely centralized by the customer,
  - purchasing is centralized by customer sites,
  - purchasing is totally decentralized by department of each site.

The six main segments shown by this analysis represent 81% of the total potential turnover. Depending on other parameters that need to be verified, they can thus be the sales priorities corresponding to the target market.

Bonoma and Shapiro segmentation

In the spirit of the preceding segmentations, Bonoma and Shapiro (1983; Berrigan and Finkbeiner, 1992), have proposed a "nested approach", based on five categories of data. Their first concern is managerial: for the company, it is better to begin the segmentation with variables that are easier to measure, and finish with those that are the most difficult. But this approach is supposed to be more pragmatic in that to obtain the most relevant information, the analysis must follow the progression of the five following categories (Figure 4.18 and Table 4.19).

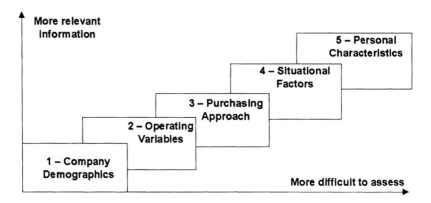

Figure 4-18. The Shapiro and Bonoma model: the nested approach

Company demographics: economic sector, size and location of the company ;

Operating variables: technology used, users and non-users of a product or a brand, purchase of associated products, customer know-how ;

Purchasing approach: organization and relationships with the buying center, hierarchical structure, centralized or decentralized purchasing, necessary bidding process for suppliers and purchase criteria ;

Situational factors: order-type, amount, degree of urgency of the order ;

Personal characteristics: affinities between sales representative and buyer, buyer motivations, individual perceptions, degree of loyalty.

Table 4-19. The succession of the five categories

The succession of the five categories are linked to the industrial buying process and the complex relationships with the different contacts in the buying center. The nested presentation makes it possible to better understand the progressive phases in building a trust-based relationship.

THE SEGMENTATION OF THE FREIGHT MARKET

By definition the freight market is global. In recent years it has become much more competitive, forcing the players involved to constantly seek increased productivity. The growth rate of the cargo market is higher than that of the passenger market[7]. However, this market is very heterogeneous ranging from the transport of small packages, flowers, chicks, to the transport of aircraft fuselage and heavy equipment. Furthermore, the cargo can be seasonal (animals, flowers, fruit), two way (a different load for the return trip) or one way (without any return load). For maximum productivity, in the latter case it is necessary to find a "triangular" solution, taking into consideration the most active destinations in terms of freight. In service terms, today the market "norm" is door-to-door delivery in a defined time ; cargo operators are much more than simple transporters, they are real service providers, participating in their customers' productivity[8].

On the cargo market, the following needs to be distinguished:

• The express package segment: frequent shipping, the weight of the package being under 32 kg (70 lbs) with homogeneous loads. Short delivery times, high costs, and big profit margins.

• The freight transport segment: shipping once or twice a week, weight greater than 227 kg (500 lbs) with variable loads, fixed delivery times, lower costs and average profit margins.

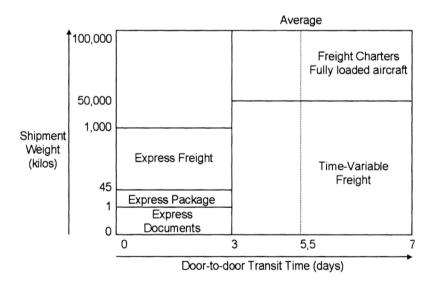

Figure 4-20. Segmentation of the cargo market as a function of the weight of freight and the delivery time

By focusing the analysis on the freight, a segmentation of the intercontinental freight market[9] can be made based on the two main criteria: the duration of the transit (rapid, or less than three days door to door, or slow, greater than 3 days) and the weight transported. For the last 20 years, on average a door-to-door transport has an average stable duration of 5-6 days.

Another segmentation of the players operating in the cargo market can be made as a function of the product range (transit time/ cost options) and the geographical range of their services (sales points).

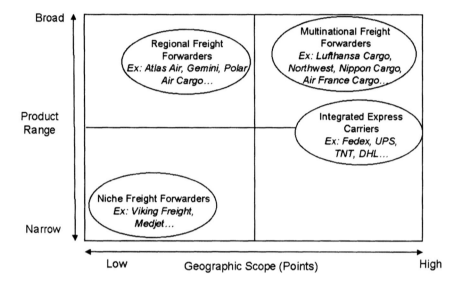

Figure 4-21. Segmentation of the cargo market as a function of coverage and product range

As far as cargo aircraft are concerned, a segmentation can be made as a function of the respective tonnage capacities.

Segment	Capacity	Types of aircraft	Accessible cargo market
1	< 30 tons	BAC One – Eleven, BAe 146, DC-9, B737, B727	22 tons
2	30-50 tons	B707, DC-8, B757, A310, A300-134	45 tons
3	50-80 tons	B767, DC-10, A300-600, L-1011	60 tons
4	> 80 tons	MD11, B747	100/120 tons (used/new)

Table 4-22. Segmentation of the cargo aircraft market

2. POSITIONING

The different segmentation methods presented provide a useful framework for strategic analysis of the markets, classifying the different segmentation factors as extensively as possible. Segmentation must be used pragmatically, taking into consideration each context and using the most fitting bottom-up or top-down methodology.

Segmentation is an essential phase in the marketing approach. Thanks to effective segmentation, the company can better target its product or service: once the segments have been defined, they have to be analyzed and evaluated in terms of their potential in volume and financial value. Targeting consists in choosing which specific segments the company is going to focus on. Then relevant positioning can be established for each segment selected, which enables the company to make the necessary decisions for each variable in the mix (Figure 4.23).

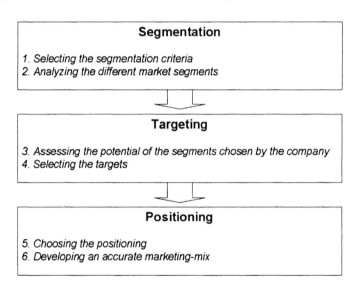

Figure 4-23. Two essential aspects of marketing: segmentation and positioning

2.1 Positioning objectives

The goal of positioning (Dubois and Nicholson, 1992) is to create a determined place and "personality" for products or services, in other words, a distinctive value in the eyes of potential customers relative to the products of direct or indirect competitors. How the market reacts to this product depends on the relevance of the positioning ; if the latter is effective, it will generate profits, creating a competitive advantage in the long term. Ideally, the positioning is immediately visible to the target populations.

The two dimensions of positioning

Positioning (Ries and Trout, 1981) occurs at the level of the product category and the company (Figure 4.24).

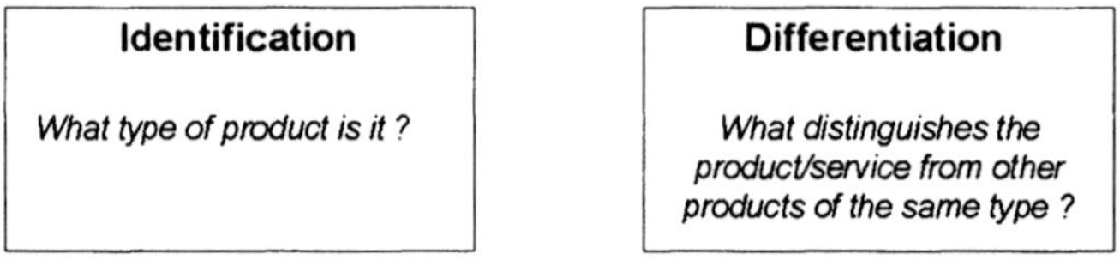

Identification	Differentiation
What type of product is it ?	*What distinguishes the product/service from other products of the same type ?*

Figure 4-24. The two dimensions of positioning

Identifying the product category is the first aspect of positioning. This is necessary when the product is not clearly defined, the goal being to inform customers who are not necessarily aware of the advantages to be gained from using the product. This is especially true for emerging market segments such as express freighters. The latter is a new category of service which must be positioned relative to postal or cargo services of traditional companies. Identification is also necessary when the product or service falls in between two categories such as earth observation satellites. First used for military applications, these satellites have increasingly been used for civil markets: follow up of natural phenomenon (floods, desertification), ecological problems (deforestation, pollution). Thus it was important to explain the interest of the different possible services to prospective customer organizations.

Differentiation by brand is the second aspect of positioning related to the products or services offered by the company. When the segment is in the mature phase, customers are generally sufficiently informed. They know the product's or service's advantages. In this case, the latter have to be positioned relative to direct customers. This is true for airlines which have very different positions so as to make clear their distinctive services in the minds of customers, thereby hoping to eventually achieve favorite airline status.

EXAMPLES OF POSITIONING: AEROMEXICO AND THAI AIRWAYS

AeroMexico is positioned with a strong local emphasis. In its campaigns, the company name is associated with Aztec secular monuments or objects. AeroMexico aims to present itself as a "must" in Central America, with 32 connections from Mexico.

In comparison, Thai Airways stresses the legendary quality of in-flight service and the natural refinement and warmth of the Thai staff.

4. Market Segmentation and Positioning

Figure 4-25. "Historical wealth" positioning of AeroMexico with emphasis on Aztec traditions

---

The essential goal of positioning is to clearly situate the product or service in the minds of customers (Trout and Rivkin, 1996) (most attractive or distinctive features). It is thus a simplified representation of the product, service or brand when it covers several different products. This vision is comparative relative to that obtained by competitors. The ideal thing is for positioning to be sufficiently distinctive ; as such it should not be too complex, rather, two or three essential characteristics should be stressed. Naturally, positioning is reductive compared to the more complex reality of the product. The four essential criteria for evaluating positioning are:

- Simplicity: One of the basic advertising principles is the Unique Selling Proposition, according to which an advertisement should only present one promise at a time. A position should be simple so as to be easily understood and memorized. For example, in preparation for the launching of the A400M military transport aircraft, Airbus Military Company positioned its offer as the European solution to European needs.

- Relevance: The idea is to choose one or several criteria which correspond to the main expectations of the determined target. In air transport, this could be flight punctuality, quality of service, or the density of the network in a particular geographic area.

- Originality: If the positioning criteria chosen by the company are the same as those of competitors, the differentiation is not functioning. The uniqueness of the positioning is an important factor in the efficiency of the advertising. This is why the criteria retained to express the differentiation are often irrational (the rational or logical criteria being the first used by competitors). Detailed analysis of the product relative to competing products reveals that in general they have many points in common. Positioning, on the other hand, is based on differences.

- Credibility: Whether they are rational or not, correspond to characteristics of the product or company, its nationality or directors, the criteria selected must be credible. An overly theoretical or exaggerated positioning will only disappoint customers.

The following examples (Table 4.26) illustrate the four required qualities for effective positioning.

	Singapore Airlines	Air France	Swissair	American Airlines
Positioning	The most modern fleet in the world	The "French touch", beauty	'Refreshing' airline	Free space
Simplicity	Whatever the aircraft taken by the passenger, the latter benefits from a recent plane	An airline which provides the customer a certain serenity: "Making the sky the best place on earth"	An airline with extremely high quality services (languages, refreshments...)	Whatever the class the passenger chooses, freedom of movement is guaranteed by more space
Relevance	A recent fleet is reassuring for passengers who are looking for a reliable company and quality services	Passengers want to spend a pleasant moment and forget their worries during the flight	Associated with the origins of the airline (the Swiss Alps, Mountain Air)	On long haul flights, passengers increasingly want to be able to stretch out and travel comfortably
Originality	Different from traditional positioning (number of flights, welcome...)	More than an airline, this positioning affirms a message of escape or getting away	The idea of freshness in contrast to traditional themes	A strong positioning which affirms the personality of the company
Credibility	The fleet is made up of richly equipped jubilee 747's and 777's and A340's.	Particular care in on-board services and a pleasing, attractive interior	The staff speaks several languages (French, German, Italian, Hindi, Chinese, Japanese)	The commitment of freedom of movement is exemplified by the new design of the B767-300 fleet.

Table 4-26. Simple, pertinent, original and credible positioning

4. Market Segmentation and Positioning 137

Figure 4-27. One aspect of American Airlines positioning : free space

Figure 4-28. One aspect of Swissair's positioning: "the world's most refreshing airline" [10]

Two other companies can be cited for their effective positioning: Virgin Atlantic and easyJet (Table 4.31). The creation of Virgin Atlantic symbolizes the spirit of the Virgin group and its founder, Richard Branson (Figure 4.29).

Figure 4-29. Richard Branson, the charismatic president of Virgin Atlantic in front of a Boeing with the company's colors[11]

In 1984, Virgin was one of the world leaders in the music industry. The young president of Virgin, following a discussion with an English-American jurist, Randolph Fields, decided to create a new airline capable of operating right from the next quarter. Today, the number two British airline for long haul flights, Virgin Atlantic humorously recalls its history on its website. " What's more, Madonna flatters us by believing she's like a Virgin ". This youthful spirit is characteristic of the company's advertising campaigns (Figure 4.30).

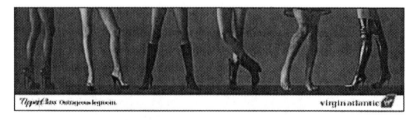

Figure 4-30. Example of Virgin Atlantic advertising[12]

In the same way, easyJet, founded 10 years later by its president Stelios Haji-Ioanou, is an atypical company (Table 4.31). When it was created, it was presented as the first company to not use the services of agencies to sell its tickets. It has continued to shake up the sector by displaying strong impact campaigns on aircraft (Figure 4.32).

4. Market Segmentation and Positioning

	Virgin Atlantic	easyJet
Positioning	Cheeky youth	Low cost and friendly
Simplicity	Identification with a young anti-conformist target incarnated by its founder Richard Branson	The message is simple: to travel as easily as possible you have to be clever
Relevance	A crossover brand (Virgin Cola, Virgin Holidays, Virgin Sun, Virgin Music...) aimed at an anti-conformist target	The target population looks for the lowest prices possible
Originality	Different relative to the promises of competitors	Positioning entirely based on prices
Credibility	Extremely high impact original advertising	No free meals or newspapers but regular cheap flights (advertising directly displayed on aircraft)

Table 4-31. The positioning of two atypical airlines

Figure 4-32. Example of ads displayed on easyJet aircraft

2.2 Setting up positioning

Positioning must be carried out as concretely as possible in terms of pricing, product, distribution and communication policies (Figure 4.33).

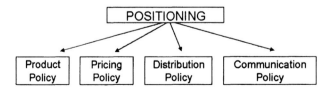

Figure 4-33. Positioning, keystone of the mix[13]

140   *Aerospace Marketing Management*

Careful applying of positioning to the concrete elements of the mix is indispensable. Most failures are not due to poor positioning but rather to lack of coordination between criteria selected and actions carried out.

Choosing positioning: When the company launches a new product or service, it has maximum leeway for positioning. In other instances, such as launching a product that has been modified or is a complement to the range, the existing positioning must be taken into consideration. By analyzing customer expectations and questioning customers about the relative importance of each expectation, a visual map can be established. To do this, each customer must grade the different suppliers on each criteria. Thus it is possible to determine the present positioning of competitors. From this, the company can use the map to visualize a new positioning opportunity.

Qualitative studies show the customer-perceived strengths and weaknesses of the product or service. The company can define to which expectations its product/service must correspond and what is the importance of these expectations. Then the company must verify the product's or service's credibility compared to competitors. Selection of outstanding features for customers follows the qualitative studies and data analysis.

The most used positioning criteria can be chosen based on:
- target customer expectations,
- positioning of competitors,
- additional product assets.

A case in point is the long distance tourism market. In general, tourists from developed countries have four main expectations (Figure 4.34).

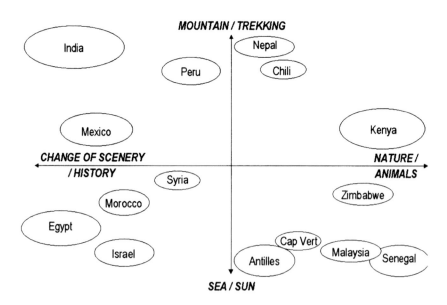

Figure 4-34. Example of visual map: positioning of countries in terms of tourism

4. Market Segmentation and Positioning

These expectations are:
- Sea / Sun,
- Change of scene / History,
- Nature / Animals,
- Mountain / Trekking.

For a study (Figure 4.34), tour operators were interviewed and their answers were used to establish the main positioning of tourist countries (excluding developing countries).

Main differentiation strategies: These depend on product characteristics, correspondence with a specific expectation (niche strategy), or a symbolic dimension of the product.

Positioning can in particular highlight:

- Certain product characteristics: performance, design, functions. In comparison to its competitors, American Airlines differentiates itself by offering more spacious seating and free use of in-flight telephones. Air Liberté, on the other hand, perfumes the cabin with a pleasant, fresh fragrance and renews the air every four minutes.

- Solutions provided by the product: productivity, practicality, easy maintenance and repairs. This can be also elements associated with the product such as Thai Airways, which has an excellent reputation thanks to the quality of its service and flight staff.

- Use of product: frequent, occasional...

- Categories of users targeted: business people, tourists, children...

- The position of the product in reference to others: the most known, the most used on the market, most, solid, least expensive...

- An innovation introducing a new category: the first air surveillance system by radar coverage (Awacs), first civil aircraft with a jet engine (B707), "the world's first twin-engined with twin aisle" (A300), the first jumbo passenger aircraft (B747)...

Strategic positioning: In terms of strategy, there are three differentiation possibilities: at the level of the central product, the tangible or actual product, and the augmented product (Figure 4.35).

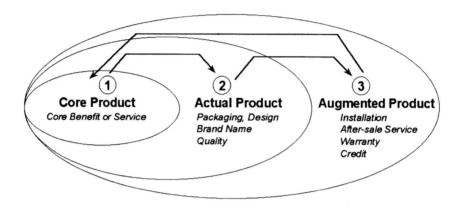

Figure 4-35. Three levels of product/service differentiation[14]

Core product (level 1) corresponds to the product's intrinsic characteristics. For a helicopter, this would mean the type of motor, fuel consumption, passenger capacity, equipment level, etc. In the beginning it is difficult to differentiate from competitors. For example, British Airways and Lufthansa occupy globally the same position on the European market. Lufthansa differentiates itself by promising reliable service and a wider offering of seats, while British Airways offers more cabin space and more comfortable business lounges.

Level 2 corresponds to the actual product, or in other words, concrete characteristics such as fuel presentation and its safety level (fuel delivered by pipeline). This level also involves irrational characteristics such as the brand.

Level 3, that of the augmented product, can be illustrated by after sales service or customer financing.

Ideally, positioning should be based on differentiation at the level of the core product. Before competitors have caught up with this level, it is wise to have already found a fresh competitive advantage at the actual product level. Once this advantage has been neutralized there remains the possibility of new differentiation at the augmented product level. The period of years over which this differentiation evolves from the first to the third level can be used to prepare and set up a new generation product. Thus it is possible in this way, using R &D, to start off once more with the new differentiation based on the core product.

This succession of differentiation levels is highly strategic. Positioning of the brand is the result of the positioning of the different products which it covers. As such, the coherence of the product portfolio, not only in terms of life cycle but also positioning of each of the products, must be carefully watched over.

In business to business, the main orientation of positioning chosen is often linked to the offer, allowing professional customers to identify the type of service provision and to distinguish it from the competition. Lufthansa Technik must therefore position its maintenance service offer in terms of those of the airplane,

4. Market Segmentation and Positioning

equipment or engine manufacturers as well as those from other airlines or independent maintenance service providers.

LUFTHANSA TECHNIK: POSITIONING BASED ON CUSTOMER SUCCESS

They have been able to communicate to customers their evolution into a customer-oriented company, supplying technical services the world over (Hamburg, Shannon, Brussels, Budapest, Miami, Dallas, Tulsa, Sun Valley, Manila, Beijing). This positioning has been achieved using different campaigns: more than just maintenance, Lufthansa Technik contributes to its customers' success.

Figure 4-36. Lufthansa Technik's positioning: above and beyond maintenance, contribute to success[15]

The strategies for modifying positioning: The company must start analyzing its offer when there is an obvious difference between the theoretical positioning, usually defined by the company, and the positioning as perceived by customers. Corrective action can be taken concerning the mix in an effort to move the perceived positioning towards that which has been previously defined, or a new positioning can be adopted. Repositioning generally takes several years. It is costly and risky and can be interpreted as an admission of failure for the original strategy chosen.

There are several strategies for modifying positioning:

- Modification of the product or service: this costly strategy allows the company to better meet customer expectations ;

- Modification of the weighting given to the different attributes: this means modifying, in the eyes of the customer, the relative importance of the different choice criteria ;

- Modification of beliefs associated with the brand: this entails carrying out an effective communications campaign designed to influence the customer's perception of the supplier brand ;

- Highlighting of an important attribute as yet unknown to customers. Boeing for example in its European campaign, put the accent on its contributions to the European economy in terms of jobs created or activities generated and sub-contracted by them within Europe. In a similar fashion, Airbus in its US campaigns presents itself not as a European but rather as an international manufacturer by citing the large number of American companies associated with its aeronautical programs.

It should be noted that companies are sometimes obliged to adopt different positioning for the same product, as a function of geographical zone. This is the true for example where they make an offer on an international scale in countries with very different buying contexts ; the type and importance of customer expectations can vary from one zone to another as a function of the level of equipment and development. The positioning adopted must take account of this. In this way a company can obtain different positioning according to whether it is looked at from the point of view of the European, Asian or American market. Nonetheless for concentrated world markets, it is usually preferable to have one positioning covering all countries in order to reinforce promises to customers by the coherence thus accrued.

NOTES

1. See Chapter 8, Pricing Policy.
2. See Chapter 5, Marketing and Sales Action Plan.
3. Source: Alcatel Space
4. See Chapter 3, Business Marketing Intelligence.
5. Sources: Jane's, Aviation Week and Air & Cosmos (1999-2000)
6. See Chapter 3, Business Marketing Intelligence.
7. See Chapter 3, Business Marketing Intelligence.
8. See Chapter 7, Marketing of Services, 5. Focus on the freight market.
9. Source: MergeGlobal, Inc. analysis.
10. Ad published in Time
11. See Virgin Website: virgin.com
12. See Virgin Website: virgin.com
13. Source: Lendrevie, J. and Lindon, D., (1997), *Mercator*, Paris, Dalloz.
14. Adapted from Kotler, P. and Armstrong, G., (1999), *Principles of Marketing*, Prentice-Hall International, New Jersey, Chapter 8, p 239.

15. Ad taken from *Air Transport World*, (1999), June, p 12-13.

REFERENCES

Beane, T. P. and Enis, B.M., (1989), Segmenting markets: state of art, *Recherche et Applications en Marketing*, vol n°3, p 25-52

Dubois, P.-L. and Nicholson, P., (1992), Le positionnement, *Encyclopédie du management*, Paris, Vuibert.

Malaval, Ph., (2001), *Marketing business to business*, 2nd Ed., Paris, Pearson Education France.

Ries, A. and Trout, J., (1981), *Positioning: The battle for your mind*, New York, McGraw-Hill.

Saporta, B., (1987), La segmentation en marketing industriel, *Recherche et Applications en Marketing*, vol n°4, p 39-51.

Shapiro, B. P. and Bonoma, T. V., (1983), *Segmenting the Industrial Market*, Lexington Books, Lexington, MA ; Berrigan, J. and Finkbeiner, C., (1992), *Segmentation Marketing*, New York, Harper Business.

Trout, J. and Rivkin, S., (1996), *Les nouvelles lois du positionnement*, Paris, Village Mondial.

Chapter 5

MARKETING AND SALES ACTION PLAN

Right from the very beginnings of marketing, planning has been identified as an important component. Taken to an extreme, it has made a nonsense of marketing. The marketing plan must be a pragmatic tool directly based on action plans.

1. THE MARKETING PLAN

" The object of a marketing plan is to put into writing the marketing policy chosen by the firm in order to ensure that the company's marketing actions are coherent. They must correspond to the sales and resource potential of the company, and to the business environment" (Dubois and Jolibert, 1992).

The marketing plan's different definitions are all inter-linked. In essence it is a document including:
- an analysis,
- the strategy and the objectives,
- the means or the action plan.

• All marketing plans contain the latter three steps which must logically follow each other. In other words, objectives are defined as a function of the analysis, and this quantified definition must be as precise as possible. The result is an estimate of the margin that the marketing department envisages earning once the objectives have been officially approved.

The third step is the means that the company is ready to deploy in order to carry out these previously defined objectives. Whether they concern the sales force, distributors, communication or the product plan, they must be budgeted, and it is up to the managing director to set the constraints involved in terms of profitability. However, the marketing plan has also gradually become an important tool for mobilizing personnel. It is highly motivating for staff to realize that a large part of the company's profits is reinvested whatever the form – sales, technology or advertising.

Companies generally make an annual marketing plan. A period of 12 months is considered necessary in marketing because it is:
- sufficiently long to be able to put things into perspective, e.g. seasonal changes,
- sufficiently short to be pragmatic, allowing concrete applications.

A yearly marketing plan is particularly recommended for companies: even if it is short, it will serve as a useful guide for the correct functioning of the company. It provides directors, managers and staff with a perspective enabling better coordination of the different actions.

• A three year plan updated annually as a function of events is used by very large companies. This is viable as long as it is not highly detailed ; marketing managers should be modest when faced with the difficulties of forecasting. Monetary, economic or political crises regularly demonstrate how unrealistic three year old statistics are, which sometimes even go as far as to predict sales per customer and per product ! It is recommended that the marketing department produce a " 3 year strategic orientation" document giving the main trends and development objectives. Whatever the case, this 3 year plan can only be produced if the yearly one has already been an integral part of company functioning for several years.

• The marketing plan is a lot of work and it often can take six months to complete. It requires numerous exchanges of information and propositions from the marketing department ; indeed the marketing plan must be the result of teamwork (Figure 5.1).

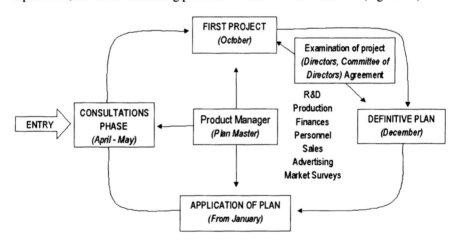

Figure 5-1. Example of setting up a marketing plan

It cannot be written by one person nor one department, however competent. In order to be realistic, participants must include operatives from sales and technical departments, managers in charge of human resources, quality and communications, etc. The product manager (if there is one) is the main driving force and should be like an editor in chief, breaking the task up into parts for the different managers. In this way, the marketing plan will stand a good chance of being accepted by the different managers and their teams. It will be a reference document for use over a whole year by the different managers. For example, when the financial director is uncertain as to the financial straits of a customer, he will be able to consult this document and get a summary of the latter's standing in the overall activity of the company or for one particular product category. Properly produced, the marketing plan can thus be used as "marketing for the marketing department" thus legitimizing the latter. If not, it will be taken as just one more useless document and the marketing department will therefore be seen as ineffective and a waste of money.

1.1 Part one: the analysis

Unlike the other steps, there are two possible plans for the analysis (Malaval, 2001). Some people prefer an initial external analysis of the environment followed by an internal one of the company. Others prefer an analysis by theme. The choice depends on the size of the company and the number of products offered.

First type of plan: external / internal analysis

This has the advantage of being systematic and avoiding omissions. It does however have the disadvantage of being redundant that is the same topic, e.g. communication, must be covered twice, first for internal, then external purposes. In the external diagnosis the full range of opportunities and threats faced by the company must be covered:
- legislation and regulations,
- the monetary, economic and political context,
- technological innovation,
- the socio-demographic evolution (trends),
- the competition: direct and indirect,
- the main distributors,
- the main influencers,
- the main suppliers.

	Opportunities	Threats
Global environment (legal, technological, sociological, etc.)		
Influencers		
Suppliers		
Distributors		
Competitors		

Table 5-2. Example of a table summarizing external diagnosis

The aim is to identify any outside factors affecting the company and its main partners in order to detect any unfavorable movements or threats, and also anything positive which could be an opportunity. It is important to summarize the main points in a table (Table 5.2). In internal diagnosis, the company's strong and weak points must be assessed. These are then analyzed under the following headings:

- products (grouped together by category),
- customers (on the one hand key accounts and on the other the rest of the customers grouped into segments),
- geographical areas,
- distribution channels,
- the sales force,
- the communication policy.

It is a question of making a diagnosis of the internal situation, and this must also be summarized in a table (Table 5.3).

	Strengths	Weaknesses
Products		
Customers		
Geographical areas		
Distribution channels		
Sales force		
Communication (branding, advertising, sales promotion)		
Pricing strategy		

Table 5-3. Example of a table summarizing internal diagnosis

The drawback of this method, the SWOT analysis (strengths, weaknesses, opportunities, threats), is its built-in redundancy, which is why a plan by themes could be preferable.

Second type of plan: by theme

In this case, each topic is analyzed from both an external and internal point of view, which avoids covering the same one twice.

Analysis of the company within its market: This part can be divided under three main headings.

- The macro environment: this includes analysis of legal, monetary, sociological, technological changes, etc. This could for example be the development of sensitivity to ecological issues (noise pollution, emissions) or social ones (impact on local, regional, national employment).
- The descriptive analysis of the market: the size of the market is studied in terms of volume and value, as well as its evolution. It is necessary to show the

5. Marketing and Sales Action Plan

composition of the market in segments which must be defined, evaluated and followed up as they change. The product/market matrix gives an overview of the company within its market by comparing for each segment the respective rates of change and their corresponding weighting in the turnover. In the example given below (Table 5.4), the daunting situation for product X can be seen: it is only making 7% of its turnover with a 10% decrease in the volume, all this in segment S1 which accounts for 25% of the market and which in addition is progressing at +12%. At the same time, company X makes almost half (48%) its turnover, up by 5%, in segment S2 which only accounts for 8% of the market and which has fallen by -20%.

Segment...	S1		S2...	
	% Breakdown	% Evolution	% Breakdown	% Evolution
Market	25	+ 12	8	- 20
Brand X	7	- 10	48	+5

Table 5-4. Product / market matrix

- Competitor positions: the competitor companies are listed with their market share by volume and value and it is thus possible to set out the density or scattering of the market. Each of the main competitors is analyzed as to the share of its turnover per segment, and as a function of how these positions are evolving.

Analysis of the mix variables relative to the competition: This part covers the following four main points.

- The product: here there is an analysis of the sales structure per product, the range as well as the prices offered, all in relation to competing brands. In this section it is recommended to analyze the conditions of sale, in order to get as close as possible to the real price set by the company relative to its competitors.
- Distribution channels and management of the sales force:
  - analysis of distribution channels: they must be described in terms of their weighting and evolution (e.g. percentage development of the Internet sales channel) ;
  - main distributors: for each channel, it is necessary to identify the main players and assess the company's place, and its evolution relative to the competition.
- Analysis of sales:
  - sales are analyzed by product and by range compared with direct competitors,
  - analysis of the sales force: here the number of sales staff representing the company must be assessed, together with their level of training and or specialization in relation to competitors as well as the marketing structure.
- Means of communication:
  - for the company as well as for the main competitors, this involves analyzing the advertising budget, the share of voice, the awareness and the image perceived,

- the media and main channels used,
- the targets: analysis of the targets whether professional or general public.

In conclusion to this first phase, the diagnosis should allow the essential facts to be visible. Product performance must be explained and the elements on which action should be taken whether price, image, sales structure, etc. must be highlighted. It is not a question of being exhaustive, merely relevant and bringing out the essential points.

1.2 The objectives

Global objectives

In marketing the global objective is usually the market share (MS).

Hypothetical evolution of the market: A hypothetical market evolution must be established, because without this, the market share objective has no real meaning.

Market share in volume and growth objective: Establishing the market share is a policy decision which cannot be set by any simulations. When the market share objective is fixed for the following year, 20% in the example below (Table 5.5), the volume of sales can then be estimated in absolute terms, allowing the rate of change relative to the previous year (i.e. 22% here) to be calculated.

Deciding about tariffs: It is here also that the company makes the decision as to whether to keep last years tariff or to modify it. If it is raised, this becomes an additional objective: apart from selling 22% more products, the sales management must also now sell them at a 5% higher price (Table 5.5). The value of the estimated sales can be calculated (volume of sales x tariff), which means that the rate of change of the sales value is for the example given, +28,3%. When there is a price reduction this is not an additional objective but rather a measure designed to make it easier for the sales staff to meet their quantitative objectives.

Initial situation:
- For year Y: MS = 18%
- Hypothesis 1: market growth + 10% pour Y+1 / Y
- Hypothesis 2: the directors are asking for 5 points progress in the MS over 3 years, i.e. for the financial year Y+1: 18 + 2 = 20%.

Years	Y	Y+1	% Change
% Market Share	18	20	+11
Market	100	110	+10
Sales Volume	18	22	+22
Pricing	-	+ 5%	
Sales Value	18	23,1	+28,3

Table 5-5. Summary to show calculation of global objectives

5. Marketing and Sales Action Plan

The final objective is usually calculated as the contribution made per product or per range (obtained from the estimated value for the turnover).

Intermediate objectives

These are sub-objectives which should be obtained in order to meet the global objective. The main sub-objectives are:
- the distribution rate or presence of the company's products,
- the brand recall,
- the penetration rate (or trial if it is a new product).

Whether global or intermediate, objectives must always be specific, i.e. quantified and with a time limit.

The diagram below (Figure 5.6) shows that global objectives must be broken down into the intermediate ones necessary for their achievement. In the same way, the intermediate objectives make it possible to recommend concrete actions. For example, take the development of an Internet site for sales of tickets (action 1) and then the decision to set up a competition in travel agencies (action 2). Both these should logically lead to an increase in the level of ticket in circulation, from 40 to 60% in the example given. In turn, the new 60% circulation level will contribute to achieving the 20% market share which is the final objective. Similarly, reinforcing the advertising campaign on TV and in magazines (action 3) should result in an increase in the level of renown (brand recall) as well as penetration of the airline company. This is the reason for action 4 which sets up a special offer " discovery special " aimed at students.

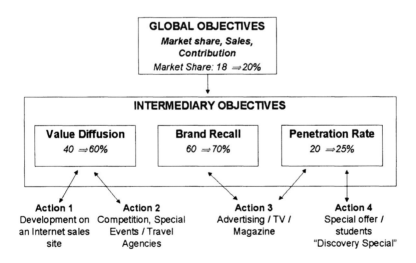

Figure 5-6. Global and intermediary objectives plus action plan[1]

1.3 The means or action plan

The means decided on by the company must be budgeted. Management will compare estimated costs to the profits forecasted in the objectives.

Global budgeting: The means are budgeted under three main headings:

- manpower: under this heading come hiring costs whether sales managers or other staff, plus training and " incentives ";
- the product plan: this is the amount which needs to be spent to launch new products but also the cost of improvements made to existing ones ;
- the communications plan: this covers spending on advertising space, agency fees, creation of new messages, etc. Sponsorship and sales promotion also come under this heading.

Presentation of the action plan: The ideal is to be highly pragmatic.

- For each action: it is recommended to follow the 5 point plan below.

- objective: this is the specific objective set for the particular action, e.g. raising the distribution or presence level of the product ;
- mechanism: the action must be described in precise detail as to how it will be carried out, e.g. the length of the operation, the amount of the bonus, etc. ;
- justification: this is explaining concisely why the recommendations for this action are relevant to achieving the objective fixed ;
- budget: all actions must be quantified as accurately as possible and cover several hypotheses when it is impossible to predict actual figures, e.g. the feedback level from a promotional offer ;
- planning: the starting and finishing dates of the operation are suggested and if necessary the reasoning behind the relative positions of two consecutive actions.

- Recap document: the different actions proposed should be summarized in one document with:

- budgetary recap: the global marketing budget is presented as " costs/action " table,
- forecasted income statement: as a function of the elements sent to the marketing department, it is a good idea to show which products are the most profitable and which on the other hand are going to be the subject of special attention:
- action plan recap: a timetable showing the logical sequence of the different actions (Figure 5.7) concludes the presentation. In the example below, the action by the sales force is planned for February and March. It logically comes before action 2, the advertising campaign in the press planned for April. If consumers want to buy the product as a result of advertising, they must be able to find it in the sales outlets. This is the sales team's job.

5. Marketing and Sales Action Plan

	January	February	March	April	May
Action 1		Sales Force Challenge			
Action 2				Advertising	

Figure 5-7. Example of a recap timetable for actions

The two diagrams below (differing only in the perspective chosen for the analysis) draw together the overall marketing plan giving details on the analysis (1), the objectives (2) and the actions plan (3). After this, the sales and marketing managers must follow up the levels of sales made and compare them to the objectives which have been fixed. Corrective actions can be taken throughout the year to make up for any eventual lateness in meeting the objectives (Figures 5.8 and 5.9).

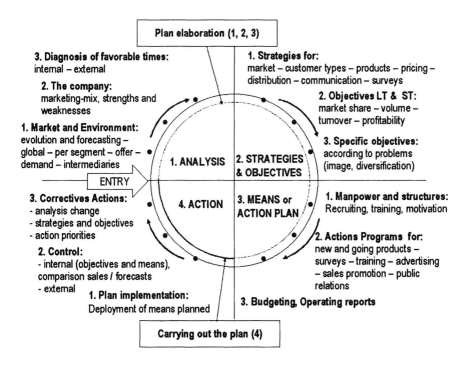

Figure 5-8. Presentation of the marketing plan: diagram to show external/internal analysis[2]

Can the objectives be renegotiated while the plan is running ? This depends to a large extent on the way a company works. It would seem preferable only to modify an objective if there is an important change in the company's environment such as a monetary crisis, a war or major catastrophe. This is because, if the objectives are modified without a really 'earth shattering reason', they lose all their credibility in the eyes of the operative managers especially those in sales.

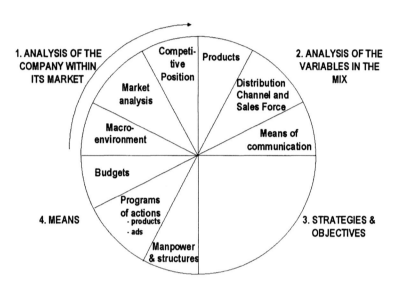

Figure 5-9. Presentation of the marketing plan: diagram to show analysis by theme

Finally, the marketing department proposes a plan to the directors. This plan is the result of transactions between the different managers of the different departments: the management departments, especially finance, want logically to increase the importance of part 2, the objectives, and decrease that of part 3, the means (which correspond to spending). The operative departments such as the sales staff but also production, want exactly the opposite, i.e. more means and more modest objectives.

It is up to marketing to play the role of referee between the different services. The main thrust is to be able to show the operatives that the more they work, the greater the profits and the more the company can reinvest in new products, publicity, etc. This reinvestment contributes to long term employment and arguments such as these are used more and more by managers to motivate their teams.

2. THE SALES ACTION PLAN

The Sales Action Plan (SAP) is the breaking up of the marketing plan into short periods and its translation in operational terms. The marketing plan is usually yearly which means that it cannot be used directly by the sales force. In addition, it concerns all the company's products, i.e. if it was given directly to the sales force,

5. Marketing and Sales Action Plan

each sales person would select his own priorities and there would be no coherent action in the field.

2.1 The objectives of the sales action plan

> Product priorities
> Customer priorities
> Essential information
> A more effective sales force
> Greater confidentiality of the marketing plan

Table 5-10. The essential objectives of the sales action plan

Establishing the "product priorities" for the period: This is necessary in order to help the sales manager to select the product to be highlighted. The products have been chosen as a function of external events: season, launch of a competing product, spread of a technological innovation ; but also as a function of internal constraints: management of the product portfolio, tariff modifications, advertising/promotional events such as exhibitions, etc. The number of products should be kept to a minimum so that the effect can be felt in the field.

Establishing the "customer priorities" for the period: Here the aim is to ensure coherence between sales operations within the zone chosen. For consumer goods as in industrial marketing, customers are increasingly consolidated via buyouts and also by the development of different forms of associations of shopkeepers, trades people, farmers, etc. Under these conditions, the same promotional operations and the same arguments must be used from one region or channel to another. " Customer priority " means a type of customer per activity sector or distribution channel, or key accounts when the customer in question is important to the company's activities.

Transmit the essential information to the sales force: For the in-house marketing department, this means sending information about launch of new products, with the codes, tariffs, advertising plan for the period, promotional plan, etc.

It is necessary to draw out the main points from studies carried out both internally and externally. They are made up of key figures, the market share and results from sales breakdown, plus their evolution and these quantitative studies concern not only the company but also the competitors. The aim is also to transmit a summary of qualitative studies, reasons for customer purchases, reasons why prospective customers do not purchase, how customers perceive the service quality, etc.

Harmonizing sales management methods: If this is not done, each sales person will only retain the arguments in the order which corresponds to his own particular concerns. This will result in a lack of coherence during negotiations carried out with a customer company present in several zones. Apart from this, a certain harmony is necessary in the different marketing department supports to be used: presentation stand, videocassettes, kits, etc.

It is also necessary to bring into line the different profiles of the people on the sales team: age, training, experience, etc. The objective is not to create a uniform team but to make a common base from where each individual can function.

Another advantage of the sales plan is to improve time management within the teams. The date for the SAP meeting is fixed at the outset. All the sales managers will take part and the management departments have all prepared for this event: everyone can leave for his sector with all the necessary information - tariffs, arguments or even models, estimates or prototypes according to the activity.

Better preserving the confidentiality of the plan: By breaking down a marketing plan into monthly periods for example, the risk of transmission to competitors is reduced. The SAP is used during sales visits and, when a document is published and presented to the customer, it can be easily acquired by competitors. By unveiling the plan on a monthly basis, the risk is reduced.

2.2 How the sales action plan is carried out

For a monthly plan, there will be 11 SAP meetings per year taking into account holidays. The meetings establish the work rhythm for the sales teams. The presentation is in two parts:

- the presentation of the plan for month M+2. In this way everyone has enough time to be able to present the agenda for the sales action plan to customers beforehand during month M+1 ;

- the review of the plan for month M+1. Although this was presented during the previous meeting, it is covered again by going through all the reactions observed and noted by the different team members over the month. Failing this, corrections must be announced such as a change in the promotional rate or a different budgetary allocation, etc.

As far as the composition of the team which sets up the Sales Action Plan, it is strongly recommended to prevent a " divorce " between the department managers and the sales team. To this end, technocratic presentations should be avoided and a mixed team should be formed from managers and operatives. According to activity sector there should be a manager in charge of communication, another for legal matters, a sales manager, etc.

There need to be sales representatives for the following two reasons:

- to act as a counterbalance to the often overly theoretical approach of the management players,
- to legitimize the SAP: since the operative sales managers take part in the decisions they can answer any eventual questions from their colleagues.

It should be noted that to be invited by the directors to participate in setting up the SAP is in itself proof of appreciation and can sometimes act as a springboard to other functions.

It is the members of this same commission who put together the " Response to Customer Objections " dossier. To choose the arguments, the team always bases itself on actual sales scenes and each person puts forward all possible objections to the arguments given. From here the team selects the best possible replies to the objections and can decide to eliminate any arguments thought to be too weak.

2.3 An example of the contents of a sales action plan

A Sales Action Plan is usually made up of the following parts:
- the exact list of the products concerned, the month's priorities,
- the sales manager's priority missions for each of the products,
- the means deployed by the company: advertising, promotion, demonstration materials, invitations to customers for an open house day, etc.
- the key points: these are the arguments which can be least objected to,
- the proof: to reinforce the impact of the arguments, these are figures whenever possible from outside organizations or benchmark customers or extracts from articles which have appeared in a well known publication.

Using his Sales Action Plan, the competent sales manager will know how to make up his Monthly Sales Plan for his " 20/80 ", i.e. the customers representing 80% of his objective (Table 5.11). By successfully meeting his sales objectives with these customers, his global objective is on the right track to success. The simplified table given below allows the objectives for a sales manager to be split up per customer (C1, C2, C3, C4 in this case) and also for each of the products in the range (P1, P2, P3, P4).

Objectives	Product 1	Product 2	Product 3	Product 4	Total per key customer
Key Customer 1					
Key Customer 2					
Key Customer 3					
Key Customer 4					
Total per product					

Table 5-11. Example of a sales plan

The diagram below (Figure 5.12) shows the cyclical nature of sales actions. Market surveys should analyze sales results in order to prepare a more efficient marketing plan for the following year.

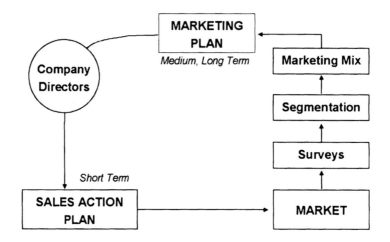

Figure 5-12. The marketing plan / sales action plan / marketing plan cycle...

NOTES

1. Source: Adapted from Merunka, D., (1992), *Décisions Marketing*, Paris, Dalloz.
2. Source: Adapted from Kotler, Ph., (1997), *Marketing management,* Paris, Pearson Education France.

REFERENCES

Dubois, P.-L. and Jolibert, A., (1992), *Le marketing, fondements et pratique*, Paris, Economica.
Malaval, Ph., (2001), *Marketing business to business*, Paris, Pearson Education France.

Chapter 6

INNOVATION AND PRODUCT MANAGEMENT

The product plays a determining role in company strategy especially in the aerospace sector. Pricing, distribution and communication strategies are all closely linked to the choices made as to the product portfolio. Grids and matrixes for analyzing the portfolio are effective tools for making sure that the company is in phase with its market and has a product range to meet objectives. However, before using them we must look at the experience curve and life cycle in the aeronautics and space sector.

1. THE LEARNING CURVE

The product or service corresponds to very different realities depending on its characteristics and technical performance, usage, solutions provided, technology and target markets. In the aeronautics and space sector, the term " product " is very wide, ranging from the complete satellite or aircraft to equipment, a component or a passenger ticket.

In spite of their differences, all of these products have in common the "experience curve" concept based on the observation that total product costs tend to decrease as the volume of production increases. This is due to time savings in manufacturing (or accumulated experience) and the decrease in total costs per unit produced. Total cost includes fixed and variable costs for a given level of production. Fixed costs do not vary with the volume of activity – investments made at launching, structural costs, etc. Variable costs evolve with the volume of production.

The following curve (Figure 6.1) shows the supply of simple components such as rivets. When the experience factor of a much more complex product such as a plane is considered, the accumulated quantities are around a thousand and the curve is much less steep.

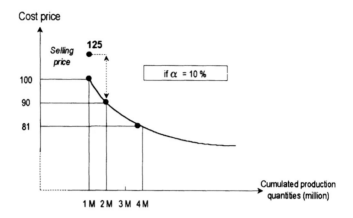

Figure 6-1. An example of an experience curve

As the company gains experience, it improves its know-how thanks to rationalization of work, better management of raw materials, stock management, faster response time to breakdowns on the production line, and the total average cost per unit goes down. This process, called the experience effect, has two origins:

- *the accounting effect*: absorption of fixed costs linked to material investments, such as production equipment and investments in research and development, design, the prototype, advertising budget, etc. The more marketing is able to prolong the product's life cycle and the bigger the accumulated quantity produced, the better the absorption of fixed costs.

- *the experience effect itself*: the more a company manufactures a given product, the more efficient the production process becomes, even if everything has been done to optimize it at the very beginning of the production. The experience effect comes both from within the company through quality circles and participative methods among others, and outside through integrating the experience of suppliers.

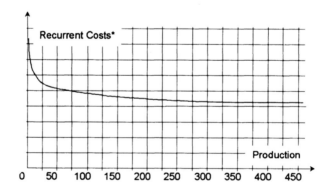

Figure 6-2. Experience curve of a civil aircraft[1,2]

6. Innovation and Product Management

The experience effect is based on the following principle: each doubling of the units produced corresponds to a cost reduction at a constant percentage (but variable depending on the product category): costs decrease by this percentage when experience doubles. Thus a pioneering, innovative company has a cost price advantage that is very difficult for the competition to match: the experience factor is a determining asset.

The experience effect was first measured in the fifties. American researchers were able to calculate a posteriori that agricultural mechanization was characterized in the twenties by experience coefficients of around 2%. Since that time the trend has been for experience factors to constantly increase, in particular in heavy investment sectors (Table 6.3). In aerospace sectors, there are services characterized by weak experience coefficients and at the other extreme, electronic components such as in-flight video equipment with very high coefficients.

Sectors	% of experience coefficient
B to B services	3 to 5
General public services	4 to 8
Integration, assembly	10 to 15
Mechanical equipment (digital machine)	20 to 25
Electronic components	40 to 50

Figure 6-3. Estimation of experience coefficients in aeronautics and space

The experience curve makes it possible to estimate the profitability threshold of the project and to forecast the evolution of the sales price. It is also possible to define the floor price to maintain a dominant position within the framework of a penetration strategy[3]. The experience curve concerns the evolution in the cost price which mainly depends on the sales volume. This is why it is important to study the product's life cycle.

2. *THE PRODUCT LIFE CYCLE*

The concept of the "life cycle" of a product can be compared to that of an individual going through the different phases of life: birth, adolescence, adulthood, old age and death. The analogy with a human being is interesting: before birth, the product must be designed and involves research costs and adjustments. When it is launched, the product is like a baby in that it is not autonomous. It must be supported by the company which can only do this if it has other products that are profitable. Adolescence is characterized by higher rates of growth but also by maximal " sustenance " in terms of marketing support. Then comes adulthood, with a slowing of the growth rate and stabilization.

However, there are two differences: the death of the product is generally provoked by the company and it is possible to influence the life curve, in particular by promoting the maturity phase.

2.1 Phases of the life cycle

Looking at the evolution in turnover generated by a product it is generally possible to distinguish the five major phases described below. The theoretical graph of the life cycle follows a bell curve. This is rarely observed in reality. Statistical smoothing is in fact often necessary to make up for the monthly fluctuations in sales due for example to the season. The curve shown corresponds to a general model (Figure 6.4):

* In the worst case, sales can drop greatly after the launching phase. This is a stillborn product;
* In the best case, marketing prolongs the most profitable phase for the company, that of maturity. Thus the product is a success and profits can be used to finance other projects.

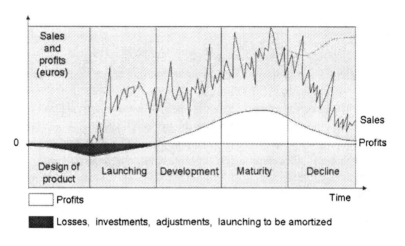

Figure 6-4. Phases of the product life cycle

Phase 1: Design This is the gestation of the product: innovation, research and evaluation of new ideas. Heavy investments are made in research and development and marketing research.

Phase 2: The launch The product is launched, improved and perfected. Profits, even if they are increasing, do not yet cover the high cost of design and research, launching and marketing. The product range is reduced and sales in volume per product in the range are weak.

Phase 3: Development This phase is marked by a high growth rate of the market ; demand is increasing. New applications result in an increase in the sales

6. Innovation and Product Management

volume. Competition is intensifying and the range is expanding. The duration of this phase can vary considerably over time: it can be very short or very long.

Phase 4: Maturity The market growth rate is slowing down, the increase in product sales is weaker. The objective is to reduce costs on major or standard products and to obtain large volumes per product in the range. Profits are at their highest during this phase.

Phase 5: Decline Growth is negative. Sales and profits decrease. Certain products are repositioned in line with new applications and are re-launched, initiating a new cycle. Other products must be eliminated. It should be noted that in general industrial products have a longer life cycle than consumer products.

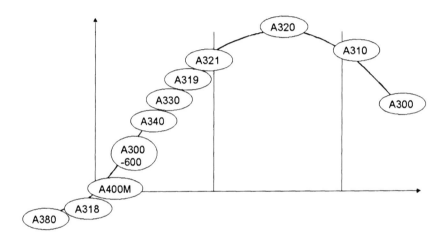

Figure 6-5. Estimation of the Airbus product portfolio as a function of the life cycle phase

2.2 Applying the life cycle concept

The classic "bell curve" obviously corresponds to a product that has been successful and which can be sold in enough examples. Consequently, there is no point in trying to show the life cycle of a product like the Concorde of which only 16 have been produced or satellites which are still characterized by tailor made projects. Nevertheless in most sectors, it is generally possible to represent the evolution in product sales of a company. The same thing can be done at the level of a product category which includes the different players in competition (Figure 6.6).

Comparing the product life cycle to the life cycle of its category shows different situations:

- The pioneering company is the first to launch the product in its category. This company's product can be in the maturity phase while the global market is still in the development phase, stimulated by the more recent product launches of competitors.

- The opposite is true with a company that launches a product late: its product is in the launching phase while the market is already mature. Obviously this is a less favorable situation.

Figure 6-6. Example of life cycle applied to the business jet category

As the life cycle of a plane can be 20-25 years, the product life cycle and the life cycle of one model should not be confused (Figure 6.7).

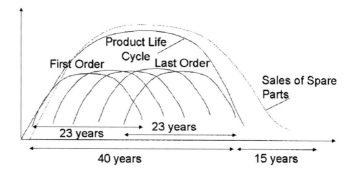

Figure 6-7. Product life cycle and market life cycle

In the above example, while the average life cycle of a plane is 23 years, the "official" life cycle of the plane as a product sold is around 40 years. But the sale of spare parts associated with this plane will continue about 15 years after the sale of the last aircraft (maintenance). As a total, the business generated by the initial product has a cycle of around 55 years.

The life cycle concept is difficult to adapt to custom-made products which do not have a product-market life cycle, but rather a cycle corresponding to the "technology that resolves the problem". This is true of important projects such as the building of an airport.

2.3 The characteristics of each life cycle phase

It is useful to study the life cycle because it can help in managing the product range and the portfolio. Each phase includes certain traditional characteristics (Table 6.8) and requires specific strategies (Table 6.9).

Characteristics	Launching	Growth	Maturity	Decline
Sales	Weak	Strong	Maximal	Declining
Growth rate	Positive	Positive	Weak	Negative
Unit cost	High	Average	Weak	Weak
Price	Pricing policy	Pricing policy	Competitive price	Lowering of price
Profits	Negative	Growing	High	Reduced
Distribution	Selective	Extensive	Very extensive	Selective
Clientele	Innovative	Early adopters	Mass market	Traditional
Competition	Limited	Growing	Stable	Declining
Norms	Little	Yes	Standard	-
Technology	Differentiated	Standardized	Substitution	-
Range	Reduced	Extension	Variety	Pruning

Table 6-8. Characteristics of phases

Looking at both the product and market life cycles helps to define product strategies. The analysis of the product life cycle is completed by that of the market, which evolves as a function of the four distinct phases: emergence, growth, maturity and decline (Table 6.9). For example the evolution of the market depends on the development of new technologies, new needs, competition, etc.).

		LAUNCH	GROWTH	MATURITY	DECLINE
MARKETING OBJECTIVES		Creating renown and encouraging sampling of the product	Increasing market share	Increasing profits and maintaining market share	Reducing expenses and "harvesting"
Market life cycle	Emergence	Opening of a new market	---	---	---
	Growth	Follower strategy or developing the range	Very favorable situation	Need to launch new products, complete the range and attack new segments	A new product should have been launched in the preceding phase, imperative to take action
	Maturity	Innovative maturity	---	---	IDEM
	Decline	Some lucrative opportunities	Some lucrative opportunities	Certain cash-cow products is it better to invest to launch a new product or maintain this product until the end ?	Ceasing investments, eventually taking the product off the market

Table 6-9. Product strategies adapted to the market context

3. MANAGING THE PRODUCT PORTFOLIO

Managing the portfolio consists in verifying that the products are coherent with the company's overall offer and in comparison to competing products (Figure 6.10 and Figure 6.11).

Figure 6-10. Airbus, Boeing and Douglas product portfolios[4]

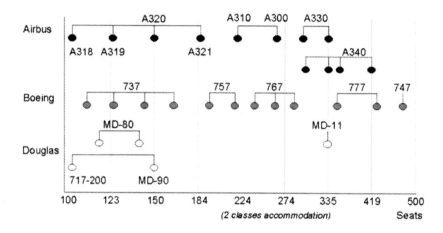

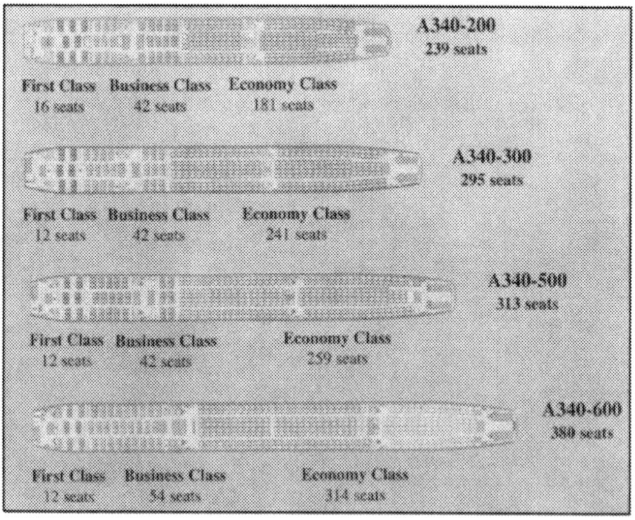

Figure 6-11. The long haul Airbus portfolio[5]

Several models have been proposed to make management of the product portfolio easier.

3.1 The BCG model

The Boston Consulting Group model is based more on the activity sector than on the product. In fact, the company has a portfolio of activities regrouping one or more "strategic segments" or "strategic business units" (SBU), each of which is associated with one or more product-market pairs. Based on the experience and life cycle concepts, the BCG model makes it possible to evaluate each strategic segment as a function of two variables:
- relative market share of the company, which is a measure of its strength in the market (its competitive position),
- the market growth rate which provides a measure of the market attractiveness.

These two variables are used to elaborate the BCG matrix (Figure 6.12), which divides activities into four categories (SBU).

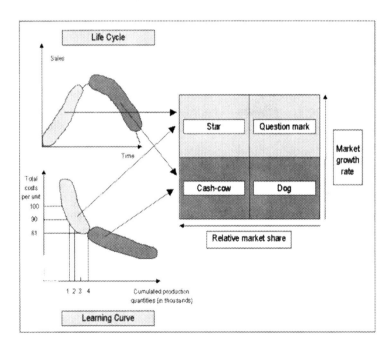

Figure 6-12. The BCG matrix relative to the life cycle and the experience curve

"Star" products: The products in this phase would seem to be in an ideal situation as their relative market share is strong and the evolution rate of their segment is above average. Thus they are in the launch or growth phases. However, from a financial perspective, they require cash flow to pay for big technical investments (increasing production capacities, quality improvement, etc.), sales (motivating the sales team, new distribution channels, etc.) and marketing (communication to increase the penetration rate). Parallel to this, recent products

benefit little from the experience effect and still have a high cost price. Accordingly, the financial analysis is generally negative for this category of products.

"Cash-cow" products: While the latter are characterized by strong relative market share, they have a weaker growth segment than the market average. Thus they correspond to mature or declining segments. In comparison to the preceding product category, cash-cow products require less investments whether from the technical point of view (daily maintenance, few changes to be made, satisfactory production equipment) in sales (satisfactory distribution, less commercial pressure) or in marketing (known product and brand, maintenance communication). Thus overall spending for cash cow products is moderate. At the same time, cost price is competitive thanks to the experience effect: the product has been produced for many years and market share is greater than that of competitors. Consequently, these products, at the model level, generate considerable cash flow.

Dogs: These are characterized by weak relative market share and a below average evolution rate for their segment. It is clear that they are not important for the company's future development and they do not merit priority investment. It is too late to win market share for dog products and it is better to save investments for future segments. Everything depends on the profitability of the products:

- If a "dog" product is bringing in money, it is important to check that its presence in the range is not harmful to the image of the other products. If this is not the case, the company should maintain this product, exploiting it without making further investments.

- However, if a "dog" product is costing the company money, it needs to be pulled.

"Dilemma" products: Also called "question mark", these products create the most difficult problem for the company: their relative market share is weak, while in the growth phase their segments represent the future of the company.

Financing requirements are very large. On a technical level, both quantitative (too limited production capacity) and qualitative (improvements necessary to make up for lag in comparison to competitors) investments are called for. From a commercial point of view, better product distribution calls for much greater spending, even more so given that better distributed competing products benefit in their sales arguments from their greater market share. As such, motivating the sales team is essential to make up for the handicap of weaker market share. As far as marketing is concerned, spending directed at the end user is very high for two reasons: the product-service category is still little or poorly known by the majority of potential customers ; in addition, when they do know it, they think first of competitors. Thus energy has to be focused on making the product category more known and on carving out a stronger position relative to competitors.

Consequently overall spending is very high for "dilemma" products, while their cost price is not at all competitive as they do not yet benefit from the experience effect. If the company's portfolio includes several "dilemma" products, it is very

6. Innovation and Product Management

important to select which of these to focus on, even if this means selling off the others.

Techniques for use and limits of the model

The BCG model stipulates that the margin made from "cash-cow" products should be able to finance investments required for the " dilemmas " so as to win market share. The most important point is to be able to count on a big share of turnover for the "cash-cow" products, if this strategy is to work realistically. Thus the BCG matrix is based on dynamic management of activities, starting with the assumption that their position is not fixed: today's "star" activities should be the "cash-cow" of tomorrow, just as the "dilemmas" are future "stars". Marketing should ensure that the global activity portfolio is balanced, thanks to synergy between the different activities, and that it is strong enough to provide for the company's future development (Figure 6.13).

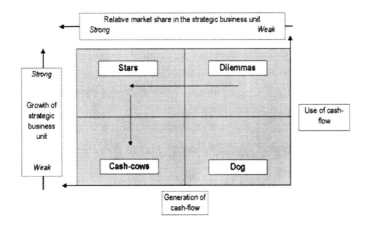

Figure 6-13. Management of an activity portfolio according to the BCG model

To actually use this management tool it is necessary to:
- situate the different company products as a function of their performance on the two axes of relative market share and percentage of market growth,
- add up the turnover from the different products in each box so as to obtain the four different turnover figures and calculate their respective share as a percentage. A balanced portfolio should have 30-50% of turnover in "cash-cows", financing the upkeep of "star" products and especially the required investment in selected "dilemmas", helping them to gain market share.

The strategic interest of the BCG model lies in the fact that it is relatively easy to use and set up, and that it takes into account financial and marketing factors. It makes it possible to take stock of and analyze the different businesses in the company, helping to determine whether the latter should be given up, maintained, exploited or developed.

However this model can be criticized:

Results, in other words the breakdown of sales into the four squares, depend on the limits chosen and in particular on the average market growth rate. But two competitors will not have the same definition of their market: the parameters differ as a function of their diversification strategy. For a given company, because of acquisitions it has made, the "business perimeter" can change from one year to another. Suppose that the market which had been defined in the past grew by + 10%. After the launch of a new activity in Internet services to companies, the new growth rate is +13%. Thus a product, whose segment has grown by +12% will be in the cash cow category (as 12 <13) and it will no longer be a star (because 12 >10). The same is true of the segmentation method chosen (macro or micro), which leads to different breakdowns of turnover by category.

Furthermore this approach does not really take into consideration all of the factors concerning the company such as its human, financial, and R&D resources, etc. Based on the experience concept, the model focuses on cost reductions, while technological innovation can sometimes entirely re-model the market; thus this tool has to be used carefully.

Another limit of the matrixes lies in their subjectivity: in effect, results depend on the calculation used for market share. A comparison between Airbus and Boeing profits illustrates this: the latter differ considerably depending on whether orders or actual deliveries are taken into consideration and of course the reference period.

AN ANALYSIS OF THE AIRBUS AND BOEING PRODUCT PORTFOLIOS

The aeronautics market, like that of satellites, is characterized by the high cost of orders. Furthermore, a delay of 3-4 years generally separates the order date from the delivery date. This is why a comparative analysis of the product portfolios of the two leading aircraft builders on the civil aviation market, Boeing and Airbus, must take into consideration different factors:

cumulated market share, which takes into account the total park installed by the two builders no matter how old the aircraft in operation,

in the example, deliveries for the year studied,

the year's orders.

A simple comparison of the total sales of the two builders since the commercial launch of the aircraft is too static and not sufficient (Figure 6.14).

6. *Innovation and Product Management* 173

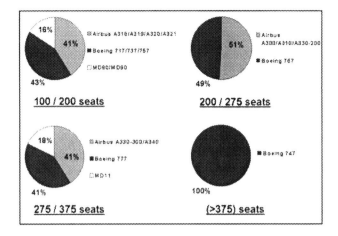

Figure 6-14. Comparison of market share calculated from total sales

The above charts compare the market share of the builders since the launch of their aircraft: 100/200 seats, 200/275 seats, 265/375 seats, Jumbo (over 375 seats). However, this comparison does not show the evolution in sales nor does it situate the products of the two competitors in a dynamic perspective. That is why a BCG analysis would be more useful.

To establish a BCG matrix, information is needed on the growth of the market and the relative market share of the products needs to be calculated. Market growth can be measured in terms of the number of deliveries and orders in a given period. Annual market results published in the specialized press[6], provide specific information on the number of aircraft ordered and delivered according to type for the two builders.

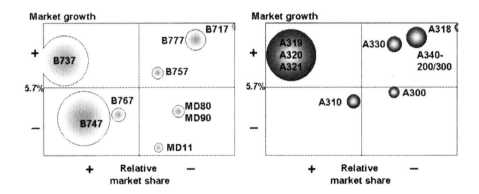

Figure 6-15. Comparative estimate of the Boeing / Airbus portfolios

The previous graph (Figure 6.15) presents BCG matrixes established by calculating the relative market share (horizontal axis) and market growth expressed in number of orders (vertical axis). The global growth rate (all products) on the market and for the period considered is 5.7%. The size of the circles expresses the volume of each of the activities in the turnover of Airbus (dark circles) and Boeing (clear circles). The biggest growth was for orders in the 100 seat, single aisle segment (A318 and B717). On the other hand, the wide-bodied segment showed a drop of around 5%, which explains why the B747-400 is in the cash cow position. Most single aisle planes are in the star product category with a strong relative market share and an above average evolution rate (B737 and A320, A321).

The global analysis shows that Boeing has a better balanced product portfolio, benefiting from the fact that the company was the first on the market. It seems clear that the B747 program in spite of modernization expenses can significantly contribute to investments required for the B777 and other programs. In addition, it would be possible to create the Airbus and Boeing matrixes taking into account deliveries from the two most recent fiscal years. A comparison of results would show a net improvement for Airbus.

3.2 The McKinsey model

The McKinsey firm has proposed a matrix (Figure 6.16) based on the value/interest of the sector for the company and its competitive position in the sector.

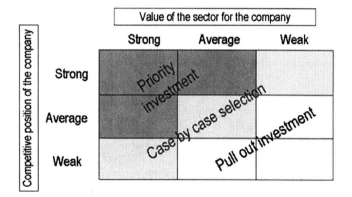

Figure 6-16. McKinsey matrix

The idea is to evaluate the opportunity for the company to invest, pull out investment or select strategic activities as a function of the context:
- external: size of the market, profitability of the sector, intensity of competition, existence of cycles, growth rate...
- internal: quality of products, relative market share, product image, experience and knowledge of the market, efficiency of sales team, price level, service...

All of these factors are assigned coefficients. The weighting obtained highlights three distinct zones:

- a priority investment zone, corresponding to an attractive sector and a strong competitive position,
- a non-priority zone, corresponding to sectors that are not very attractive and a weak competitive position; this means that it is time to recuperate any profits without re-investment. If the activity is losing money, cutting off investment must be considered,
- an intermediate zone where the company must make a case by case decision about each activity to invest or pull out.

While this tool is more complete it is less used owing to its complexity and possible bias in the weighting of criteria.

3.3 The Little model

Little developed another matrix which takes into consideration the maturity of the sector and the competitive position (Figure 6.17). The maturity of the sector is based on the same process as that of the product's life cycle: launch, growth, maturity and aging. Each phase has distinct characteristics whether these are the intensity of the competition, financial factors or the most efficient marketing actions.

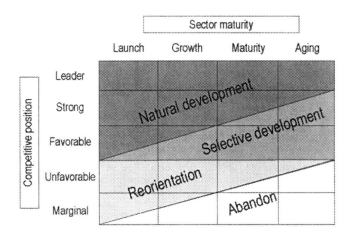

Figure 6-17. The Little matrix

For each strategic segment, the key success factors are determined which, once they have been weighted, will make it possible to evaluate the company's strengths. Thus by comparing the latter to the factors necessary for success and examining them in light of the strengths of competitors, it is possible to determine the positions of the company in the different strategic segments. The company can thus situate itself as a function of the particular market phase and its strengths, making clear the right strategy: abandon, reorientation, selective development or natural development.

3.4 Marketing and management of the product portfolios

Managing the product portfolio must take into consideration success factors, parameters internal and external to the company, the life cycle of products and the market. The marketing department defines the major orientations of product policy, better managing the portfolio by focusing on the coherence and balance of the offer, playing on the four parts of the mix (Table 6.18).

Statu quo, or "no change": This generally applies to a developing market where product sales are already greater than those of competitors. The aim is to keep track of the product without taking any particular action.

The launch of new products: This is strategically important, the goal being to prepare tomorrow's assets. Improving and modifying existing products helps the company to preserve its competitiveness on the market.

Completing existing products with others: Extending the range allows the company to benefit from the experience effect and offer a wider choice of solutions to its customers. Re-positioning the product is often the decision made to make up for a market loss on a growing market.

	Launching	Development	Maturity	Decline
Product	Product adaptation to customers and sales force advice	Product improvements after customers reactions	Product adaptation with the features of best seller competitors	No expensive change
Price	Choice of a strategy: - skimming - penetration - technological substitution	Strategy adaptation versus competitors reaction	Price reduction when pricing competition or technological evolution	No change or price reduction
Distribution	Channel choice: - direct - wholesaler - combination of both	No choice change but extension to new channels to better cover the market	New extensions if necessary to maximize the product availability	Selection of cheapest channels
Communication	Promotion actions, sampling, product trial and corporate advertising	Strengthening product actions and institutional campaigns	Corporate campaign and reduction of communication expenditure	Little or no action

Table 6-18. Summary of actions for each major phase in the life of the product

By reducing the costs of its products: Number of parts constituting the product, simplification of the structure and maintenance, improvement of quality (process, products) : the company can also optimize its range.

Abandoning a product: This is necessary when the "dog" product costs more than it brings to the company in terms of profit, image, entrance on a market...However it is accompanied by the development of new products and consideration of the already existing park, possible guarantees underway... A product is often abandoned when, on a matured market, products are undifferentiated. Elimination of a product is based both on marketing and financial considerations. Contribution to growth, effects on the profits and sales of other products, consequences on relationships with clientele, capacity for a new product to efficiently replace the product concerned.

THE BOMBARDIER NICHE STRATEGY (Les Echos, 1999)

With close to 56,000 employees, the Canadian company founded by the Quebecois J. Armand Bombardier, has two main activities : rail equipment and aeronautics. In this activity, Bombardier Aeronautics in just ten years has become the third constructor in the world behind Boeing and Airbus and the world leader in regional transport aircraft in front of the Brazilian company, Embraer. With its twin engine Global Express, its range of LearJet aircraft and especially its regional 50-70 seat-jets (CRJ 200 and CRJ 700), Bombardier has developed thanks to a niche strategy. Given its expertise and the results of market analyses, the group has been able to offer adapted and differentiated aircraft. Its range includes :
- business jets : Bombardier Continental Business Jet, Global Express, Challenger 604, Special Edition Canadair, Corporate Jetliner, LearJet (31A, 45 and 60),
- regional aircraft : CRJ 200, CRJ 700 and CRJ 900 and Q series (derived from the Dash 8 range),
- water aircraft : CL-215, CL-215T and Canadair 415.

Figure 6-19. The Canadair 415: the latest plane in the fire fighting range[7]

4. MANAGING THE PRODUCT RANGE

The product range consists of all those products linked by their usage and functioning which are commercialized and distributed in the same way. The product range thus includes the products themselves and the product lines which regroup variations around the basic product.

4.1 The characteristics of the range

Three dimensions characterize the range :
- the width corresponds to the number of product lines commercialized by the company,
- the depth corresponds to the number of distinct products included in the product line. The deeper the range, the more it expresses the will of the company to answer different customer expectations,
- the length depends on the width and the depth, signifying the total products sold by the company. The range can be short, including a limited number of products (Table 6.20) or long, with many products (Table 6.21). The choice of how wide a range or how many products to offer is strategic: if the range is too limited it is perhaps not as profitable as it could be, if it is too extensive it can increase costs.

ADVANTAGES	DRAWBACKS
1- Focusing investments on a small number of products • commercial efficiency of salespeople who know their products very well • production efficiency through technical specialization 2- Each product is supposed to have high volume sales and a competitive cost price 3- Easier sales administration 4- More efficient logistics with limited costs	1- Little choice for customers. They risk hesitating with a competitor's product instead of choosing among the company's products 2- More difficult to use a direct distribution channel 3- Higher risk to be hit by a competitor attack 4- More vulnerable to face a technological evolution 5- More difficult to renew an important product

Table 6-20. Main advantages and disadvantages of short ranges

Often companies specialized in high technology or up-market products targeted at a very specialized clientele have a short range ; this is also true of companies producing standardized products, thus benefiting from economies of scale.

6. Innovation and Product Management

ADVANTAGES	DRAWBACKS
1- Better to adapt a product for each market segment	1- Necessary to invest in a lot of products: difficult to focus on a selection
2- Within a line, a synergy can be found for the products around the core one	2- Market managers and sales representatives may know their products less
3- Better reactivity to answer the customers needs and to face the competitors technological reactions	3- The sales administration is more complex
	4- The logistics costs are higher
	5- Risk of cannibalism between the line products

Table 6-21. Main advantages and disadvantages of wide ranges

Another important dimension to consider when managing the product range is its coherence. This involves evaluating the homogeneity of the different product lines as a function of their final usage, production imperatives, and the distribution network. The more coherent the range, the better the reputation of the company in its field of expertise. Aircraft builders have generally gone for wide ranges. This is true of Airbus and the A340 range (from 239-380 seats).

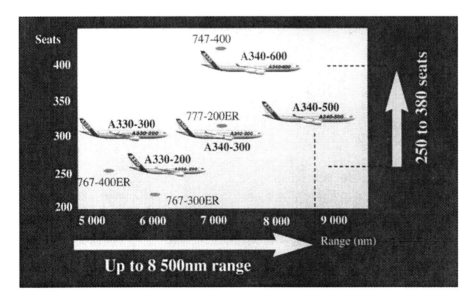

Figure 6-22. The Airbus A340 range

THE CFM INTERNATIONAL ENGINE RANGE

CMF International, a joint venture[8] between Snecma Engines and General Electric, offers a wide range of engines based on technical characteristics (thrust capacity) and customer needs (Table 6.23). CFM is the leader in single aisle planes with a cumulated market share of 82% since 1984. The extent of the range is one of the key factors of its success.

DEPTH	CFM Engines for Military Applications		CFM Engines for Airbus			CFM Engines for Boeing		
	56-2A	56-2B	56-5A	56-5B	56-5C	56-2C	56-3	56-7
	B. E3 B. KE3 B. E6	B. KC135R B. C135FR B. RC135R	A1 A320 A3 A320 A4 A319 A5 A319	B1 A321 B2 A321 B3 A321 B4 A320 B5 A319 B6 A319 B7 A318 B9 A318	C2 A340 C3 A340 C4 A340	DC8-71 DC8-72 DC8-73	B1 737-300/500 B2 737-300/400 C1 737-300/400/500	B18 737-600 B20 737-600/700 B22 737-600/700 B24 737-700/800/900 B26 737-800/900 B27 737-800/900

◄──────────── WIDTH ────────────►

Table 6-23. The range of engines offered by CFM

The range was developed from the CFM56-2 engine, which was first used for the DC8 series at the end of the 1970's. The CFM56 was initially designed as an engine with modular architecture, simple and upgradable, capable of integrating new components and equipment in step with technological innovations.

Figure 6-24. The CFM56-3 engine, one of the flagship engines in the CFM56 range, more than 4,400 sold[9]

6. Innovation and Product Management

This design logic (the same architecture with specific motors depending on the applications) has allowed the company to produce CFM56, the most reliable and durable engines on the over 100 seat aircraft market. It is today the world's most used engine on the civil airplanes market. The four sizes of engines answer the needs of the different aircraft, with thrust from 18,500-34,000 pounds (Figure 6.25).

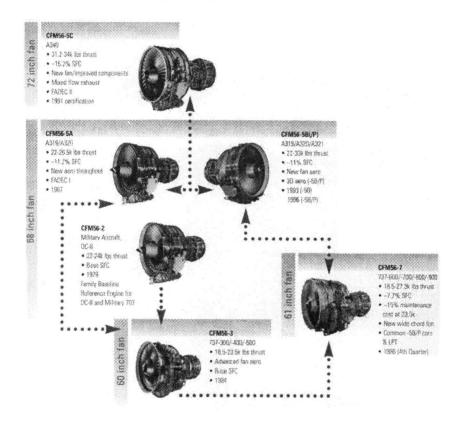

Figure 6-25. The CFM56 family tree: one architecture for specific and adapted engines[10]

In general, the product ranges of companies expand over the years. At the same time, spending increases: production costs, launch, storage, transport, communication, etc. Companies must make choices so as to maintain those products that are the most profitable and the most adapted to their strategy orientations, thereby keeping their product offer continually updated. They can:
- widen the range by attempting to penetrate new markets,
- reinforce the range by launching new products,
- modernize by "re-lifting" the current range: for example by integrating the latest equipment,
- rationalizing by pruning away "dog" products that wastefully consume valuable resources.

The marketing manager must help to balance two trends: widening the range to better meet demand and at the same time looking for competitive cost pricing at the best quality. Using a marketing-based analysis managers can decide which products to extend and which to give up. Managing the range is therefore part of the company's global strategy.

The "behavior" of ranges on the market, as well as that of other competing products (substitution) must be carefully watched. Managers should keep track of the sales, profits and positioning of all products, analyzing each in comparison to its competitors. They should watch the life cycle and be ready to intervene in case of poor performance.

The aim is to dynamically manage all the products and services offered by the company: new products will replace old ones and others will be re-launched. Certain short life cycle products play a tactical role in the range, according to a niche marketing policy aimed at hurting the competition or responding to their attacks. Other products called "regulators" serve to minimize the effects of sales fluctuations of the product leader and absorb part of the fixed costs. Product leaders, on the other hand, have the best margins and the largest market share ; they are the founders of the range: their life cycle influences that of other products. Thus the global products or services offered by the company must be balanced in terms of profitability and satisfaction of customer expectations.

5. MANAGING INNOVATION

Analyzing the company's product portfolio and range can reveal weaknesses in comparison to competitors and thus the need to launch a new product. The company can settle for launching a product similar to those that already exist, in which case there is no real innovation. A truly innovative product on the market can have:

- new functions: for example, the Landscape Video Camera system by Latécoère which allows the passenger to see the passing landscape on an individual screen,
- improved options to meet the same need: relaxation in the example of 180° seats from Weber Aircraft.

LATÉCOÈRE'S INNOVATION: ON-BOARD VIDEO SYSTEMS

Latécoère has known how to innovate on the equipment market and has become the world leader in on-board video (Figure 6.26). In particular, it produces a video system for passengers on the A320-A321 which allows them to watch the landscape that is being flown over. This system only functions at cruising altitude as takeoff and landing are considered to be too potentially tension producing.

6. Innovation and Product Management 183

In addition, Latécoère provides a camera system used for ground transport (Taxi Aids Camera System, Tacs) in particular on A340-500/600 aircraft. The company has also developed a cargo hold video surveillance system (A300-600ST) and an in-flight video refueling monitoring system for the KC135.

Figure 6-26. Latécoère: the Taxi Aid Camera System (TACS)[11]

Companies in the aeronautics sector are able to offer new products thanks to a strategy of innovation. This means developing a new product whose design, once it has been validated, should make it possible to offer later versions by improving the product's performance and functions in accordance with its future uses. This is especially true of aircraft builders, which after the design and production process, offer a finalized version of their plane. The latter can then be improved, or extensively modified depending on customer requirements. This is also true of motor manufacturers such as CFM or Rolls-Royce.

In addition to innovations which directly concern the products, companies in the aeronautics sector are constantly working for improvements and innovations upstream in terms of product design and production. This is how the assembly of the first 767-400ER Boeing benefited from innovations in production (Figure 6.27). The precision of the cockpit's assembly of the plane built by Boeing in Wichita and of the front section of fuselage made in Japan by Kawasaki Heavy Industries (KHI) is due to the use of new three dimensional digital design tools such as the Catia software from Dassault Systèmes. Using a unique design database, Catia defines the plans for parts to be produced anywhere in the world, as well as all the tools and structures necessary for their production and assembly.

Figure 6-27. The 767-400ER cockpit and fuselage assembly[12]

The improvements made to the product are a renewal process. They can often be seen when new versions with modified characteristics are launched. Real innovation is marked by a break with previous products and results in the launch of a so called new generation.

THE DIFFERENT GENERATIONS OF A PRODUCT: MILITARY AIRCRAFT

The concept of generation is linked to chronology and to the concept of a technological leap in one or several major sub systems of the airborne arms system such as the cells, radar, avionics or the engine. As a general rule, a new generation is identified by the appearance of an entirely new aircraft which, in most cases, justifies a new name or the re-use of the previous one in modified form (" F/A-18 Hornet " / " F/A-18 Super Hornet "). The table given below shows the present and future generation changes in the military aircraft sector (Table 6.28).

	Predecessor	Present Aircraft	Successor
USA	F-4 Phantom F-104 Starfighter A-7 Corsair	F-15 Eagle F-16 Falcon F/A-18 Hornet	F-22 Raptor JSF F/A-18 Super Hornet
France	Mirage F1 Jaguar	Mirage 2000	Rafale
United Kingdom	Buccaneer F-4 Lightning Jaguar	Tornado	Eurofighter 2000
Sweden	J35 Draken	JA37 Viggen	JAS39 Gripen
Russia	Mig 21 Mig 23 Mig 25	Mig 29 Su 27	- -

Table 6-28. Examples of generations of military aircraft

6. Innovation and Product Management

Figure 6-29. A new generation of military aircraft: the Rafale[13]

Within one particular generation (iso-generation), there can be technical and/or technological evolutions, function changes (new products, improvements, etc.) and performance changes. These evolutions pave the way for the new generation (Table 6.30).

	Basic Aircraft (original name)	Iso-generation Evolutions (unchanged name)
USA	F-15 A F-16 A/B F/A-18 A/B	F-15 C/E F-16 A/B/C/D F/A-18 C/D
France	Mirage 2000C	Mirage 2000D, Mirage 2000-5/9
Russia	Mig 29A Su 27	Mig 29C, Mig 29SMT Su 30/32/33/34/35/37

Table 6-30. Examples of the evolution of military aircraft within the same generation

---

Ideas for new products designed by the company generally have two main origins (Figure 6.31):
- a market that generates a new need, this is the "pull" strategy ;
- innovations from the company's R&D department. The " push " strategy is based on the idea that applications of interest to the customer must be found.

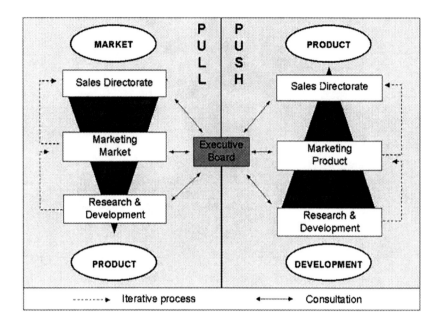

Figure 6-31. Two different impetus for innovation and the same global process

5.1 In-house innovations: the "push" strategy

There are many sources for new product ideas within the company coming from production, sales, engineering, marketing as well as research teams. In order to generate relevant new ideas different creative techniques are used such as brainstorming, benchmarking, etc. This internal process designed to encourage teams to develop new ideas and to make them commercially viable is called "intrapreneuring".

The main driving force for innovation within the company is still the Research and Development department whose task is to look for new ideas (Figure 6.32). For each product studied and without inhibiting the creative process, the R&D department, must take into account:

- its technical-economic characteristics, meaning the technology to be used, the investment necessary and the cost of launching,
- its sales characteristics: advantages, outlets, positioning and profitability,
- the company's product policy: management of the actual portfolio, patent applications, associated distribution, etc.,
- human resources needed for the project: skills, sufficient teams, etc.,
- the company's strategic orientations.

6. *Innovation and Product Management* 187

Figure 6-32. The push strategy

To guide research towards the most sales effective solutions, the departments covering marketing-sales, purchasing, quality and finance are consulted together with the general management. Technological prowess is not enough; the new product is only worth what it offers the user and has no intrinsic value. The industrial customer is buying a solution to a problem, functions and services. The new product must take account of these imperatives: costs, maintenance, actual and future equipment, output, etc.

WEBER AIRCRAFT: INNOVATION AT THE HEART OF THE OFFER

Founded in 1941, this American company, a subsidiary of the Zodiac group, boasts the major airlines among its customers. It has a complete product range covering (Figure 6.33):
- passenger seats (tourism, business, first class) ;
- special seats for the crew and pilots;
- cabin equipment such as galleys (on-board kitchen modules, coffee machines, ovens, etc.), toilet units and the " airstairs " (built-in stairs).

Thanks to these innovative products, Weber has been able to capitalize on its long experience and it was the first offer a product complying with the TS0 C127 / FAR 25 norms. It also produces first class seats which convert into comfortable beds (with actual separations between them).

Figure 6-33. An example of seats developed by Weber Aircraft[14]

The innovation based approach is mainly of the push type. For example the arrival of ultra-thin, high resolution Mitsubishi TV screens induced Weber Aircraft to offer them for first class passengers on B777 and A340 aircraft. Their marketing department, having discovered the technology in the domestic equipment sector, then persuaded in-house teams to design hi tech seats corresponding to this target. Passengers in this class expect a level of comfort and quality equal or better to that in their own homes. And so the armrest of the seat has sophisticated multimedia equipment such as a top of the range, retractable LCD screen for watching TV films, a telephone and all the necessary connections for computers or video games.

With an eye to increasing the effectiveness of its product offer while at the same time improving customer satisfaction, Weber has developed a complementary activity ; they offer a personalized maintenance and repair service for airline companies.

In-house based innovation is a rigorous and interactive process: solutions must be abandoned and others worked on, etc. The more the company progresses down this road, the more manpower, time and money will be devoted to it. Possible solutions must be selected extremely carefully and the project followed through with rigorous attention to detail to reduce the risk of failure on launching as much as possible. One of the main pre-requisites for the success of this type of venture is the mixture of the different internal and external know-how within the same project. Efforts and teams in isolation are to be avoided.

The diagrams below (Figure 6.34) are a humorous look at the dangers of allowing a single team to design an aircraft.

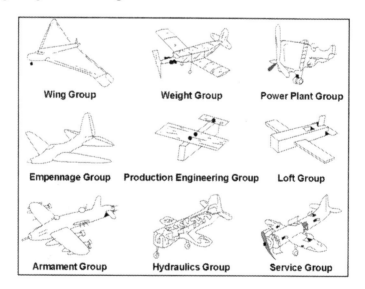

Figure 6-34. Caricatures of innovation with different project leaders[15]

5.2 Innovations from outside: the "pull" strategy

Innovations from outside are those suggested or dictated by the market: the demand "pulls out" the innovation (Figure 6.35).The company's marketing department must detect opportunities for the development of new products and transmit these to R&D.

Figure 6-35. The pull strategy

The market players are professionals, capable of describing precisely what they need and what type of material, product and function they require. Often an industrial customer mentions to sales staff, specific problems encountered with a particular piece of equipment supplied by them or a competitor. From then on, the objective must be to identify the exact nature of dissatisfaction in order to point research and technical solutions in the right direction, studying difficulties for set-up, maintenance, production capacity flexibility, quality, waste, eventual compatibility, etc. There must be close cooperation linking the company to its suppliers and to the customer or prospective buyer, in order to design new solutions in step with everybody's real concerns.

Several outside sources can play a part: customers or prospective ones, influencers, distributors or public organizations. Contacts and sales relations with distributor's agents, concessionaires and sales technicians plus reading matter, congresses, visits to international exhibitions, are all potential sources of information. In business to business, studies, customer observation and even the use of "pilot customers" often turn out to be much richer than traditional creativity techniques. In this way, for the launch of the A380 project customers were involved and met regularly to take part in its development (Emirates, Virgin, etc.). They were part of the project very early on especially concerning choice of options[16]. Their commitment in terms of orders, allow the industrial launch of the aircraft.

In the military sector, since 1946, the Martin-Baker company has been developing a coherent range of crew evacuation systems second to none in confidence ratings (Figure 6.36).

Figure 6-36. The Martin-Baker America ejector seat systems[17]

Their ejector seats are used in particular to equip planes from the US Navy with whom they work particularly concerning the design and set up of the products. The Mk 16 seat is the fruit of years of collaboration with the customer and in this way is the only one capable of safely ejecting a man or a woman weighing between 45 and 111 kg (best human safety record with 99,4% successful ejections, i.e. 6810 lives in fifty years).

The " pull " method of innovation is frequently encountered in the aeronautics and space sector. Industrial partners working together in a common program must meet an exacting specification sheet. The technical and functional characteristics of a project are such that companies are obliged to innovate to satisfy demands. Rolls-Royce for example, in the former A3XX project relied on the flexibility of its range of engines (same fan and core). However the development of the Trent 900 needed cutting edge technology to meet the specific needs of the new ultra large Airbus A380.

6. Innovation and Product Management

THE EVOLVING RANGE OF ROLLS-ROYCE TRENT ENGINES

The range of Rolls-Royce Trent engines has been built up from the RB211, the first to use the more efficient and even stronger three shaft architecture. This has allowed the company to produce an evolving family of engines with thrusts from 50,000 to100,000 lb. (Figure 6.37).

Figure 6-37. The Trent 900 engine developed for the A380[18]

5.3 Product development phases

The developmental process of a new product goes through several stages and includes numerous disciplines and functions. This iterative and interactive development between the different departments of the company, reduces the risk of failure. It consists of designing and launching a product for profitable sale based on an idea or a concept. The idea for a new product can come from (Figure 6.38):

- market studies of the type used for consumer goods,
- tenders of specific customer needs,
- needs of the State (arms, satellites, etc.): " imposed products ",
- " unbridled research " carried out internally and in all directions,
- specific research, i.e. directed,

- economic, environmental, competitor and technological scanning (allowing the company to offer similar products which are based upon, adapted or entirely new, etc.).

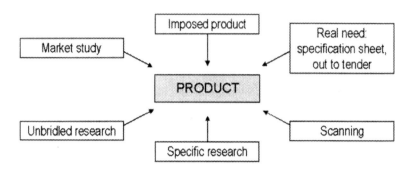

Figure 6-38. A new product's different sources

The process of innovation is initiated by strategically defining the main lines of research, i.e. the direction in which research and the search for creative ideas will be carried out. The process of innovation leading to the creation of new products has six main stages (Figures 6.39 and 6.40).

1. Looking for new ideas both internally and externally

2. The analysis, evaluation and selection of ideas using screening methods to reduce the number of ideas according to criteria such as technical or commercial feasibility, suitability for customers, distribution networks, means of production, etc.

3. Evaluation of the sales potential (analysis of market opportunities, the technical, technological, economic, legal and political environment, etc.). Apart from estimating the market using simulation tools, financial estimations must be made (sales, margins, cash flow, investments needed and the return on these for each product).

4. The development of the product taking into account market demands and production constraints (tools, reduction in the number of parts, etc.), supplies (collaboration with suppliers, quality standards, costs, deliveries, etc.), financial resources (design to cost: minimization of costs, maximization of performance) and the organization of the company (mobilization of teams, delays, etc.). This phase ends with the production of the first prototype.

5. Evaluation and verification tests: a fresh analysis of the product's profitability, production of new prototypes with lab, on-site and customer factory-based tests, appraisal of inadequacies and pre-series production tests again using certain special customers during exhibitions or demonstrations. This phase ends with industrialization of the project.

6. The actual sales launch must quickly follow the pre-series, since competitors have been alerted: the marketing is based on coherent preparation (information to

influencers supplied beforehand, communication, distribution, sales team and product support service activities). The airlines' aircraft purchasing intentions are often decisive for manufacturers. For the latter, such commitments reassure the financial markets as to the project's viability. Apart from their " shop window" function, these announcements coming from companies which are well known on their market, give the manufacturers the benefit of the " domino effect ": their international competitors are prompted to imitate these early adopters to be on equal terms.

Figure 6-39. The developmental process for a new product

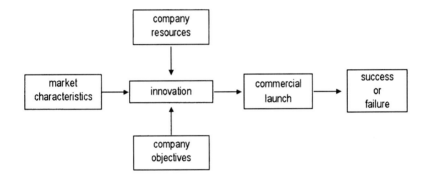

Figure 6-40. From innovation to commercial launch

5.4 *The conditions for successful development*

The successful development of a new product depends on the resources allocated and how motivated the innovating teams are. Internally, due consideration must also be given to all the constraints linked to strategy, production, purchasing and quality. Externally, the constraints of the market, the competition and technology must be analyzed.

The teams are composed of people with varying profiles and complementary skills and in particular design, production and quality engineers, designers, marketing managers, buyers, certain managers from customer companies and suppliers, etc. These cross functional teams limit conflicts between different departments who admittedly have the same general objective but sometimes divergent views as to how to achieve this.

According to its characteristics, size, resources, degree of innovation, risk, tendency to innovate, the company can choose between the four main types of organization in current use:

- the *functional structure*: this gives functions such as marketing or R&D the responsibility for coordinating the work, the characteristics of which have been defined and agreed right from the outset for all the other functions involved in the project. The development stage will accordingly be long with the dividing up of tasks needing special attention: each function checks its phase in the project before passing it onto the next one. This structure can be adopted in the case of stable technology and competition and where modified products are being developed.

- the *lightweight structure*: this is a cross functional coordination committee made up of a liaison manager for each function, thus facilitating the circulation of information. Each person's role evolves with the development of the project and there is no individual identified as its protagonist. Each department participates in decisions concerning the product, the price, the technical support, the distribution and its network, the sales team, the communication, etc.

- the *heavyweight structure*: this is a task force focused on the project with a person designated to be in charge.

- the *independent structure*: this is almost a company within a company and is especially adapted for highly innovative situations. It is made up of a leader who supervises the developmental stages of the project, defines the objectives and sets out work procedures. He is entirely responsible for the results. The team is composed of people seconded from the various company departments and working full time on the project. Actual production is left in the final instance to the company's production department.

The organization chosen for developing the product is very often linked to that of the marketing department, according to whether it is structured by product category (product manager), by type of customer (market manager) or by key accounts (main accounts manager)[19] . Whatever the organization, the aim is the same:

-to bring together the different skills in the best possible way to succeed,

- to separate the process of innovation from the pressures and constraints of everyday activities.

The marketing department's task is to coordinate the overall development process from suggestions for lines of research from the market, right up to the validation by this same market of the resultant innovations. The marketing department must be careful that the innovations are appropriate to market expectations and also that the company can ensure a good quality service for the product: availability of after-sales technical support, actual physical distribution, etc. By listening to the market and to R&D, the marketing department arbitrates and

6. Innovation and Product Management

coordinates, thus favoring the circulation of information. The whole point is to favor the harmonious development of new products. Experience shows that a conflictual gestation period sows the seeds for an eventual commercial failure of the product.

An innovation's success does not only depend on the conditions under which it saw the light or how much was invested in it. Thus, the launch of an entirely new product entails by definition greater risks than a modified or repositioned one. Similarly, any easily copied or unpatented innovation has a high failure risk for the company. Causes of failure of new products are more to do with the market than the technology used: they come from competitor reactions (price wars, buying out of patents), overestimation of potential, too high prices, technical problems linked to the product or badly directed marketing efforts.

In view of the investments needed for designing new products and processes, companies try to guard against such eventualities. This is why, before putting anything onto the market, the company carries out numerous evaluations concerning not only the innovation itself but also the market potential (demand, competition, general environment). Similarly, it is in the company's interest to organize as often as possible meetings about the project with all those concerned.

The conditions for success of new products is based on an osmosis between the company's resources, good project coordination, precise definition and rigorous follow up. The technical excellence of the product, its objective and perceived capacity to answer customer demands, all these are conditions which contribute to its eventual success. This successful policy for innovation can also be used by the company in its advertising campaigns to show its willingness to develop products adapted to its customer's expectations (Figure 6.41).

There is a strong correlation between success and the technical excellence associated with deep marketing and strategic thought concerning the innovation. The latter, the fruit of the interface between technical and sales staff, must exactly express spoken or latent needs to succeed. More than just an innovative product, the company must reason in terms of an innovative system to provide optimum service to customers. Indeed, the mere integration of innovation into a product is not enough to ensure its success, competitors will do their utmost to catch up as quickly as possible. The service offered to the customer must be improved to give him the greatest possible added value.

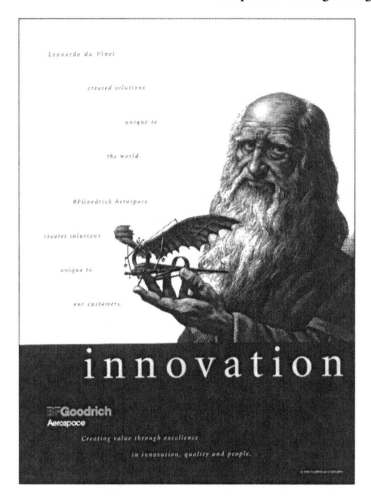

Figure 6-41. Innovation, the BFGoodrich main line of communication[20]

INNOVATION AT SPOT IMAGE: SPOT THEMA

Spot Image, the world leader in space satellite imaging is a good illustration of the innovation process. At the end of the nineties, market forecasts showed the need for new services and in parallel, there was growing competition with the appearance of new, and up to then unauthorized technology (transfer of military observation applications). The company was worried by this situation because the Spot satellite family that they were using could not incorporate such technology (radar and optical resolution to less than a meter).

A study of market requirements and an estimate of the future demand allowed Spot Image to get a better idea of the global market, to better identify the target markets and evaluate its future market share. Based on these studies and analyses, a new product was quite quickly

6. Innovation and Product Management

designed by a team made up of experts from the company and from its customers. Close and effective cooperation between the team members allowed the project to advance rapidly. The project, known as Spot Thema, is an urban scale ground cover data base using innovative mapping systems. It is a digital vector product which can be directly integrated into the users geographical IT system.

Spot Thema is based on recent pictures and on a data base started in 1986 which permits a dynamic analysis to be made of the evolution of the areas, and it also uses new image enhancement processes. It provides the most up to date information possible and permits study at various levels. Spot Thema allows customers to:

- obtain knowledge about and follow urban and peri-urban territory (mapping ground occupation, evolution of urban cover, space consumption monitoring),
- facilitate urban planning (development of road and transport infrastructures, business parks, risk management, etc.),
- model and analyze territorial spaces (countryside analysis, travel maps, etc.).

Professionals concerned with urban planning and development (town planning agencies, public organizations, councils and town committees) thus have access to dynamic information about urban spaces (Figure 6.42).

 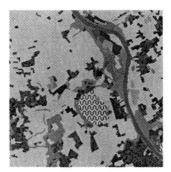

Figure 6-42. Spot Thema: innovation for customers

This innovative approach has allowed development of products and services and their adaptation in terms of customer requirements and environment (tools, architecture, geographical IT systems). Customers can use the database without requiring any special knowledge about space or satellites. The service quality and added value for the customer is thus increased, and the latter can benefit from a product which has no market equivalent and which has already been validated by other big name customers from the Spot Image market.

6. INNOVATION, THE KEY TO DEVELOPMENT OF THE A380 WIDE-BODY JET

6.1 Analysis of the market and the range

The A3XX project was born out of an analysis of the market and the product portfolio[21]. In view of the continuing and predicted growth in air traffic (+5% per year), air movement saturation (routes, airports, etc.) and environmental factors, Airbus began to study the possibility of developing a new, ultra large capacity aircraft, the A380[22] (Figure 6.43).

Figure 6-43. The Airbus jumbo aircraft project: A3XX

On a world scale, 50% of traffic in seats per kilometer takes place on only 6% of the routes linking 2% of the airports; consequently the latter are approaching saturation. This is true for the five main Japanese airports where almost 400 Boeing 747's take off daily. For the world's high density routes such as Tokyo-Sapporo, Tokyo-Fukuoka, Tokyo-Osaka, it is no longer possible to meet the real frequency demand using aircraft with the capacity of a B 747 (in size, productivity and range). And this is how company and airport manager's expectations for a new ultra-large carrier which would eventually reduce the number of take-offs and landings was brought to light. The market analysis also shows that almost all the major airlines in the world use the Boeing 747 in its various versions. Therefore there is a large, latent, potential demand for ultra-large aircraft over the next 20 years which market forecasts estimate to be between 1,000 and 3,000 units[23].

A study of the Airbus range reinforced this decision to develop aircraft in the over 500 passenger category[24] (Figure 6.44). Presently, only Boeing has such an aircraft, the B 747, and its most recent version the B 747-400.

6. Innovation and Product Management

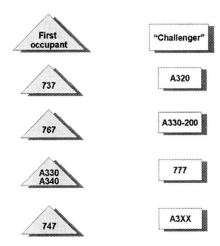

Figure 6-44. Comparison of the offers from Boeing and Airbus at the moment when the A3XX was being planned[25]

6.2 Taking into account customer's expectations

This situation led Airbus to offer an alternative to the B747 monopoly, giving the airlines a choice of aircraft. This decision aims at extending the Airbus range so that customers will be able to choose from a complete family. Finally, this new aircraft must be designed to offer the airlines a veritable tool for their development in terms of running costs, profitability, etc. The task is to offer a new solution capable of giving customers enhanced value. This is why it is necessary to involve the latter as far upstream as possible in the project in order for their exact requirements to be taken into account and to define the specification sheet for the future aircraft with them. The needs analysis has been carried out within the " A3XX club ".

This club is made up of the main companies interested in the project. Their involvement allow to better define specifications for the A3XX family:
 - a much higher seating capacity (from about 550 to 700),
 - variable operating ranges (short haul, especially for Japanese companies, medium and long haul and combi plus cargo versions).

In concrete terms it is a question of developing an aircraft which will make a sizable reduction in the cash operating cost (COC), i.e. 15-17% compared to the existing market offer (Figures 6.45 and 6.46). This can be achieved by size (increase in the number of passengers carried), and by using new technology and optimizing existing ones (type A340). The design of the new aircraft thus takes the market into account (economic constraints) and the required technology (anticipation of innovations and improvements).

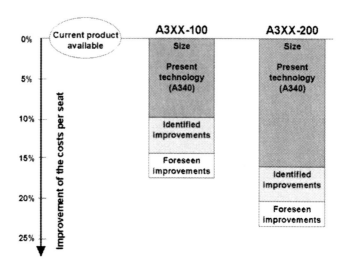

Figure 6-45. Productivity improvement thanks to the design of the aircraft[26]

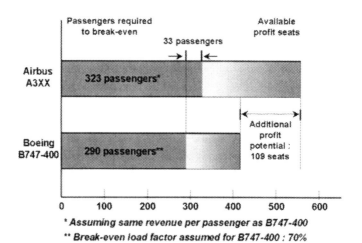

Figure 6-46. The proposal for a more easily achievable profitability threshold[27]

6.3   Producing a solution

The design office have studied the different possibilities for extending the capacity using recent wide bodies. The project for a much lengthened version is unrealistic because of difficulties for on ground maneuvering, which would make

6. Innovation and Product Management

the aircraft incompatible with most airport equipment. Technical constraints have ruled out flattened oval section versions.

In order to optimize costs, the design of the aircraft must not introduce a major technological leap but rather use a traditional configuration (low wings, wing mounted engines, traditional tail fin). The innovation lies in the three level, ovoid section fuselage (the hold and the two passenger decks are connected by a large straight staircase in front and a spiral one at the rear).

Little by little, the idea of having a third deck running the whole length of the fuselage with two levels reserved for passengers above the hold, has found favor with the engineers. These two cabins will thus avoid a "transporting the masses" effect. The aircraft is much bigger inside for passenger comfort, mobility and access. The external dimensions can be contained within a square of side 80m in order to minimize investments necessary to accept the plane in the main airports of the world (Figure 6.47). Above 80m, there would need to be a complete revision of airport infrastructures (parking slots, terminal taxiways, corners to be modified, etc.).

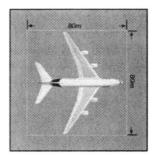

Figure 6-47. The external dimensions of the aircraft take into account airport infrastructures[28]

6.4 Innovation in the cockpit

The cockpit is situated at a level between the main and upper decks in order to reduce drag and aerodynamic noise, while at the same time facilitating crew access to the two passenger decks (Dupont, 2000). It is also designed to require only one type qualification for the whole future A380 family (Figure 6.48). Furthermore, it retains the layout of the A340, A330 and A320 in order to allow Cross Crew Qualification[29].

Figure 6-48. What the cockpit of the A380 will look like[30]

The cockpit will be entirely automated with electrical controls as in its predecessors; it will also be highly user friendly with eight rectangular screens (instead of six square ones) with a 75% increase in display surface compared with previous generations. The two extra screens have the control and display modules for the flight management system (FMS) plus the messages received from data links. The pilot will be able to manage from a single screen all the dialogue with air traffic control from clearance right up to execution by reprogramming the route followed by the FMS. Furthermore, a keyboard built into the retractable tray in front of the pilot allows in-flight messages to be sent via air - ground data links or consultation and updating of the on-board information system. The latter, using an Ethernet network (100Mbit/s), will store airport maps, permit flight preparation, calculate the aircraft's center of gravity, update the flight plan and link with maintenance crews. The latter will be able to download onto their laptops all the diagnostic information retained in the on-board maintenance system. An electronic pointer will carry out the same function as a mouse except that it will be able to move over all the screens on the dashboard. Using the rectangular format, the pilot will be able to obtain a navigation display (ND) for the trajectory, including contours of the terrain being flown over and the minimum safe altitude published in the aeronautical documents (Daniel, 2000).

6.5 *Production innovation to reduce costs*

The main innovations destined to reduce running costs concern the aerodynamics, the mass and maintenance. New calculation methods and wind tunnel experiments have enabled the drag to be reduced while at the same time preserving simplicity and developmental capacity within the family, without changing the wing structure. A large decrease in mass (or an increase in the maximum range) has been obtained by moving the center of gravity backwards allowing the tail fin surface area to be decreased, as well as by using new materials. The latter are for example:

6. *Innovation and Product Management* 203

- weldable aluminum alloys such as 6013 from Alcoa or 6056 from Pechiney (Figure 6.49),
- Glare a hybrid material made up of thin layers of aluminum and pre-impregnated glass fiber,
- carbon fiber reinforced plastic (CFRP) for sub assemblies with high loads.

In addition, the lower panel stringers will be laser welded and not riveted, giving a gain in weight and above all a reduction in cost.

Figure 6-49. An advertisement for Pechiney Aerospace, supplier of alloy and aluminum[31] : *"to understand all the benefits provided by aluminum, it is necessary to get high. 10,000 meters at least"*

6.6 The involvement of industrial partners

Apart from the customers, industrial partners have been very rapidly associated with the project, e. g. AIDC, Belairbus, Eurocopter, Finavitec, GKN Aerospace, Hurel-Dubois, Latécoère, Saab, Stork Aviation, etc. Engine manufacturers have been approached for studies on auxiliary power units (Pratt&Whitney) as well as for the actual aircraft engines. A memorandum of understanding (MoU) has been signed with Rolls-Royce for its Trent 900 engine as well as for the GP7200 from Engine Alliance (alliance between Pratt & Whitney and General Electric). An entirely new

20 wheel landing gear must also be designed (Messier-Dowty and BFGoodrich are in the running).

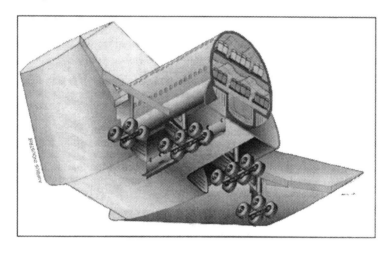

Figure 6-50. The projected landing gear for the A380[32]

The aircraft must in addition, be as well adapted as possible to the existing airport infrastructures. Almost 50 world airports have been associated with the development of the product. In general, these airports do not have the possibility of building extensions and so the A380 must conform to present structures. Preparation for industrial manufacture necessitates the use of different partners and the drawing up of a timetable to monitor progress (Figure 6.51). There is also the question of anticipating production methods and assembly (place, equipment, manpower, etc.).

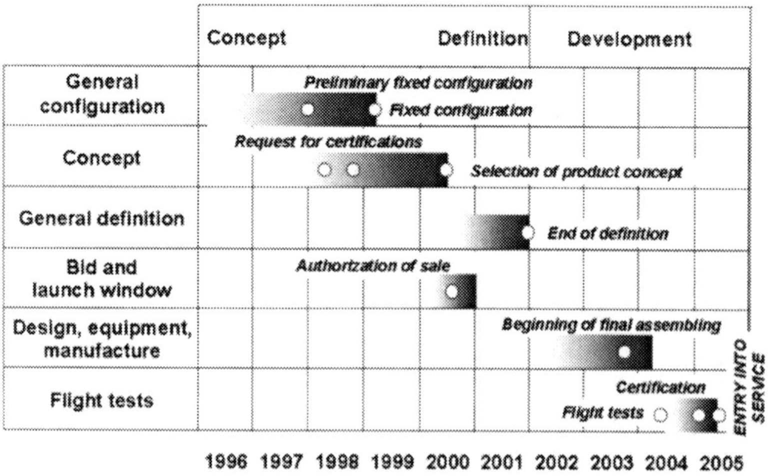

Figure 6-51. From the design to entering service[33]

6.7 Cabin fittings

Designers have been working on the cabin fittings right from the beginning of the project and to do this, the opinion of future passengers and users has been taken into account[34]. In this way, in view of the layout of the aircraft, design studies have been carried out for the transfer of cooked dishes from the hold to the two cabin levels (Figure 6.52). The idea has been to optimize the available space and maximize that devoted to passenger and cargo transport (freight, luggage, etc.). By installing lifts for delivering the dishes, the crew's work is simplified and the loading/unloading of food will be improved.

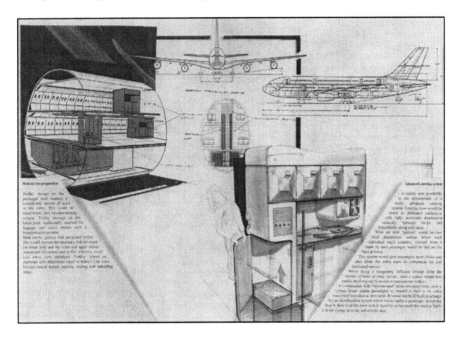

Figure 6-52. The transfer of cooked dishes from the hold[35]

The interior of the aircraft (double deck and double aisle) is such that airlines have several possibilities to personalize the layout and thus differentiate their service offer. An open space can be flexibly fitted out according to the wishes of the company, with the possibility for people to move around during the flight towards a self-service zone and to talk freely with other passengers. This space can be used to install a conference or medical center (Figure 6.53).

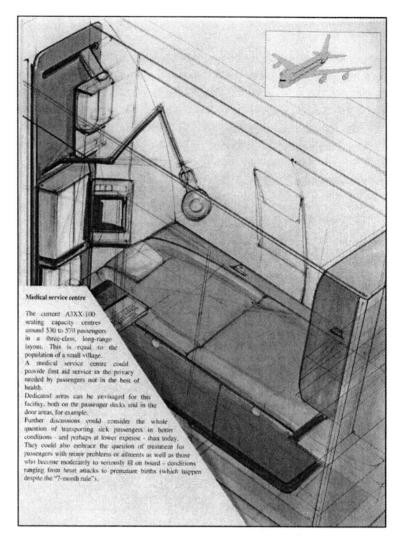

Figure 6-53. Possible fitting out of a medical care center[36]

The designers have also thought about the possibility of having a fitness center at the rear of each deck (Figure 6.54). This area will allow passengers to do a little bit of exercise so as not to suffer the effects of sitting down for long periods. Exercise permits better vascular circulation, makes the passengers feel better and allows them to arrive at their destination in top form.

6. Innovation and Product Management

Figure 6-54. The possibility of having a fitness center[37]

6.8 Future versions of the A380

In anticipation of future evolution of the new generation aircraft, Airbus has defined the different versions of the A380 from the initial A380-100 taking 555 passengers in three classes with a range of 14,200 km (7,650 nm), to the A380-200 capable of carrying 656 passengers over the same range of 14,200 km, right up to the A380-F, the cargo version with a capacity of 150 tons over 10,630 km (5,725 nm). An A380-100R version is also planned, taking 555 passengers over a range of 16,200 km as is a shortened version, the A380-50R carrying 481 passengers over the same range. A dedicated Japanese market version, the A380-100S will provide a solution to the need for frequent flights. Combi versions are also being studied:
- 473 seats, 37,9 tons of cargo over 13,500 km (7,270 nm),
- 421 seats, 51.3 tons of cargo over 12,940 km (6970 nm).

These versions of the A380 will allow a complete family of products to be built, covering the different types of needs for ultra large capacity aircraft (Table 6.55 and Table 6.56).

	A380-100	A380-100HGW	A380-100R	A380-200
Wingspan	79,8m	79,8m	79,8m	79,8m
Length	73m	73m	73m	79,4m
Height	24,1m	24,1m	24,1m	24,1m
Capacity	555 pax	555 pax	555 pax	656 pax
Freight (Load Devices 3 Containers / pallets)	36 + 2/12	36 + 2/12	36 + 2/12	36 + 2/12
Maximum take off weight	548 t	560 t	590 t	590 t
Maximum landing weight	383 t	383 t	390 t	412 t
Maximum unfuelled weight	358 t	358 t	365 t	385 t
Maximum empty weight	275 t	275 t	282 t	292 t
Maximum payload	83 t	83 t	83 t	93 t
Fuel capacity	325000 l	325000 l	370000 l	370000 l
Engine thrust	302 kN	311 kN	333 kN	333 kN
Range	14400 km	15100 km	16200 km	14200 km

Figure 6-55. Characteristics of passenger versions[38]

	A380-F	A380-100R C11
Upper deck	17 pallets	397 pax
Main deck	28 pallets	11 pallets
Hold (LD3 containers / Pallets)	36 LD3 / 12 pallets + 2 LD3	36 LD3 / 12 pallets + 2 LD3
Maximum take off weight	583 t	590 t
Maximum landing weight	427 t	407 t
Maximum unfuelled weight	399 t	382 t
Maximum empty weight	249 t	276 t
Maximum payload	150 t	106 t
Fuel capacity	325000 l	370000 l
Engine thrust	333 kN	333 kN
Range	10410 km	12540 km

Figure 6-56. Characteristics of freight and combi versions[39]

6. Innovation and Product Management

Figure 6-57. Drawing of Airbus A380

NOTES

1. Sources: Airbus - Type of aircraft: non revealed.
2. Recurrent costs are all those associated with the mass production phase. Non-recurrent costs are those linked to the product development phase.
3. See Chapter 8, Pricing policy.
4. Source: Airbus
5. Source : Documentation Airbus
6. See *Air & Cosmos, Airline Business...*
7. See www.aerospace.bombardier.com
8. See Chapter 15, Alliance Strategies.
9. Source: CFM International
10. Source: CFM International
11. Source: Latécoère
12. Photo Boeing published in *Air&Cosmos Aviation International*, (1999), Friday 16 April, n°1699, p 19.
13. Illustration from Dassault annual report.
14. Source: Weber Aircraft.
15. Source: adapted from Aerospatiale-EADS documents, 1991.
16. See infra and Chapter 3, Business marketing intelligence.
17. Advertisement from *Aviation Week*, (1999), 1 November, p 55.
18. Advertisement published in *Air & Cosmos* (2000), n°1764, 29 September, p. V.
19. See Chapter 9, Selecting distribution channels and sales team management.
20. Advertisement published in *Aviation Week*, (1999), 1 November, p 58.
21. Source: Airbus / Press Office, Airbus A3XX AI/L 810.0120/00, website www.airbus.com and the different dossiers published in the trade press.
22. Thanks to a sufficient number of firm orders received, the A3XX was officially launched as the A380 on 19 December 2000.
23. See Global Market Forecast, Airbus.
24. See supra, 3. Managing the product portfolio, figure 6.10.
25. Source: Airbus
26. Source: Airbus
27. Source: Documentation Airbus
28. Source: Airbus
29. See Chapter 14, Building loyalty: Maintenance, Training and Compensations
30. Published in *Air & Cosmos*, (2000), n°1764, 29 September, p XX.
31. Illustration published in *Air & Cosmos*, (2000), n°1764, 29 September, p XVII.

32. Illustration published in *Air & Cosmos*, (2000), n°1764, 29 September, p XVI.
33. Source: *Air & Cosmos*, (2000) Op. Cit.
34. See Chapter 3, Business Marketing Intelligence
35. Source: Airbus.
36. Source: Airbus.
37. Source: Airbus.
38. Source: Airbus in *Air & Cosmos*, September 2000, n°1764.
39. Source: Airbus in *Air & Cosmos*, Op. Cit.

REFERENCES

Daniel, J.-P., (2000), Cockpit of the XXIth Century, *Revue Aérospatiale*, p 28-31, March, n°166.
Dupont, J. , (2000), Special Dossier A3XX, *Air & Cosmos*, 29 September, p III-XXXI, n°1764.
Les Echos, (1999), Bombardier, champion du transport aérien régional, 28 June, p 70.

Chapter 7

MARKETING OF SERVICES

Marketing was initially developed with products in mind, which explains why it has been applied relatively late in the service sector both in Europe and the United States. Over the last few years, this trend has definitely changed both for consumer and business services. In particular the growth of services linked to the Internet has accelerated the development of the marketing of services.

1. THE CHARACTERISTICS OF SERVICES

In comparison to products, services (Eiglier and Langeard, 1987) are essentially characterized by their intangibility, their inability to be stored and their indivisible and variable nature.

1.1 Intangibility

Before it is performed, a service is in fact difficult to demonstrate. The salesperson can try to describe it, citing customer references, but it is impossible to show ahead of time the real results applied to the customer's specific case, whether in the field of daily services (restaurants, telecommunications, health), financial services (banking, insurance, consulting), transport, tourism, etc. Thus in comparison to a product, a service is intangible. However, there are very few "pure" products or services (Malaval, 2001). In fact, there is a continuity between products and services:
    - either because the service is an extension of the product, i.e. after sales, maintenance, or repair...;

- or because the service itself includes tangible elements after being completed: the range of dishes offered in a plane, regularity of functioning of equipment/machine tooling, a telephone or electricity bill...

In fact, a service consists of a combination of tangible and intangible elements (Flipo, 1989). Accordingly, services have been classified as a function of their tangible or intangible dominant factor: hotel and restaurant services are considered as more tangible while banking, insurance and training are all considered as mainly intangible services (Figure 7.1).

By analyzing these combinations, it can be observed that the more intangible a service is, the more difficult it is to use standard marketing tools initially developed for products. In communication, for example, when there is no tangible proof of the product's superiority, word of mouth is extremely important.

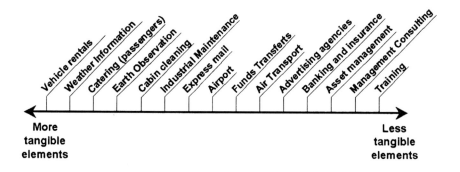

Figure 7-1. A classification of services according to their degree of tangibility (consumer and business markets)

Customers will be sensitive to different outside signs. For example, in the case of a maintenance check-up, the customer is reassured by the coherence projected by the condition of the personnel's equipment, his car, clothing, the clarity of his reports, etc. In the case of automated maintenance, one of the possible signs could be the transmission of data, maintenance visit statistics, etc. This is even more important given that customers rarely have the technical skills or means necessary to fully evaluate the service performed.

ON-LINE INFORMATION: A SERVICE TO INCREASE MAINTENANCE EFFICIENCY

One of the main concerns of airlines is reducing maintenance costs and increasing the life time of aircraft (time of use). It is possible to meet these requirements by providing technical documents on-line from the aircraft builder and from the different suppliers of the aircraft. Boeing has done this with its own myboeingfleet.com website created for airlines. However,

7. Marketing of Services

the giant parts manufacturers such as Honeywell and UTC have preferred to create their own sites to directly reach final customers.

Airbus has developed its own Internet portal, Airbus On-Line services (AOLS), on which it proposes constructor documentation, spare parts optimization inventory systems, and intelligent maintenance and engineering systems. To make it user-friendly, the site has links with the Idis portal which is aiming to become the leading site for technical documentation, regrouping as many components suppliers and plane makers as possible. Through this on-line information, constructors and components manufacturers provide support to their products and services and bring added value to their customers.

Compared to a product, a service is intangible and for the customer it involves greater risk. To minimize this risk, the customer looks for a concrete basis of evaluation. Consequently, it is important to provide a framework for the service with tangible indicators that can be evaluated before and after its completion (Figure 7.2).

Figure 7-2. Making the service tangible

Cleaning companies illustrate this "tangibilization". After the cabin has been cleaned, the head of the service company submits a signed document to the airline. This document specifies any problems observed, such as a damaged passenger tray that has been repaired or a reading lamp that has been replaced. In this way, the airline has a written record of the service performed.

1.2 Perishability and stock-impossibility

Perishability refers to the fact that generally services cannot be stocked. An example that is valid both for the consumer market and business to business is the number of seats available in a plane: empty seats at takeoff represent a revenue loss which cannot be made up on following flights. This explains the development of yield management[1] which consists in offering reduced prices on seats still unsold a few hours before takeoff. Unlike when managing a product, it is not possible to move stocks around. Holding capacity limits sales but also determines the cost price.

Services are even more difficult to manage owing to the fact that demand is difficult to project and often fluctuates significantly (strong fluctuations in demand for energy or air transport).

For airlines, peaks in demand coincide with the vacation calendars of developed countries combined with fluctuations linked to professional events such as the main trade shows and congresses. To reduce these differences in demand the airlines have a sophisticated pricing policy[2] inciting customers to spread out their departure dates for long haul flights and modify the departure times of their domestic flights.

1.3 Inseparability

Inseparability is a concept which was initially developed to characterize consumer services. However it also applies to industrial services in that the customer evaluates the latter in their entirety and is therefore judging a complete entity. Thus for an airline, this would be all of the in-flight services offered such as the quality of the meals served, the pleasantness of the crew, the choice of films, etc. The latter are as important in the passenger's evaluation as the comfort of the seat and the punctuality of the flight. This service will be judged not only on the basis of material elements but also on less rational factors such as ambiance and the feeling of having received personal attention. It is the customer's global perception of service which will be taken into consideration.

1.4 Variability

Services are variable because unlike a product they are difficult to standardize. In fact one can almost say that the quality changes each time that the services are provided. The human factor is very important in the performance of a service and the more it comes into play, the less uniform the service will be for it is linked to the individual carrying it out. For example, in a restaurant a meal will be more or less appreciated depending on the talent of the cook and the attitude of the waiter. In order to guarantee a consistent standard of quality it is important to develop quality control and invest in procedures to calibrate the different service phases. As an example one can cite auditing firms and more generally all service companies that seek to obtain an ISO certification. Thus one of the present development strategies of airports is to improve the quality of baggage transfer and follow-up. Increasingly, baggage is handled by an automatic sorting system capable of functioning in a minimum of time to meet the constraints of check-in and connecting flights. Airport parts manufacturers offer equipment designed to improve the handling of baggage in terms of sorting, safety control, and tracking. This standardization improves the quality of services provided by airports to passengers, ensuring maximum comfort and safety. The second source of variability comes from the customer:

- The same flight could be evaluated differently depending on the passenger's state of mind, which in turn depends largely on his reasons for travelling. The evaluation will be more favorable if the trip is associated with a positive professional or family event.

- Furthermore, the passenger's appreciation also depends on his involvement during the flight. An active passenger who asks the stewardess to hang up a jacket for example, or give him a magazine will have a better perception of the flight than the passive passenger.

Of course this has to be qualified depending on the service sector. In a field like collective food services, suppliers such as Sky Chefs, Gate Gourmet, Servair, Select Service Partner (Compass Group) supply turnkey solutions and the customer can be relatively passive. However, for a service like training, for example with Alyzia Airport Services or the National Air Traffic Services, which work with airport operators and airlines, participant involvement is essential: the same trainer teaching the same course can have very different results depending on participant involvement.

We can highlight here the importance of direct contact between the service company employee and his customer. The image that this employee projects is always crucial: if he provides poor service or is negligent, this can destroy all the goodwill that the company has been building upstream. The risk is minimized in business to business when the same person is assigned to deal with the customer company ; this employee thus ensures regular follow-up and is aware of the customer's satisfaction or needs expressed the last time the service was performed.

Overall, a service is characterized by a combination of tangible and intangible elements (Figure 7.3). It is perishable, inseparable and variable. The customer pays for the use of a service without owning it ; a priori nothing remains after its use. All of these characteristics together should help the service company to focus on its customer's real expectations.

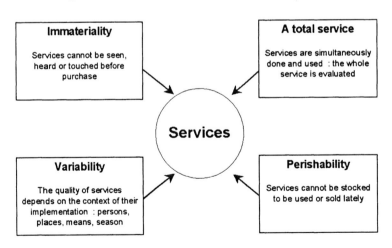

Figure 7-3. The four characteristics of services[3]

A client decides to buy a service based on the advantage that it represents. The first step in the creation of a service or in its evaluation lies in defining the concept of customer benefit, in other words, the main advantage that the customer will gain from the service. Close observation of customer expectation is crucial in developing a strong service concept.

The next step is for the service provider to define the concept by studying in detail what services he will have to provide to satisfy the customer benefit. The coherence of the latter will determine the overall service and how it is perceived by the customer. Finally, the service has to be actually set up, ensuring its tangibility. All of the services defined need to be organized so that they correspond exactly to customer expectations.

SPOT IMAGE: A SERVICE WITH MULTIPLE APPLICATIONS

A world leader in its market, Spot Image analyzes, processes and commercializes data from Spot earth observation satellites. Extremely varied and reliable, the data is used in numerous sectors: environment, agriculture, geology, surveillance, urban planning, cartography... Space observation, apart from its value for scientific research, is extremely important for all aspects of managing the earth's resources. In particular, the Spot system makes it possible to collect local or regional data on ground utilization, vulnerable zones, deforestation, erosion, desertification, urban areas, and the environmental impact of construction projects.

Using satellite imagery, and in particular with the development of the Geospot range (Spotview, SpotMap), Spot Image offers personalized solutions adapted to diverse customer needs in multiple domains such as:
  - the exploitation and management of natural resources: geology, forests, sea life,
  - agriculture: irrigation, desertification, rural planning,
  - prevention and management of natural risks: floods, volcanic eruptions,
  - civil engineering projects,
  - urban planning and management[4],
  - environmental protection.

The Geospot range, which provides both standard and tailor made solutions, is continually being expanded with new applications. Starting with the Geospot range, Spot Image offers ready-to-use geographical data " products " such as mapping of protected zones, defense areas, forestry zones, studies on territorial planning... Applying a service strategy, Spot Image has developed know-how in carrying out turnkey projects, or providing complete service solutions from the supply of images to the training of future users. This is true of the ALIS project (Agricultural Land Information System) an aid for agricultural resource management in Egypt.

The Ministry of Agriculture and the Soil and Water Research Institute (SWRI), under the aegis of research in food self-sufficiency, wanted a permanent tool which would measure the evolution over time of areas cultivated with the main crops, keep track of uncontrolled

7. Marketing of Services

urbanization of agricultural lands, and analyze the potential of new agricultural lands. Supplying a turnkey solution to solve the initial problem required training local specialists and transferring know-how to ensure the continuity and operation of the system. The project was based on 6 key steps:

- the acquisition and supply of images,
- the projection of Geospot space maps,
- the identification of land usage,
- agricultural statistics and associated maps,
- setting up a computer network for the production and exploitation of results,
- technology transfer and team training.

			Spot Image: ALIS
Level 1 ↓	Customer Benefit Concept	What is the customer really looking for ?	The self-sufficiency of Egyptian agriculture
Level 2 ↓	Transpose Customer Expectations Into a Service	What benefits are the service company offering ?	An aid for the management of agricultural resources
Level 3 ↓	Precise Definition of the Service Proposed	What services should be proposed ? • tangible and intangible • type of services • levels of service	• supply of images • space maps • ground occupation inventory • agricultural statistics and associated maps
Level 4	Setting up the Service	What means ? • skills of staff employed • technical characteristics of equipment	• training of local teams • setting up a computer network

Table 7-4. Example of a service concept: Spot Image

Spot Image provided a complete service to its customer – the customer benefit being food self-sufficiency for Egyptian agriculture – from the acquisition of images and the completion of ground occupation maps and agricultural statistics, to the setting up of a computer network and a specialized training program. The service provided was based on the provision of technical assistance, but also on technology transfer, with the aim of building a long term working relationship between Spot Image and its Egyptian customer. In short, Spot Image helped to improve management of Egyptian agriculture within the ALIS project.

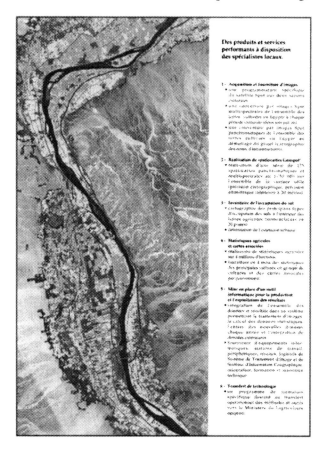

Figure 7-5. Spot Image: from the supply of satellite images to the training of local teams

2. DIFFERENT CATEGORIES OF SERVICES

Services are extremely heterogeneous from those aimed at consumers to those targeting businesses.

For example the field of air transport services (Figure 7.6) ranges from those provided by airport operators (air traffic, baggage and freight, passenger information, control and security, access, parking...) to airline services for passengers (transport, welcome, information...) as well as the services of airline partner companies (catering, maintenance, cleaning...).

Some services are just between professionals such as the catering company Servair and the airline (1). Another service category concerns the general public, i.e. the on-board menus (2) served by Air France.

7. *Marketing of Services* 219

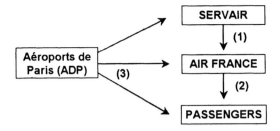

Figure 7-6. Examples of services in the air transport sector

Then there are those services which target both professionals and the general public (3). A case in point is the airport company ADP (Aéroports de Paris) which provides the airlines with an infrastructure adapted to the size of the aircraft, refueling facilities, waiting rooms and boarding bridges, and above all technical assistance and adequate organization for flight operations. ADP also offer services to passengers between flights and the people who accompany them such as car parks, restaurants, various practical commodities (information areas, Internet access, baggage checks, shops, restaurants, etc.) and the appropriate religious facilities (Figure 7.7).

Figure 7-7. ADP's double target: passengers and airlines

Table 7.8 shows a selection of services according to target: professionals, general public or mixed.

Services for the General Public	Services for Professional	Mixed Services
Airlines e.g. United Airlines, British Airways, Thai Airways, Air France...: sales of tickets	Alyzia Airport Services: ground assistance, security checks, rent of equipment, shops and car park management, technical support and maintenance, advice and training	Aviapartner: passenger assistance, aircraft runway assistance, freight and postal processing
Airlines: - supplying varied and adapted on-board meals - supplying news and leisure magazines (films, music) - taking charge of unaccompanied minors and people with restricted mobility	Servisair: runway assistance, freight assistance, traffic and operations, VIP rooms, on-line maintenance, duty free shops, supervision, business aircraft assistance	Airports such as JFK, Abu Dhabi International Airport, Inchon International Airport (Korea): passenger services (religious facilities) and cargo transport (hub)
Internet sites for choosing and reserving tickets	Fedex: Fret Express	Weather services
Leisure tourism: Club Med	AirLiance: supplying spare parts for airline planes	Car rental services
Films, on-board TV: AirTv	Lido (Lufthansa Aeronautical Services): Flight Management System	

Table 7-8. Some services in the aeronautics and space sector

3. PROFESSIONAL SERVICES

Professional services cover all those activities of which more than 50% of the business is for companies; these include services for the daily functioning of companies (IT, energy, banking, insurance, etc.) as well as those which contribute to the service delivered to the end customer (catering, health services, on-board video).

The development of industrial services is the result of three main factors:

- the growth of industrial production which needs more and more services, i.e. transport, advertising, information processing, etc. ;

- the outsourcing of activities which are not within the company's competence: catering, engineering, delivery, etc. ;

- the appearance of new services stimulating demand: e.g. security systems.

The trend in airports is to have detection systems for explosives in luggage using tomography and X-ray imaging such as those developed by Heimann Systems, PerkinElmer Instruments or Rapiscan Security Products. Also a process known as Quadropole Resonance (QR) allows the presence of explosives on a person to be detected. Ion Track Instruments have developed a very precise system (Itemiser)

7. Marketing of Services 221

which can detect the presence of infinitesimal amounts of explosives (including in suspension) and by supplying such systems to airports, these companies ensure optimum passenger safety (Figure 7.9).

Figure 7-9. Ion Track Instruments: detection systems serving airports and passengers[5]

Services to companies include consulting and studies, advertising and communication, IT services and telecommunications, passenger and goods transport, legal activities, architectural and engineering activities, personnel selection and subcontracting, real estate and equipment rental services for companies.

While this list is long it is not all inclusive. Services can also be classified according to whether they are required by law, and according to the type of function they are linked to.

3.1 Services which are required by law and regulations

This is any demand for external services which are required by law:
- auditors: the accounts of limited companies and partnerships limited by shares plus all other types of company of a certain size are legally required to be checked and certified ;

- architects: contracting authorities are required by law to use the services of an architect for all work covered by a building permit.

3.2 More general services linked to management and strategy

These are all the services which a company may use to optimize management, i.e. banking, insurance as well as specialized consulting services in marketing, human resources management and strategy plus the external or contractual checking and drawing up of annual statements by chartered accountants, required by law. For example, companies in the aeronautics sector increasingly use on-line services in an effort to improve their purchasing management efficiency.

3.3 The aeronautical " marketplaces ": a new type of service from MyAircraft to AirNewco

Marketplaces have been created to bring together buyers and sellers with a common interest linked to their activity and they can be created by an independent player or market maker who thus runs the group. The ownership of the actual site is generally shared equally between several partners for greater credibility and operational neutrality. Because of the specificities of the aerospace sector, the main market places used are vertical (i.e. specific to the sector). However, for products and services with a lower priority (e.g. office supplies) aeronautics companies sometimes use horizontal market places (covering several sectors) such as buying-partner.com for general purchases. For buyers, the main advantage of this type of model is access to a wide range of suppliers, with the best prices and reduced transaction costs. For sellers, the market place facilitates sales to appropriate customers, unearths new markets and reduces distribution costs.

Internet B to B exchange sites have multiplied with the association of a few players from the same industrial sector and an IT provider (Oracle, Commerce One, Ariba, etc.). These promise to bring together buyers and suppliers thanks to many services: catalogues, comparative prices, bidding, tenders.

Independent marketplaces

Several independent market places have been developed (Partsbase, AviationX, Skyfish, SynerDeal, etc.) in order to allow customers to improve their purchasing efficiency. Thus SynerDeal has been created to reduce the purchasing costs of aeronautical industrial companies through the use of Internet. The goal is to reduce costs for standard goods such as raw materials, industrial parts, machining, tooling, electronic components, etc. This platform identifies the new suppliers, helps customers put together their tenders and ensures the appropriateness of supplier's replies.

7. Marketing of Services

The partsbase.com site is a dedicated market place for the aeronautics, space and defense sector. Here, sellers can move surplus stock or slow-selling products by auction and buyers can acquire what they need at much more competitive prices. Over 2300 types of products are offered in the following categories:

- aircraft,
- supplies,
- avionics,
- ground support,
- equipment,
- direction finding and guidance instruments,
- helicopter equipment,
- chemical supplies,
- hydraulic components,
- computers,
- interior equipment,
- connectors,
- landing gear,
- control systems,
- mechanical components.

Figure 7-10. Partsbase marketplace

The RightFreight hub has been made in the freight transport sector to supply purchase management solutions. It also aims to reduce costs, speed up the decision making process, facilitate information exchange and improve customer service. Using the data from the customer's management system (ERP) via the RightFreight connector, the RightFreight Contract Manager sends out information concerning the shipment (nature of the product, quantity, characteristics, price). The offers of service are then examined and the transporter chosen (using the customer's criteria) can then be retained by the RightFreight Booking Manager. Invitations to tender can also be sent out via the RightFreight Shipper Exchange and the RightFreight Document Center supplies, fills in and automatically sends all the necessary shipping documents to the customer's ERP. The site allows users to pinpoint current shipments using a tracking system: RightFreight In-Transit Visibility. It also allows total shipping costs including taxes and the various outlays necessary to be calculated plus shippers' performance to be compared.

Figure 7-11. The RightFreight marketplace

Airline sites

Faced with the development of specialist sites, the airlines (which have sometimes already joined together within large alliances[6]) have decided to create market places. For cargo there has been the GF-X (global freight exchange) project announcement and the extension of passenger alliances to cover freight. For air transport, the airlines have developed several operational platforms.

The aeronautic site Aeroxchange groups together Air Canada, All Nippon Airways (ANA), America West Airlines, Cathay Pacific, Fedex, Japan Airlines, Lufthansa, Northwest Airlines, Scandinavian Airline System (SAS), Singapore Airlines, Air New Zealand, Austrian Airlines and KLM. In this way these companies aim to improve their operating efficiency and that of suppliers (administrative costs, purchasing costs). This marketplace offers a wide variety of products and services destined for the aeronautics industry and sets up common standardized procedures for the different airlines.

Figure 7-12. The Aeroxchange marketplace

Another market place, AirNewco, aimed at suppliers' customers (the airlines) was set up by the airlines to develop an exchange network designed to harmonize the complex logistics chains in the sector. AirNewco's founding members were American Airlines, British Airways, Continental Airlines, Delta Airlines, and United Airlines. More recent arrivals are the freight carrier United Parcel Service (UPS) and the European companies Air France, SAirGroup and Iberia. This Washington based marketplace aims to reduce buyers' and suppliers' costs on the aeronautics market.

Figure 7-13. The AirNewco marketplace

Aeronautical supplier sites

In a similar fashion, the rise in power of independent sites has encouraged the OEM's to invest more in the Internet and offer their own marketplaces. In this way there is EverythingAircraft for example which is the result of collaboration between Boeing, Lockheed, Raytheon and BAe Systems.

The supplier orientated MyAircraft platform was announced at the Singapore air show. Developed by the parts manufacturers United Technologies and Honeywell (to which BFGoodrich has been added) plus i2 Technologies (a specialist in the publication of e-business solutions, notably the TradeMatrix technical platform), MyAircraft is supposed to be a dedicated electronic marketplace for business

7. Marketing of Services

between aeronautical professionals, an estimated annual market of some $500 billion according to its creators[7].

Open to all industries within the aeronautics sector, the site initially aims to connect Honeywell and United Technologies to their trading partners: Pratt & Whitney for aircraft engines, Sikorsky for helicopters, Hamilton Sundstrand for aerospace systems, etc. . The aeronautics players (manufacturers of control systems, electronic devices and avionics equipment, etc.) can meet on this site to collaborate and complete transactions. In this way purchases and sales can be dynamic, i.e. they will integrate the notion of automatic supply or order flow. Participants on this market place can optimize their supply chain thanks to forecasting and restocking functions. The site also offers products and services for airline companies and business or military aircraft operators. The real power of the marketplace lies in strength in numbers ; buyers and sellers united in the common goal of improving efficiency. The fall in costs thus concerns not only the company and its supply chain but also that of the suppliers and sales teams.

Figure 7-14. The MyAircraft marketplace

The market place Aerospan.com focuses on the air transport market, with price negotiation, signature of deals, simple and fast access to useful information and real-time market follow up.

Figure 7-15. The Aerospan marketplace

At the outset the result of a joint venture between AAR and Sita (www.sita.int), it has been owned since by Sita, the world leader in integrated telecommunication and information systems for the air transport industry.

Suppliers	Purchasers
Surplus inventory reduction	Procurement cost savings
Sales efficiency / market access	Inventory reduction
Increased product exposure	Increased delivery reliability
Market trading efficiency	Market trading efficiency
Improved market intelligence	Secure, encrypted global trading network
Secure, encrypted global trading network	Online documentation
Innovative trading method	Global search mechanism
Industry news, regulatory information, white papers and feature articles	Industry news, regulatory information, white papers and feature articles

Table 7-16. Main advantages for the purchasers and suppliers

Recent trends

The whole of the aeronautics industry is involved in the development of marketplaces. The airports in London (BAA), Paris (ADP) and Copenhagen, Dallas-Fort Worth, Houston, Pittsburgh, Indianapolis and Melbourne have announced their intention to develop together a " WorldAirports " website for purchases in common. The airports can in this way put out to tender, group up their purchases and get better rates from their suppliers.

Following on from the numerous aeronautical website creations, there is now a consolidation phase with natural selection operating in favor of the most solid and coherent projects. Thus AirNewco and MyAircraft have decided to merge to give one vast electronic market (The Guardian, 2000; Beauclair, 2000), expected to generate 10 to 15% of business between the different civil and military aeronautics players. It is in the airline companies interests to work with the suppliers on an independent platform which is open to all participants from the sector. This new platform must cover five main areas designed to optimize the logistics chain:
- maintenance and engineering,
- fuel and associated services
- catering and on-board services,
- airport assistance service,
- general purchases.

The key to success lies in making a critical mass in terms of procurement, in order to reduce stocks by about 6%. The site aims to connect up buyers and sellers by offering services such as on-line catalogues, bidding, tools for stock and logistics chain management, and help with transactions. It also wants to be a "community website". However it does not want to compete with parts manufacturer's e-business, which, like in General Electric, goes through individual, on-line procurement channels.

3.4 Services linked to the production process

These are all the services directly involved in the production process, like the hiring of machine tools, production materials or IT production maintenance, etc. The use of these services can be due to a sudden increase in activity or a breakdown, but often they are the result of a decision to outsource certain functions.

3.5 Sales related services

These are services associated with the product which aim to reinforce customer satisfaction. Examples are freight companies specializing in urgent, prompt deliveries (Fedex, TNT, UPS, DHL) or in a particular technical characteristic (cold, etc.). Similarly some Internet reservation sites or travel agents offer to oversee business ticket reservations[8]. Original and different services can be developed to differentiate the offer and give greater customer satisfaction. A case in point is Air France who set up a partnership with the sea transport company Emeraude Lines to organize, during the summer months, a daily catamaran service between Nice Airport, Cannes and the port of Saint-Tropez (around 1h 15m journey time). This permitted passengers to discover the French Riviera in an original way (Figure 7.17).

Figure 7-17. Nice – Cannes – Saint-Tropez by catamaran[9]

3.6 Technical and commercial global services

Services very often contribute to a company's production process as well as improving customer satisfaction. In the aeronautics sector, for the A330 and A340, Airbus have developed an in-flight information service for the technical and commercial crew as well as passengers. This system connects the plane to the ground communications network, whether in flight or on the ground. In order to bring the maintenance, repair and overhaul personnel into closer contact, the work environment is recreated in the air. The crew and especially the technical teams can take advantage of better operational conditions thanks to electronic documentation, an electronic log, real time weather forecasting, route and airport information plus e-mail communications. This also lets passengers watch live TV programs from their seats, connect up to the Internet and use the other facilities available (games, laptop connection, etc.).

One of the aims of components suppliers such as Honeywell or Tenzing is to offer an Internet service (e-mails and surfing) to passengers, in conjunction with the manufacturers or directly with the airlines. Companies such as Air Canada, Singapore Airlines, Cathay Pacific and Swissair have already tested such a service on different aircraft (747-400, A319, etc.).

Global services also include those which participate indirectly in the company's production by ensuring efficient everyday functioning. Industrial cleaning, working clothes (Kimberly-Clark, etc.), catering (Sodexho, Compass-Eurest, Servair, etc.) and accommodation (Accor, Marriott, Holiday Inn, etc..) are often outsourced because they are not strategically important for the company, cost less or are of higher quality externally.

SERVAIR: A GLOBAL SERVICE FULFILLING THE EXPECTATIONS OF AIRLINES AND THEIR PASSENGERS

The cleanliness of the cabin and the quality of the meal trays are high on the list of priorities for airline passengers. In the light of these expectations, Servair[10], world number three for in-flight catering, offers a complete service to almost 130 airlines, from meal planning to final cabin cleaning. In fact they are the only company in the world to offer:
- food production,
- catering (meal planning, preparation of dishes, making up the meal trays and loading them on-board),
- preparation and handling (management and stocking of hotel products, loading/unloading aircraft with the various products: pillows, blankets, newspapers, soap, etc.),
- aircraft cleaning,
- engineering and training consulting.

Servair is present in over 22 sites throughout the world and serves companies such as Air France, Continental, British airways, SAS, Sabena, Thai Airways, KLM, Lufthansa, Emirates, Air Mauritius, All Nippon Airways.

A high degree of organization is required from reception of supplies, preparation of the trays, loading of the trolleys together with the drinks and on-board sales, preparation of all the various types of aircraft through to the recuperation and cleaning of the trolleys (about 700 aircraft cleaned daily). To ensure a high class service there must never be a break in the supply of any product. This is where logistics comes in, dealing with the volumes and diversity of meals served (500 different menus per day) and the accredited suppliers. In turn

7. Marketing of Services

logistics relies on a transit facility which means a fleet of 200 refrigerated vehicles transporting over 200,000 tons over millions of kilometers each year. Provision must be made for the variety of materials necessary for each type of aircraft, the number and diversity of passengers, the different classes which vary from one company and/or flight to another, the range of special menus (crew, vegetarian, diabetics, kosher, etc.).

Food	Tons	Food	Tons
Lemons	292	Cherry tomatoes	24
Lolo rossa lettuce	90	Ham	20
Grapes	80	Pastrami	20
Foie gras	60	Live lobsters	13
Pressed duck	50	Bayonne ham	10
Smoked salmon	40	Caviar	5

Table 7-18. Essential logistics to ensure a wide variety of products: some examples of annual consumption

Apart from the logistics, the main preoccupation is with how the food is conserved. More than 20,000 analyses are made each year to ensure that food hygiene norms are respected. With this in mind, Servair has developed a unidirectional flow system: no food (raw ingredients and prepared products) can return upstream in the chain. Products served follow the Hazard Analysis Critical Control Point (HACCP) food hygiene protocol and refrigerated trucks ensure delivery to the aircraft 45 minutes before take-off.

Figure 7-19. Servair refrigerated trucks ensure delivery to the aircraft 45 minutes before take-off[11]

In order to ensure this service quality, the company has organized its different activities as follows:

- a food production site, ISO 9002, whose job is to ensure large scale cooking of the dishes for shipment to other members in the group,
- a site in charge of meals for long haul flights plus the production of thousands of meal trays prepared daily using original recipes (special dishes, exotic cooking),
- a site in charge of meals for medium haul flights, specializing in cold meals,
- a site for hot and cold meals for short haul flights,
- a site specializing in meals for business aircraft: Jet Chef runs a top of the range service (e.g. for VIP's) plus catering outside the airline sector.

Figure 7-20. An example of a meal tray from among the 500 regional, vegetarian or exotic menus[12]

Behind this efficient industrial organization are the individual trades, teams of professionals, cooks, pastry cooks, pork butchers, butchers all concerned with the taste of local products and respect for culinary traditions, e.g. there are 65 references for bread ! Similarly, regional dishes served for 3 months need several months preparatory work by chefs upstream. What Servair aims to do is reconcile customer expectations (gastronomy, personalized recipes, international and/or local menus, etc.) with the airline's demands (time constraints, meal trays, hundreds of thousands of meals served on-board every day, volume production and respect for hygiene rules, etc.).

4. *CONSUMER SERVICES: TRANSPORT AND TOURISM*

Tourism is a growth industry not only in developed but also in developing countries. In countries with planned economies, the transport and tourism sectors have long been state controlled. Here, employees respect their hierarchy rather than what is in the customer's (or " user's ") interest. The opening up of the economy and the arrival of competition has allowed these companies to progressively adopt a marketing approach giving the customer pride of place in decision making. Employees have begun to understand that the future of their jobs depends first and foremost on customer continuity. In an increasingly open sky market, airlines quickly set up loyalty programs in order to retain their market share and develop their activity.

QUALIFLYER, AIRLINES SERVING CUSTOMERS

Qualiflyer is an alliance between the airline companies[13], Swissair, AOM, AirEurope, VolareAirlines, Air Liberté, Portugalia Airlines-PGA, Air Littoral, Crossair, Lot-Polish Airlines, Sabena and Air Portugal-Tap. Under the Qualiflyer loyalty program, the passenger can benefit from a multi-hub network giving access to over 200 destinations in Europe and 330 in the world. He can also accumulate miles and exchange them for a wide range of bonuses: diamonds, sugared almonds, bouquets of flowers, a stay in a luxury hotel, a day as a pilot on a Swissair, Sabena or Crossair flight simulator, donation of miles to a worthy cause, etc.

Miles can be accumulated if the passenger uses certain partner hotels (Holiday Inn, Le Méridien, Marriott, etc.), rents a car (Avis, Europcar, Hertz, Sixt), pays by credit card (American Express, Diners Club, Visa, Eurocard-Mastercard) or phones using Era GSM, GlobalOne, Marconi, Mobistar, Mobilkom, Proximus Sunrise or Swisscom. When he joins the loyalty program, the passenger must give certain information: name, age, address, telephone/fax number, e-mail, preferred language for correspondence and interests. He can at the same time express his preferences for seating (window or aisle) and meals (vegetarian, Asian, kosher, traditional Hindu, pork free, special diabetics, salt free). In this way the companies within the Qualiflyer program can offer a service which is tailor made in terms of seating and food, increasing the satisfaction of the customer who in turn feels respected and listened to.

Airline companies wishing to strengthen their customer base have increasingly adopted the one to one approach (Table 7.21) from business to business marketing techniques. Here each customer is unique thanks to optimal use of mega databases followed by data mining and the setting up of Customer Relationship Management. Knowing the exact expectations and characteristics of customers allows a personalized offer to be made with messages, methods of distribution and appropriate methods of payment, ultimately improving customer satisfaction.

Mass Marketing	One-to-one Marketing
Average customer	Individual customer
Anonymous customer	Profiled customer
Standard product	Customized market offering
Mass production	Customized production
Mass distribution	Personalized distribution
Mass advertising	Individual message
Mass promotion	Personalized incentives
One-way message	Two-way message
Share of market	Share of customer
Wide target : all customers	Profitable niche
Conquering customers	Customer retention

Table 7-21. The main characteristics of traditional and one-to-one marketing[14]

One of the keys to success in service marketing consists of accurately targeting the contact personnel, i.e. those in direct contact with customers, who answer their questions and perform the expected services. Cases in point are hotel receptionists or stewards and hostesses on board an aircraft who could for example be in charge of an unaccompanied minors.

TAKING CARE OF UNACCOMPANIED MINORS

For children, taking the plane is always an adventure. For the airline companies, the transport of unaccompanied minors (UM) is a special activity needing care and attention and the number of UM's per flight is limited in the interests of security. There is no extra charge for transporting a UM unless an accompanying adult is necessary, as is the case for children under four. Conditions vary according to company, British Airways takes charge of children from five up to eighteen years old, Air France from 4 to 12 years (Blue Planet). When the reservation is made, a personalized dossier is established (child's name, age and address plus a dossier concerning the adult who will be meeting them, telephone number, relationship to child, etc.). On the day of departure a clearly visible badge identifies the child for company and other airport personnel, and a ground hostess takes care of the child who, on boarding, is entrusted to the cabin crew. At the end of the flight, another hostess comes to pick up the child and lead them to the adult (who must give proof of identity) who is meeting them. If the latter is not there, the child will remain under the responsibility of the airline with a hostess permanently in charge. This rigorous procedure is well established and airlines carry numerous children each year especially during the school holidays.

Marketing must ensure the quality of the service and continuously being aware of customer perception. This means in particular mobilizing the personnel with customer contact (Figure 7.22).

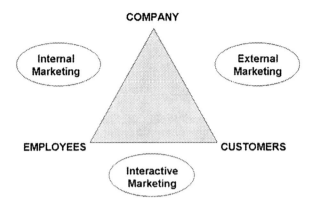

Figure 7-22. The three components of marketing adapted to services (Gronroos, 1984)

External marketing: this is all the advertising or promotion operations aimed at the customer. With long-term perspectives in mind, the airlines are more interested in keeping their existing customers than in winning over new ones, whether business class or tourists. Hence the advertising and promotional fidelity campaigns offering miles as a function of distance traveled.

Internal marketing: in the first instance this means remembering the importance of those on the front line in contact with customers. Does their profile suit this type of job, do they enjoy serving and being in direct contact? Once this is confirmed, the organization must do everything to ensure that those working closely with customers can do so with no problems and under good conditions. Marketing and human resources management work together here with the object being to improve performance and satisfaction on internal and external targets.

Interactive marketing: customer satisfaction not only depends on service quality but also on the way it is done. This perceived service quality often depends on the quality of the seller-buyer relationship.

AN EXAMPLE OF HEALTH SERVICE: MEDJET INTERNATIONAL

Founded in 1983 under the name Jetco, the MEDjet International company specializes in long distance medical transport (over 20 hours flight time). It is the only company offering such a service in the United States and covers the world with a fleet of special CAMTS (Commission on Accreditation of Medical Transport Systems) approved aircraft. These are Learjet 35A-RX and 36A-RX (Figure 7.23).

Figure 7-23. The MEDjet company[15]

The company offers a very high quality of service by taking care of the patient during transport to any destination in the world. In this way they have already carried out 8,500 missions in more than 100 different countries.

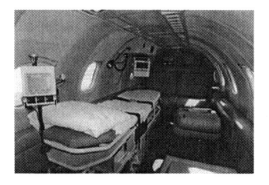

Figure 7-24. MEDjet's specialized services with aircraft adapted for transporting patients[16]

Each aircraft is specially fitted out for transporting patients. The company has been the first to use a civil aircraft equipped with a supply of liquid medical oxygen. For each flight there is a specific team of the required medical personnel who accompany and look after the patient in-flight. Each aircraft can be configured to take up to two patients and a four man medical team. The latter are made up of at least one doctor, a nurse and an intensive care specialist. As a function of the state of the patient, this team can be completed by another nurse and a casualty or other specialist doctor. MEDjet also has specialist teams for transporting new born babies.

The cultural context and the notion of service

Marketing of services, even more than marketing of products, depends upon the country's cultural and even the historical context. In countries which have been under various oppressive regimes, the idea of service can be unconsciously

7. Marketing of Services

associated with servility, and this is why even today it is more difficult to get good quality service in some countries. Even in tourist sites and hotels it can be a bad thing to be seen doing too much for the customer. Conversely in most Asian countries, being of service fits more easily into daily life, and this means that service companies here will be more effective. These cultural differences are rooted in the status originally given to the function of selling in each country. Often, merchandising in a production economy came second to the more noble function of production and it is now the marketing department's job to explain the technicality and important role of the sales function in companies.

THE EXEMPLARY SERVICE QUALITY OF THAI AIRWAYS

The service culture: the Royal Orchid Service

Thai Airways has based its development on customer oriented service known as The Royal Orchid Service. In keeping with the refined culture of Thailand, the company aims to satisfy all its customers who fly on board its aircraft, whether they be in First Class, Business class (Royal Executive Class) or Economy Class and regular satisfaction studies made with passengers allow continual improvements to be made.

Figure 7-25. Service at the very heart of Thai culture[17]

Because of the increasing number of flights to China the company has for example set up crew partnerships with Air China and China Southwest (learning the language and having in-flight Chinese personnel, etc.). This means that those on board now speak Mandarin with the Chinese passengers who are obviously delighted. Similarly, in order to make the passenger's life easier the immigration papers are given out by ground staff before boarding for short haul flights to neighboring countries.

This focusing on the customer can also be seen in the welcome messages in four languages (German, French, Spanish, Italian), the music and the actual videos which work immediately on boarding and on all flights. On internal flights, magazines and newspapers are given out on boarding the plane. In addition, almost all of the company's aircraft are equipped with seats having built in video screens. A large number of regularly renewed film programs are available (always with several exclusive ones) including classics, dance and musicals and sports reports.

The quality of service for First and Business classes

In keeping with its reputation, Thai Airways has a business class (Royal Executive) which is very often equivalent to many First Classes with 120° reclining seats and an individual video. On the menu passengers can choose from an exceptional range of dishes and an excellent selection of wines or refreshments. On intercontinental flights or those which are direct between Thailand and Japan, China or Korea, a traveling kit is also provided. In the airport the Royal Executive class lounges are available (telephone, fax, Internet, television, newspapers, office, etc.).

Figure 7-26. The Royal First Class and Royal Executive Class services[18]

7. Marketing of Services

The Royal First Class service provided in First Class is a reflection of centuries of culture and traditions. Thus, on board a 747-400, passengers have couchettes reclining to 180° and can stretch out their legs with ease over the 2.15m space available. At the touch of a finger the passenger can turn on the video screen and adjust his individual lighting and at meal times, can enjoy refined cooking served on a table with a cloth, porcelain china and silverware. The wine cellar on board offers famous names such as Dom Pérignon champagne and also fine wines chosen by prestigious sommeliers. On flights between Bangkok and Europe the Souper Royal (royal supper) is served before the nights rest. In Bangkok airport, the Royal First Class salons allow passengers to telephone, connect to Internet, send a fax, watch the TV or go through the days newspapers. There is also an office available as well as small meeting rooms and a changing room. Staff are on hand to serve refreshments and snacks.

Thai Airways, the incarnation of Thai service

Well known for its abundant natural resources as well as its culture and exotic food, Thailand is to a certain extent mirrored by the Thai Airways company. The latter bend over backwards to apply these qualities in their minutest detail to the services offered to passengers on board their aircraft.

In this way the company has a great deal of experience helping handicapped passengers and has introduced new wheelchairs which are available on all its aircraft. Obviously this is much appreciated during cultural or sporting events where disabled people are participating (as in the FESPIC games).

Similarly, the company has participated to the full in the " Amazing Thailand " international promotional campaign with, for example, the national dish " Thai Somtam " being served with success in First Class on flights from Europe. In addition, an introduction to the dishes of the 4 main Thai regions " Amazing Thai Taste " has been added to the menu in First and Royal Executive Classes on European flights. Desserts served on board have been decorated in line with the festivals of the moment e.g. a sugar Chang Chai-Yo, the mascot for the 13th Asian games has been made and served during the games. Special menus are offered over the holiday period, during special events like the New Year or for international occasions such as the Oktoberfest. For Japan, special operations have been organized with a Japanese chef on board for Thai flights coming from or going to this country. Finally, still within the " Amazing Thailand " campaign, Thai Airways have set up a special area where the crew can exchange information with tourists, travel agents, those in charge of the Tourism Authority of Thailand (TAT) and manufacturers of Thai products.

As a general rule, the Thai Airways personnel are encouraged to participate in cultural activities. A case in point is the " Smile for the Father " campaign organized jointly by the company and the Foundation of Medical Research and Handicapped Persons with Princess Chulabhorn as patron. Similarly, a classical Thai music ensemble has been created by volunteers from the company to welcome the " Amazing Thailand " operation in the Royal Orchid salon of Bangkok International Airport, and since then they have played for other charitable activities organized by the company.

The subtle association of Thai culture and cooking of Thailand is deeply rooted in the service on-board, which is why Thai Airways is world renowned and has been honored by numerous and varied international institutions.

5. FOCUS ON THE FREIGHT MARKET

The cargo transport market includes activities by sea and air. The former are more concerned with bulk cargos (cereals, oil, cars, etc.) transported for several weeks. The latter concern the dispatch of various loads meeting much shorter deadlines (Figure 7.27).

The air cargo market representing 1% of world trade in volume, covers the sending of freight (goods) and mail (parcels, express packets). The United States is the main market covering 60% of all freight worldwide when all types of transport are taken into consideration.

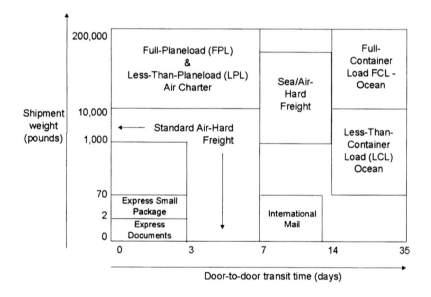

Figure 7-27. The cargo transport market: air and sea

In the air, the exchanges are much more balanced as can be seen from the diagram below (Figure 7.28) North America / Asia-Pacific exchanges come first just ahead of Europe / Asia-Pacific.

7. Marketing of Services

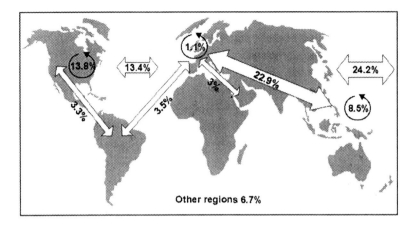

Figure 7-28. World distribution of air cargo traffic flow

Freight is very seasonal and directional (one way transport of goods, from one place to another). Even though historically the first commercial cargo flights were just before the First World War and the first regular cargo service was established in 1926, it was only after the Second World War (Malkin, 2000) that the cargo market really came to life. The airlines sought to extensively develop their services on the freight market in order to meet the growing production company (shippers) exports. Thus the cargo market on an industrial scale is a recent activity which, since the fifties, has enjoyed an annual growth rate of about 6-7%, higher than that of passenger traffic (Figure 7.29). However, it still depends upon this passenger market in that this fleet is also used in part for freight transport.

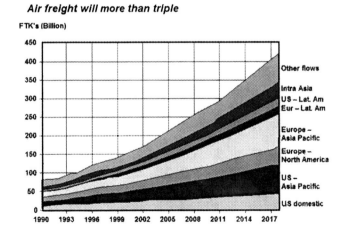

Figure 7-29. Projection for the evolution of air cargo transport[19]

The cargo market groups together different players:

- the customers who are sending (shippers): industrial production companies, exporters/importers, trading companies, etc.
- the carriers: airline companies, cargo companies,
- the forwarding agents: distributors / logistics experts who ensure the delivery of shipments.

5.1 The evolution of the freight market: expansion of the "integrated" carriers

For many years the cargo market was characterized by the perishability and urgency of goods transported as well as high revenue per ton transported. Constructors progressively adapted to airline expectations by offering them aircraft with a higher capacity such as the Boeing 747 cargo (100 tons payload) (Vareilles, 1999).

At the same time in order to increase their sales and stay focused on their main task (passenger transport), the airlines have encouraged an indirect distribution network, the forwarding agents.

The freight forwarders

Apart from their function as distributors, forwarders have gradually expanded to cover shipment bulking /un-bulking, shipment preparation in line with administrative, customs and health regulations, reception and delivery of shipments (after customs formalities) to importing companies.

To get attractive rates, forwarders tended to group together shipments from different shippers. At the same time, small shipments invoiced at a higher rate earned more in terms of actual weight transported. To maximize their margin, the agents offered industrial customers a combination of these two types of shipment with door to door service. Almost 85% of freight traffic goes through forwarders and the 15 largest ones on the international stage represent 49% of freight tonnage. As for the airlines, their task is reduced to an airport to airport (ATA) service, which is part of the forwarder final service.

For the vast majority of airlines, freight comes second to " passenger " activity. However for certain airlines such as Lufthansa, Korean Air, Air France, Singapore Airlines or Cathay Pacific freight has become an activity in its own right with specific equipment (cargo and combi aircraft). The service provided is then better adapted to the market in terms of timetable and transport capacity (bulkier or more dangerous goods than those carried in the holds on passenger flights).

7. Marketing of Services

Integrated carriers

Companies like UPS, DHL, Federal Express or TNT, carrying small parcels and documents are recent arrivals to the cargo market, initially at the regional level and now internationally (Figure 7.30).

Figure 7-30. An integrated player: Fedex

Concentrating on small parcels and shipments, these players known as the integrators have set up a new service quality standard:
- tariffs adapted to small shipments,
- infrastructures and procedures permitting a quality, door to door service,
- a service working hand in hand with just-in-time production and supply chain management,
- improved information management: shipment tracking.

The integrated carrier have had considerable success (25% annual growth rate) taking over 6% of the world market and almost 60% of the American market in just a few years. Projections (Boeing, World Air Cargo Forecast) give them 40% of the world market in about 15 years. Memphis airport has become the world number one for freight and highlights the force of the integrators: it is in particular the site of Fedex's head office (Table 7.31).

Airport	Abbrev.	Traffic (millions of tons)
Memphis	MEM	1,93
Los Angeles	LAX	1,72
Miami	MIA	1,71
Hong-Kong	HKG	1,59
Tokyo Narita	NRT	1,56
Frankfurt	FRA	1,53
New York	JFK	1,51
Louisville	SDF	1,37
Seoul	SEL	1,36

Figure 7-31. World main cargo hubs: 2/3 of the market [20]

Customers want to be able to rapidly obtain any type of product from anywhere in the world at any time. They don't want to have to manage stock. Keywords are service flexibility, reactivity and quality, and these demands present a great opportunity for the cargo market (Table 7.32).

1.	Federal Express	11.	Cathay Pacific
2.	Lufthansa Cargo	12.	American Airlines
3.	United Parcel Service	13.	Northwest Airlines
4.	Korean Air	14.	China Airlines
5.	Singapore Airlines	15.	Delta Airlines
6.	United Airlines	16.	Cargolux
7.	Air France Cargo	17.	Eva Air
8.	British Airways	18.	Swissair
9.	Japan Airlines	19.	Nippon Cargo Airlines
10.	KLM	20.	Martinair Holland

Figure 7-32. The main cargo transporters (revenue/ton/km, RTK) [21]

In this way the market has evolved with the globalization of industrial and distribution activities, the improvements in productivity and the growth of competition. From being mainly deregulated and multiple, the industry gradually has become focused and little by little new players have arrived: low cost companies and postal services.

The low cost companies: The first step has been the replacement by the airlines of their long haul fleets, 747-200 and DC10's with Boeing 747-400 and Airbus A340 having a longer range. The new players then bought these old aircraft cheaply and transformed them into cargo planes able to function at very competitive rates.

The postal services: The express transporters have today been gradually joined by the postal services from various countries (United States, Germany, Holland, etc.). Indeed some forwarders and integrators have reinforced their services by proposing logistics such as customer stock management.

5.2 The reaction of the cargo airline companies

To face up to the competition from the integrated carriers and new players, the cargo airlines have developed different types of distinctive services. This could be guaranteed delivery deadlines or special transport conditions for specific products such as live animals (race-horses), perishable or expensive products or dangerous materials.

These improvements to service quality involve investments for modernizing and enlarging ground installations, for obtaining more efficient IT tools (e-commerce, EDI, tracking, analysis/treatment of service anomalies, etc.). Thus Air France Cargo (www.afcargo.com) in collaboration with the Traxon IT communications network (www.traxon.com) offer new services to customers thanks to Internet technologies (real-time follow up of shipments using mobile phones, shipment arrival times, etc.).

Moreover, horizontal alliances between airlines already allied for passenger traffic[22] have begun to develop and also vertical partnerships (between shipping agents and airline companies). In fact for even greater control over the key factors for market success, speed, frequency and cost, companies must add another ingredient, the network. By becoming logistical partners for their industrial customers, the cargo companies are well placed to give satisfaction. Links with the hubs and presence on the main routes are essential factors for the success of these service providers. The SkyTeam Cargo Alliance founded between the partners Aeromexico Cargo, Air France Cargo, Delta Airlines Logistics and Korean Air Cargo offers shippers access to:

- a global network based around 12 main cargo hubs across the world,
- harmonious service (perfect coordination of shipments and deliveries for all the SkyTeam destinations),
- a fleet of 1,070 aircraft with 6,810 flights weekly to 411 destinations in 100 countries.

Furthermore, the development of Internet related activities can have an effect upon the global evolution of the freight market (modification of the competition and of negotiations with the companies, offers of services to end consumers). However it is still, difficult to measure the real impact[23]. The initial dot-coms such as QuoteShip.com, Freightgate, RightFreight.com or GlobalFreight Exchange (GF-X) have very quickly offered wide ranging services from the traditional reservation of a transport service to shipment follow up, right down to on-line transactions. They now find themselves in competition with carriers and forwarders offering services on the Internet. Having at the outset offered their services as transporters for goods bought on B2C sites like Amazon.com, they then turned to the B2B market place (car parts, chemistry, industrial machines, etc.) to play the same role.

The consolidation of cargo activities on the Internet is progressing, joining dot-coms to transporters and/or forwarders. The forecast is thus for a period of great change in the cargo market, modifying the influence of the various players present.

NOTES

1. See Chapter 8, Pricing Policy.
2. See Chapter 8, Pricing Policy.
3. Adapted from Kotler, Ph., (1999), *Principles of Marketing*, N.J., Prentice-Hall.
4. See Chapter 6, Innovation and Product Management.
5. Advertisement published in *Jane's Airport*, (2000), October, vol 12, Issue 8, p 48.
6. See Chapter 15, Alliance Strategies
7. *Les Echos-Aéronautique Business* (2000), Supply chain sur Internet pour l'industrie aéronautique, 30 March.
8. See Chapter 8, Pricing Policy.
9. Illustration Aphoram/Le Bavar published in *Air France Magazine*, (2000), n°38, June, p 20.
10. Source: *Air France Magazine*, (1999), n°31, p 13-20.
11. Servair / Gosset illustration published in *Air France Magazine*, (1999), n°31, p 15.
12. Illustration by L. Vidal published in *Air France Magazine*, (1999), n°31, p 14.
13. See Chapter 15, Alliance Strategies.
14. Adapted from Peppers, D and Rogers, M., (1993), *The One-to-One Future*, New York, Doubleday/Currency.
15. Illustration taken from www.medjet.com
16. Illustration extracted from www.medjet.com
17. Source: Thai Airways
18. Source: Thai Airways
19. Source: *Global Market Forecast*, Airbus, 2000.
    FTK: Freight Ton-Kilometers ; RPK: Revenue Passenger Kilometers.
20. Source: Air Transport Intelligence, KLM study.
21. Source: Airline Business (1999).
22. See Chapter 15, Alliance Strategies.
23. The 2000 MergeGlobal Air Cargo World Forecast.

REFERENCES

Beauclair, N., (2000), Une place de marché mondiale par Internet, *Air & Cosmos*, n°1769, November, p 14.
Eiglier, P. and Langeard, E., (1987), *Servuction, marketing of services*, Paris, McGraw-Hill.
Flipo, J.-P., (1989), Marketing des services: un mix d'intangible et de tangible, *Revue Française du Marketing*, n°121, p 21-29.
Gronroos, C., (1984), A Service Quality Model and its Marketing Implications, *European Journal of Marketing*, n°4, p 36-44.
Malaval, Ph., (2001), *Marketing business to business*, 2nd. Ed., Paris, Pearson Education.
Malkin, R., (2000), An Air Cargo Century, *Air Cargo World*, www.aircargoworld.com.
The Guardian, (2000), E-market takes to the air, 27 October.
Vareilles, J., (1999), Le marketing: une démarche nouvelle pour les compagnies aériennes dans le transport du fret, *Revue Française du Marketing*, n°173-174, 3-4, p 209-218.

Chapter 8

PRICING POLICY

Price, which is often a determining factor for customers choosing among different products or services, is a key variable in the marketing mix. When establishing a sales price external constraints (regulations, market price, customer behavior) as well as the constraints associated with in-house strategic objectives (costs, profitability, policy, etc.) need to be considered. While there are different methods for determining the price, the final decision depends on the strategy of the company. Depending on the market response, prices often have to be adjusted. This is in part the case of airlines when they use yield management.

1. *FACTORS INVOLVED IN PRICING*

Even more than other variables in the mix, pricing is very closely linked to all other all aspects of company policy. As soon as the content of the product or service is modified, its value for the customer changes, thus calling for a price adjustment. At the same time, manufacturing costs as well as changes in the distribution channels used modify the cost price and in turn the sales price. At the same time, the price is closely linked to the product image and to the communication policy (price includes the communication costs).

The concept of pricing is essential, whatever strategy the company chooses:
• Superiority based on "product leadership" (Treacy and Wiersema 1995), which consists in focusing on providing the highest performance product possible,
• Superiority based on "customer intimacy", which consists in providing personalized service to each customer,
• Superiority based on more efficient, rationalized internal functioning or "operational excellence", which allows very competitive cost prices.

The price is fundamental in the three situations:

• Attaining product or service superiority means considerable investment in R&D. Before justifying a higher price than competitors, this strategy involves a higher cost price, if only due to absorption of R&D costs.

• Offering a personalized product or service to customers also means extra costs for the company, whether in terms of elaboration of a specialized product or the provision of additional customer follow-up. However, a strategy of maximum customer satisfaction can result in a higher customer loyalty rate.

• Savings on internal operational costs of course have a direct influence on cost price: for a comparable result, the company commits less resources. This can also allow the company to lower the sales price thus benefiting the customer.

However, these traditional strategic models need to be considered in light of the great diversity of industrial markets. Disparities can especially be seen in the capacity to differentiate products or services offered: the freedom the company has in pricing is very different depending on whether it deals with entering goods, heavy equipment, intermediary equipment goods or industrial services. The different methods for determining prices are thus more or less applicable depending on the characteristics of each market. In project marketing[1], when a bid is answered companies tend to focus on technical rather than cost-based solutions but obviously not exclusively. Thus the aim is to combine strategies, rather than to apply one global model.

In general, pricing policy is influenced by a number of external constraints such as customer demand, intensity of competition, the market context, currency fluctuations and the regulatory framework. It also depends on internal constraints including the cost structure, priorities in terms of product and customer portfolios, distribution channels, etc. (Figure 8.1).

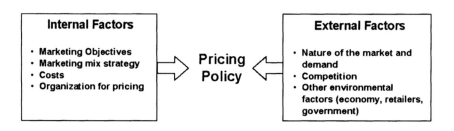

Figure 8-1. Factors influencing the pricing policy (Kotler and al., 1999)

1.1 External constraints

In business to business, in particular in the space and aeronautics sector, it should be kept in mind that with his purchase the customer assembles, builds, and elaborates his own product or service. To determine the sales price, it is thus

8. Pricing Policy

essential to first analyze what the purchase represents for the customer, the benefits he expects to gain from it and the costs involved.

Customer requirements

The derived demand effect: It is essential to estimate the relative weight of the product purchased in relation to the total cost price of the goods produced by the customer. In fact, depending on this factor, an increase in the price of supplies will not always have the same effect on the total cost price. For example, an increase in the price of polytetrafluoroethylene has greater consequences for a manufacturer of adhesives than for a producer of fiber optics while for the latter, the relative weight of the polymer remains low in the cost price. Thus, the negotiation will not be carried out in the same context for the two cases.

To summarize, it can be said that for a customer, price sensitivity depends on the weight of the purchase in the final product. With this in mind, it is important to verify whether the customer can pass on a price increase into the final product.

Benefits expected by the customer: These can be classified in different ways. Fundamentally, customer expectations about product advantages are based on two objectives: saving money or having a more attractive product. Benefits can be classified according to whether they concern products, the production process, buying procedure or the status of the purchaser or his company.

Expected product benefits can be very varied. For example, the addition of a particular desired ingredient such as Nutrasweet® to meals served on a plane could better satisfy certain passengers' dietary expectations. This is also the case with a change in cabin design, such as the addition of attractive Sogerma bar units to the A340.

Production benefits could be greater personnel safety, for example the Mison® protective gas from Aga which improves working conditions for welders by reducing the ozone formed from electric welding. This could also be anything that facilitates the production process by reducing the number of movements that an operator must carry out or making them less tiresome using ergonomic equipment or lighter materials. Benefits can also derive from increased product reliability such as through reducing the number of components. Customer expected benefits about the product are related to financial, commercial and social interests.

The way the buyer manages the different risks[2] in part determines his buying behavior he buys: buyers are confronted with different risks. Thus the aim for the company is to define the product or service in such a way as to reduce this perceived risk, for example through providing a return guarantee, or an ad campaign based on the material's high reliability or a certification... It is especially important for industrial brands[3] to reduce perceived risk and to help customer companies at the commercial, technical and general operational level. Certain expected benefits are easily identifiable by the sales representative while others are more difficult to perceive. Many buyers do not reveal all of the benefits they stand to gain from a product or service, thereby conserving a stronger position in the negotiation.

Costs linked to the purchase: To understand the buyer's perspective, it is important to consider not only the price of the product or service but all the costs associated with the purchase such as those linked to reception and installation of equipment, necessary changes made to the production site, training etc. Additional financial costs, for example if the customer has to take out a loan to purchase the materials, must also be considered so as to propose the best financial solution.

Taking into consideration operational costs: For an engine maker, it is very important to evaluate the operational costs of the engine when defining the sales price. Maintenance costs are linked to the number of cycles carried out by the engine (Figure 8.2). A cycle includes takeoff (the most demanding phase for the engine), ascension, the trip itself and landing. The cost of life limited parts depends entirely on the number of cycles completed. Fuel costs depend on the distance traveled.

Long haul carriers are characterized by a low number of cycles with the trip representing the essential part of the mission completed: fuel is thus the main cost. However, short or medium haul carriers are characterized by a large number of cycles and thus takeoffs, while trips involve short distances. In this case, engine maintenance costs and the cost of spare parts are the most important.

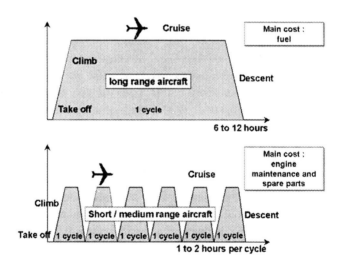

Figure 8-2. Breakdown of operational costs according to the type of flight carried out by the aircraft

Intensity of competition

Price cannot be determined without looking at the competition. The higher the number of competitors on the market, the more price pressure there is and the more difficult the negotiation will be. When there are many companies in the running, there is a good chance that one of them will adopt an aggressive strategy so as to set or modify the positions acquired. However, as long as there are only 3-4 players in competition they can be tempted to maintain the status quo. Taken to an extreme, this situation can even result in illegal price fixing.

For price determination, when the products involved are perceived as equally worthy, the competitive proposal, or lowest price, sets the top limit of the negotiation. When there is a quasi monopoly, the company must determine its prices especially carefully. While it may benefit from a privileged position, the company must check what are the alternative solutions open to the customer. In this case, it is the indirect competitor of a product, service or technology which sets the ceiling.

Competition intensity has an essential role in the balance of power between the buyer and the seller (Figure 8.3). The framework for the negotiation can be determined based on an objective, mutual evaluation of the buyer and seller using weighted criteria such as relative market share (the respective weight of each in purchases and sales of the category concerned). By applying the two final scores to the axes, for example the score of 35% obtained by the seller and that of 58% obtained by the buyer, it is possible to situate the negotiation in the quadrant, "price dictated by the buyer".

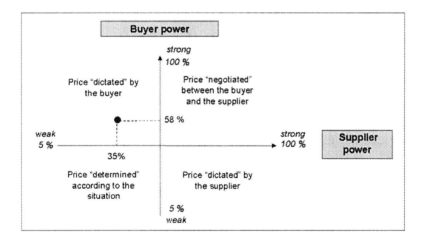

Figure 8-3. Different ways of determining the industrial sales price

The market context

The economic outlook and currency fluctuations also influence price determination. When there is an economic slowdown, the companies in a given sector suffer the consequences and some of them may be in a situation of economic under-activity. To keep the company running and avoid layoffs, they can make very aggressive proposals to ensure orders, later changing this strategy when possible.

In the case of big currency fluctuations, a company operating in its own country only suffers the variations in prices of imported materials and equipment. However, a company that exports or answers an international bid is subject to the currency fluctuations of the contractor.

If for example the customer country experiences a 20% devaluation of its currency, the financial proposition of the foreign country will be 20% over evaluated in comparison to a local competitor. To avoid this type of risk, many bids stipulate that the proposal must be made in a stable currency such as the dollar or the euro. This nevertheless does not eliminate the risk of fluctuations between these currencies and the company's national currency. Thus when determining prices the company must consider international monetary fluctuations.

The regulatory framework

To determine its prices, the company must also take into consideration current regulations. It is illegal for competitors to reach price fixing agreements in order to divide up the market ; it is also illegal to sell at a loss but it is difficult for governments to prove that actual dumping is taking place. In addition, companies are not allowed to use discriminatory practices, in other words to not treat all customers equally for the same supplies or service. The company must also sometimes deal with the protectionism of the country depending on the category of products considered (in particular military equipment).

1.2 Internal constraints

The company's pricing policy also depends on internal constraints stemming from management of cost structure, product and customer portfolios and distribution channels.

Cost structure

The first internal constraint is the cost structure. Each product has a cost price that needs to be analyzed and followed up. The cost price of a product is made up of the technical cost price plus a percentage of fixed structural costs. The technical cost price is itself made up of variable costs such as labor, components, raw materials, packaging and fixed costs corresponding to the absorption of fixed production expenses such as salaries, amortization of production means, etc.

Whatever the method of price determination chosen, the company must ensure that its fixed costs are covered: not only the fixed costs of production but also all

8. *Pricing Policy* 251

those of operational departments, from R&D to marketing. Without fulfilling this basic requirement, the company will lose money.

THE BUSINESS PLAN OR FORECASTING FINANCIAL PROFITABILITY

An aeronautics program is a long-term project: it can easily take 40 years to get from the initial plans to sales and after-sales (spare components) including design, development and industrialization.

Given the financial and technical magnitude of launching an aeronautics program, many different players are generally involved:
- industrial companies: one or more builders, suppliers, sub-contractors, engine makers, components suppliers,
- organizations: certifying organizations, key political decision makers,
- banks and financial institutions,
- the airports (to ensure compatibility of equipment)
- launch customers involved in design upstream of the project (options, expectations, etc),
- lobbies and the media,
- competitors.

The decision to launch a program essentially depends on the potential profitability of the project. The business plan is the tool used to estimate the financial profitability of investing in a program. This involves simulating for the total duration of the project (30-50 years to come) the evolution of different parameters and taking into consideration different hypotheses:
- planning with main dates from the authorization to offer to production authorization including development, the production rate schedule and closure of financing,
- estimation of the number of aircraft and the versions to be produced relative to the competition,
- sales price of aircraft and conditions of pre-payment before delivery (20-40% of total),
- non-recurring costs associated with the development phase and recurrent costs associated with production,
- general economic forecasts: currency rates, evolution in price indexes and costs,
- worksharing,
- investments.

• *Not-recurrent costs include*: design (specialized or not), aerodynamic tests, structural tests, system tests, simulations, equipment development (model, preparation, machining) development of the aircraft, flight tests, modifications, ground equipment, spare components, documentation, management programs, supply, other diverse costs and refurbishing.

• *Recurrent costs include* those associated with production and in particular salaries, equipment, components and the engine (between 25-30% of the cost).

To these must be added costs from insurance, continual technical support, transport, product customization, additional services, modifications, specialized products, etc.

The results of a business plan can be evaluated using:

- Internal rentability rate (IRR) which corresponds to the return rate on capital used. This is linked to the risk involved with the project,
- Net present value (NPV) which corresponds to cash-flow generated by a given interest rate,
- Maximum exposure which corresponds to the maximum amount to be spent to finance the project,
- Break-even threshold which corresponds to the point at which profits are equal to expenses.

Based on the forecasts and parameters selected, it is possible to visualize the expected results (Figure 8.4).

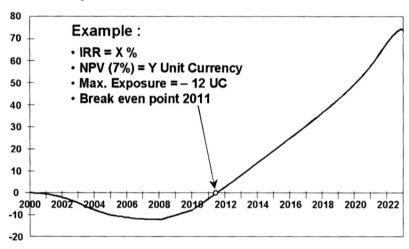

Figure 8-4. An example of a simulated business plan for the launching of a new aircraft

A business plan is necessarily drawn up with a long-term perspective: the production of an aircraft involves customer follow-up for 30-50 years (customer orientation). This customer support includes maintenance, technical publications, training, special ground equipment, computer systems, packaging, moving of equipment, storage and transport of spare components and sub-assemblies.

Managing the product portfolio

Depending on their age, products can be situated in different life cycles[4]. Thus they benefit to different degrees from the experience effect: for example, a new product is sold at a cost price that is not yet competitive and consequently requires important sales backup as well as communication. For a mature product, promotional efforts are not as important and the cost price benefits from the already amortized fixed costs. To select priority products and price levels, the company must have calculated respective market shares and especially the contribution of each

8. Pricing Policy 253

product to present and potential profits. As far as determining the sales price, it is especially important to choose a strategy at the time of launching, because the price will largely determine the positioning, that is difficult to modify latter on. Managing the product portfolio has little relevance for service companies or those evolving in project marketing. In the latter cases it is more important to focus on managing the customer portfolio.

Managing different types of customers

Sales price varies as a function of the sales volume. Thus customers can be segmented according to the volume of business that they generate, establishing the "20/80" turnover or the exact percentage of customers who make up 80% of turnover ; the ABC rule can also be used, according to which customer A represents 65% of turnover, customer B 25% and C the remaining 10%.

However, this commercial analysis is a bit random given that it is based on sales figures that are perhaps far from the real potential of each customer. The ideal is to establish the same classification as a function of realistic estimated potential and especially as a function of the potential profits represented by the different customers.

This quantitative approach needs to be qualified by identifying key customers who are so prestigious that the simple fact of working with them provides a sales pitch to be used with future customers. Reference customers are renowned market leaders or simply leaders in their geographic area. The latter require special follow-up and costly attention (financial analysis of the results, customer quality requirement).

Customers can also be classified as a function of their relative seniority: recent customers, confirmed customers, potential customers, and former customers. In general, it costs more to win over a new customer than to keep an old one, and this should always be kept in mind when establishing the pricing policy. If an old customer observes that he is not treated as well as a new one, he will feel that his loyalty has been poorly rewarded and look for new suppliers.

Making sure that the distribution channel is satisfied

When a company opts for an indirect distribution channel[5], it must watch that prices destined for the final customer remain competitive, while at the same time providing satisfactory financial rewards for the distributor. In this case, distributors can act as partners, respecting recommended prices and passing on valuable grassroots information on customers and the competition.

2. PRICING APPROACHES

Whatever the methods of determining price that the company decides to use, the average price level practiced should lie in a range defined by the cost price on the one hand and by the competitors' price on the other. However, one of the essential goals of marketing is to render the products or services offered by a company incomparable: a more attractive offer in terms of product, service and communication can win a market even if the price is higher then that of competitors. It is important to look at models for establishing prices, including the most common cost-based and customer perceived value approaches (Figure 8.5), as well as the particular case of markets obtained through bidding. There are many more or less complex models for establishing prices.

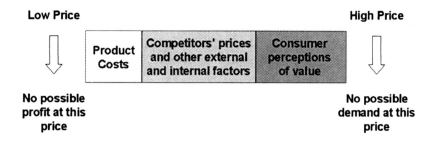

Figure 8-5. The key variables in establishing a price[6]

2.1 Cost-based pricing

This consists in determining the price based on the cost as well as the different margins: those of the company and those of the distributor. This method is simple and reliable as it is based on known elements. However, it has a major disadvantage: the company is not certain that the final price corresponds to customer expectations.

While this is strictly an accounting method, it can be modulated depending on the breakdown of the cost price, the relative weight of fixed costs and variable costs. It is essential for the company to cover all of its fixed costs. If this has not been achieved, the company is losing money. Then, any additional sales contribute to profits. It is in the company's interest to extend its cost analysis to the global cost: in this way, it can control its industrial costs and when establishing a price, take into consideration invisible and later costs (usage, support, maintenance) linked to the manufacture of the product (Figure 8.6).

8. Pricing Policy

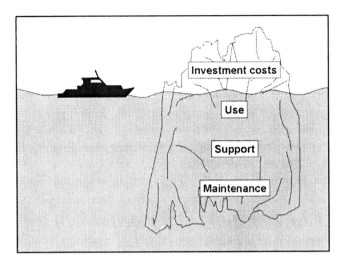

Figure 8-6. Determining prices: the global cost takes into consideration invisible costs

In difficult periods, certain companies are tempted to use direct costing to establish prices. The idea is to only take into consideration the variable cost of production which artificially lowers the price. The trading profit, equal to the difference between the sale price and the direct cost, must then cover fixed costs and make a margin. In this situation, success depends upon the real volume of sales. This approach is very dangerous since it can result in losses if sales are insufficient. It can only be used carefully in two types of cases:

• in relation to product portfolio management, when it is important to defend a product that is considered a priority but is in a delicate situation: a new product for which sales have not taken off or a product which is undergoing a very aggressive attack from a competitor.

• In relation to the customer portfolio when entering into a partnership with a new priority customer. The offer proposed represents a kind of first price: the company hopes to prove its serious intentions through this first exchange, later developing business which will recuperate the commercial investment and make a profitable margin.

The marginal cost, whether used for a product or a customer, should only concern a very limited percentage of turnover. If not, it puts at risk the income statement. In addition, this practice is problematic in that it can be considered as selling at a loss. The company in this case can be sued by a competitor for unfair competition.

The cost-based method of pricing should not be confused with direct costing: it is based on the total cost including in particular the generated fixed costs. The company has room to manoeuvre depending upon its margin. The cost-based approach must take into consideration the production capacity of the firm. If a new market is obtained, it could require additional investment so as to increase capacity.

Under these conditions, the cost price is modified by the addition of fixed costs that have to be absorbed. The degree of fluctuations has a negative effect on the cost price: the company must increase production capacity which is only used a couple of weeks a year. Providers of electricity or airlines are subject to major hourly fluctuations in demand, which is why companies in these sectors adopt an attractive pricing policy which in turn influences demand. By proposing reduced prices for slow periods, they can better control cost prices by reducing demand (passenger loading rate for aircraft, for example). To make their services more profitable, the airlines have developed yield management which requires the use of complex models[7].

EXAMPLE OF A MODULATED PRICING POLICY FOR DOMESTIC AIR FRANCE FLIGHTS

For domestic flights, there are two levels of prices offered:
- Full price red flights that correspond to rush hours from 6:00 to 8:30 in the morning and from 6:00 to 9:00 in the evening. The only reduction possible is through subscription and customers are mainly managers and executives.
- Blue flights corresponding to slow hours when there is less traffic. By offering up to a 40% reduction, the airline incites passengers to shift their departures and arrivals. These passengers mainly include people who are paying for their own tickets such as retired and young people as well as families.

2.2 Value-based pricing

This approach is called the "psychological price method" for the consumer market and the "customer perceived value approach" in the business to business context. Instead of starting from the cost price, the process is reversed. You start with the sales price that the customer is willing to pay as a function of already established specifications. Then the cost price is calculated corresponding to these specifications, taking into account the sales forecasts for the absorption of fixed expenses.

When selling to individual consumers: the psychological price method

In the following example (Figure 8.7), three sales prices were proposed to the customer (P1, P2, P3), and for each the purchasing intention was recorded. The maximum quantity Q1 corresponds to the lowest price P1, while the quantity Q3 corresponds to the highest price P3.

8. Pricing Policy

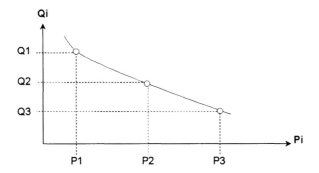

Figure 8-7. The relationship between prices proposed and expected quantities

Thus for each sales price, Pi, the company was able to calculate the purchasing intentions Qi and a cost price corresponding to CPi. It is thus possible to calculate the gross margin GMi = (Pi - CPi) x Qi. In the example above (Figure 8.7), it can be seen that GM2 is the solution that maximizes the margin, which corresponds to intentions Q2 and the sales price P2.

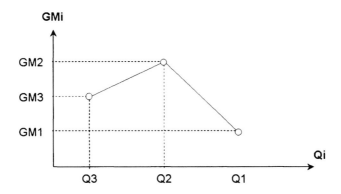

Figure 8-8. The choice of the P2 price with the aim of maximizing the margin

Using this approach, the company is certain that the sales price envisaged corresponds to real demand. Quite often, projects are put on hold because the mix suggested is not profitable enough. This method is increasingly used for the sale of services and products to the consumer public.

Another method (Figure 8.9) consists in asking consumers what is the maximum price they would be willing to pay and the minimum price below which they would have doubts about quality. In this way, the company optimizes its price proposal by selecting the price that is considered as acceptable by a maximum number of consumers interviewed.

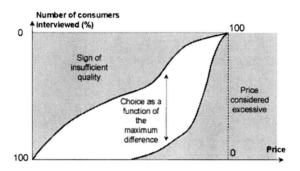

Figure 8-9. The psychological price

The company must also consider the psychological dimensions of the price. The evaluation made by the consumer consists in comparing price ranges within which he is willing to purchase the product. Psychologically acceptable prices are situated between two limits: a bottom limit which depends on the quality effect and an upper limit which depends upon a revenue effect. If the price belongs to this "acceptable" level of reference the purchase will take place.

The psychological price varies between one individual and another depending on consumer category, which makes this method a useful tool to segment demand. But it should not be forgotten that the psychological analysis of prices is solely based on declared intentions rather than behavior. This can lead to systematic under-evaluation of the price that the consumer claims he is willing to pay.

Business to business sales: perceived value approach

In business to business sales the main difficulty is estimating the level of perceived value (Kortge and Okonkwo, 1993), which requires in-depth communication with the different members of the customer's buying center. By analyzing customer satisfaction as well as unfulfilled expectations, it is possible to determine the price that customers are willing to pay for the desired solution. This estimation is essential as it evaluates both the customer and the competing proposals that he has already integrated. The estimation of the perceived value must be validated by the company's internal departments (marketing, methods, research and development) which verify the cost price approach.

The customer perceived value approach is similar to the psychological price method for the consumer market. Instead of starting at the cost price, the process is reversed. You start at the sales price that the customer is willing to pay as a function of the already established specification. Then the cost price corresponding to the specifications is calculated, taking into consideration the sales forecast for the absorption of fixed costs. The company validates only the viability of the project.

The launching approval is only given as a function of the forecasted profits and the importance of the customer.

The main difficulty is to estimate the level of perceived value, which requires extensive meetings with the different members of the customer's buying center. By analysing their satisfaction or dissatisfaction with past purchases from the company or from competitors, the objective is to establish a price that they are willing to pay for the desired solution. This evaluation is essential since it takes into consideration both the evaluation made by the customer and the competing proposals. The estimation of the perceived value must be approved by the different company departments – marketing, methods, research & development – which verify the cost price approach. The ideal for the company is to use both the customer perceived value and the cost-based approach. In this way the different external and internal factors of price determination are taken into consideration.

HELICOPTER ENGINE MANUFACTURERS: TAKING INTO CONSIDERATION THE MARKET AND AFTER SALES IN THE PRICING POLICY

The retrofit and the upgrade

On the helicopter market, retrofitting of engines has developed in light of economic constraints. This satisfies the expectations of three players:

- the helicopter operators: they want to improve their general performance with less costly operating conditions, better consideration of operating regulations and certifications (JARop's and FARop's) and the adaptation of the engine to specific environments (off-shore activities...). The engine should correspond not only to market expectations but also should be accompanied by services (spare parts, maintenance, etc.).
- the manufacturers of helicopters who integrate the engines: they can thus sell new aircraft providing an additional service to the customer;
- the engine manufacturers, who can in this way increase their market share.

Retrofitting is a complementary activity for the manufacturers of helicopters and engines. This trend is less profitable than the upgrading of engines which consists in improving the park/fleet of the big operators such as armies. Upgrading involves replacing parts (rotors...) and improving the operating of the helicopter (additional power) at a lower cost than the purchase of a new engine (almost 50% less).

Taking costs and pricing into consideration

An engine manufacturer must take three types of costs and prices into consideration: production costs, OEM prices and spare parts prices (Figure 8.10).

- Production costs: they depend to a large extent on the efficiency of the manufacturer (production tools, manufacturing process, etc.). Depending on where the factory is, salaries can represent up to 60% of the costs regardless of the type of engine built - large or small.

260   *Aerospace Marketing Management*

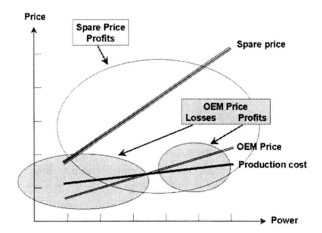

Figure 8-10. Pricing policy equilibrium to ensure profitability for engine manufacturer

• The OEM pricing policy: Taken that the engine represents between 10 and 20% of the price of the helicopter, the OEM price is the subject of intense negotiations. The pricing policy applies to the whole family of engines, giving good industrial visibility (annual projection). The price offered by the engine builder must be in phase with the price proposed by OEM to stick close to the market price. The engine builder must take into consideration the time taken to manufacture an engine (about an 18 month cycle), because unlike spare parts there is no stock of engines.

The operators, who are the helicopter manufacturer's customers, pay a great deal of attention to the intrinsic, engine operating costs, which to a large extent govern the end customer price. The engine affects operator costs 75% of the time: fuel, insurance, maintenance, depreciation of the craft (Figure 8.11).

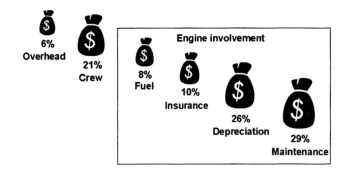

Figure 8-11. Breakdown of the operating costs and the implication of the engine

For the operators, maintenance costs evolve in the following manner (Figure 8.12). After purchasing (0) and during the period of utilization, the operator must carry out a certain

8. Pricing Policy

number of first and second level maintenance operations such as replacing filters, gaskets and small parts (every 40 to 60 flying hours). After an average of 3,000 hours TBO (time between overhaul) the aircraft is taken out of service and a general, reconditioning overhaul is carried out. The latter is costly (almost 80% of the cost price) but allows the aircraft to be used for a further 3,000 hours. Additional services may be offered at the time of the overhaul, guarantees, supply of engine parts, access to a 24h hotline, etc. A Support By Hour (SBH) rate can be calculated from the total maintenance, costs and use. At the second overhaul, costs are higher since several major components with a longer life cycle, need to be replaced.

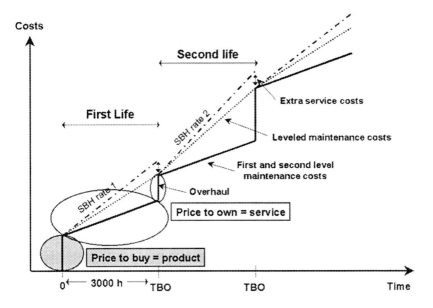

Figure 8-12. Maintenance costs from the operator's point of view

- The pricing policy applied to spare parts: this is directly linked to the relationship with the end customer. The tariff is fixed using a price list in a catalogue and is applied to a certain number of engines. The engine builder must evaluate risks to be better able to control delivery delays. The pricing policy contributes to the return on investments of the program.

The engine builder therefore needs to find the break-even-point between the price applied on the different types of engines and the spare parts. The aim is to ensure a big enough volume of sales and enough profits. Activities can be broken down into:

- a large number of small engines which in the beginning represent a business loss but which generate a large turnover in spare parts ;
- an average number of large engines which allow profits to be made right at the beginning and which subsequently represent an average turnover in spare parts.

2.3 Bidding

The procedure of putting a market out to tender is mainly used today by municipal bodies, and public and private companies when it is an important market such as the construction of a factory extension, an airport or a satellite. It corresponds to the selection of a group of suppliers in the wake of the publication of the call for tender setting out the specification and expected result.

The price in the call for tender

The process of putting out to tender using the principal of the "best tender" represents the majority of purchases in public markets. It differs from the "lowest bidder" selection which is an adjudication process less and less used today. The main characteristic is uniqueness. Each project is special because of the technical specifications imposed, constraints fixed by the customer and the duration of work involved. In actual fact, the price is put together in a specific fashion under the responsibility of the project leader who coordinates the estimates from the different departments concerned. At this stage, the company works out the cost price of the operation and the difficulty lies in knowing what price to ask in order to be selected.

The best approach is not that based on the perceived value because the contractor cannot give price indications. Anyway a project is too complex, made up of different parts. More over, each company has room to maneuver on the technical content while simultaneously integrating the specification's constraints.

The costing method is also a non-starter because of uncertainty regarding the global project. The company relies mainly on a record of past projects whether they have been involved or not. To this end they need to build up a database in order to have information about a sufficient number of projects which are comparable in type of technical complexity, physical context – climatic and geological conditions on-site – and relative to the overall cost of the transaction. For any particular call for tender, according to country, the company therefore knows:

  - the price which it proposed, whether it won or lost the market ;
  - the price of the winning competitor's proposal which is published.

In certain countries such as the United States, it is also possible to find out the prices proposed by the losers. By comparing the levels of competitor's prices for different projects in the database with their own estimates, the company can work out the probable price put forward by competitors for a new market.

A selective participation in calls for tender

In the American context, the company can gather together the competitor's proposals – winners and losers – after the market has been attributed. Models have been developed to work out the probability of winning a market as a function of the different proposals submitted.

Such quantified approaches are probably not realistic in every country due to the confidentiality of those proposals not chosen, with the exception of markets made by adjudication. Here the use of probability can only be made by large national and international groups, which can build up a sufficiently large database.

Empirically speaking, estimating the right proposal depends most of all on market scanning organized by the sales team, and in particular, on all the contacts made upstream: working meetings organized with the contracting authority's technical team, with the prime contractor and with the engineering and architectural design office.

The number of tenders which are won is directly related to the number of proposals submitted, i.e. the number of competitions that the company decides to participate in. However, a proposal is expensive since it presupposes, apart from the work upstream, a large amount of preparatory work on the dossier, initial studies and the final putting together of the proposal. The company must therefore select where it is going to participate in order to eliminate those projects where it has only a small chance of success, e.g. where one or more competitors are favorites right from the start because of previous work for the same main contractor or due to recognized know-how concerning a specific task.

3. *PRICING STRATEGIES*

Having examined the different ways of determining prices, it is clear that the price has to absorb the different company costs. Apart from the accounting aspect, the price variable plays a strategic role because it has an effect on the sales volume, and on the actual image of the products concerned. Indeed, the company can make prices evolve over the years as a function of its own objectives and within the market context. Pricing is based to a large extent on a skimming strategy or a penetration strategy. A third flexible strategy has gradually developed above all in project marketing. In the services sector over the last few years, another strategy has come to the fore, that of yield management.

3.1 *The skimming strategy*

Skimming consists of fixing an initially high price with the intention of lowering it gradually afterwards. The first advantage of this strategy is that it guarantees a high unit margin as soon as the first products are sold. Conversely, the risk of having a high price is low sales, which goes against the objective of maximizing short term profits. Progressively (more or less) the sale price comes down at the same time as the cost price: mathematically, the cost price decreases with the total sales whereas the reduction in sale price is decided by company policy. It should be noted that when the inflation rate is taken into account, an actual decrease in sale price can sometimes be achieved by apparently maintaining the same price level.

The main advantage of skimming is to strengthen the positioning of a product in the top of the range (Table 8.13). This strategy, which must be backed up by product superiority, will only give sales to a limited number of customers. A successful skimming strategy can in this way excite the envy of other potential users and this envy can also be amplified by advertising. The downside is that competition is attracted and stimulated by the high price and may even bring in new players.

Skimming can only be used when the product is launched. During the subsequent phases of the life cycle, a price reduction is always possible, however the opposite is not the case: it would be difficult to use skimming in the mature phase of a product.

	Advantages	Drawbacks
Skimming	- Allows a premium positioning - High profit margin per unit from beginning - Later price reduction follows volume cost reductions (experience)	- Risk of lower sales at the beginning - The high profit margin is attractive: incentive for competitors and new in-comers
Penetration	- Sales supposed to be more important (low price) - Experience curve: allows competitive cost pricing more quickly - Discourages new competitors	- Low profit margin at the beginning; risk of financial losses - Not appropriate for a premium positioning

Figure 8-13. Comparison of skimming and penetration strategies

3.2 The penetration strategy

There is some symmetry between the penetration strategy and skimming. It consists of launching a product at a low price in order to maximize the market share and thus the sales volume. It is an aggressive strategy designed among other things, to discourage competitors not attracted by a low unit value. This decision binds the company to the long term in that they cannot hope (or only with difficulty) to increase the sale price because of its actual low end positioning. Low cost airlines like Buzz (a subsidiary of KLM) are a case in point. They aim to attract customers with extremely competitive prices by reducing all the extras, e.g. on-board service is optional (drinks, meals, etc.).

From a financial standpoint, a penetration strategy consists of anticipating the fall in the cost price: the low price should stimulate a large volume of sales which will in turn give access to more competitive cost prices. It is therefore a risky strategy which can lead to high losses if the sales volume does not attain the expected level due to a counter-attack by a competitor for example.

The penetration strategy also risks tarnishing the image of the product, particularly for complex goods: it is difficult to give credibility to a product's high quality when the price is much lower than the competitor's. Because of the necessity for a high volume of sales, penetration does not work well with non-standardized activities where each product has its own technical specifications. It can on the other hand work with widely distributed industrial services, e.g. tools or IT supplies.

Penetration strategy can be applied as soon as the product is launched and unlike skimming, it can be adopted during the life cycle of the product or even after a skimming phase. This tactic presents the risk of displeasing customers who have paid the full price. In fact, it can happen where a technological innovation has been rapidly generalized, adopted by competitors and become 'banal'. In the mean time the company has perhaps come up with another technological innovation.

In business to business, the strategies employed are much closer to skimming than penetration.

CHARTER ON-LINE BUSINESS AIRCRAFT

The airline ticketing reservation market has been shaken up by the revolution of on-line reservation. User-friendly, new sites with clever names have appeared, offering travelers their tickets at the click of a mouse. On-line reservation is presented as simpler and faster than the traditional phone call to a ticket agent. Charter on-line has gradually developed in the United States on the growing business aviation market, side by side with the established charter companies and those offering fractional ownership such as NetJets, part of the Executive Jets group. The charter on-line companies have developed in response to growing dissatisfaction, on the part of businessmen, with the airlines and airport infrastructures (lateness, missed appointments, loss of time, indirect flights). Taking advantage of this window in the market, they have diversified their reservation service with specific offers such as charter reservation, fractional ownership, bidding, long distance business charters, programmed public charters and corporate shuttles. The idea is to make reservation on a business aircraft as simple as buying a traditional ticket. Among the offers is FlightTime (flighttime.com) who have signed a large corporate shuttle contract with Procter & Gamble for their "Global Program". FlightTime manages and coordinates an A320-200 with 56 seats, just like a private airline, flying for Procter & Gamble four times a week between Cincinnati and Brussels.

Apart from this corporate shuttle approach, new on-line companies offer individual seats for sale on regular business flights, for a price equivalent to a full tariff on a regular airline service, these are the "public charter sites". A case in point is Flightserv.com which offers places on light and mid-size business aircraft between DeKalb Peachtree airport, Atlanta and Teterborough, New York. Similarly, Flyindigo.com offers places on a Falcon 20 between Midway, Chicago and Teterborough, New York. The transJet.com site offers charter services between Fort Lauderdale, Florida and Los Angeles with Hawker 600's. Programmed flights on the business aviation market, is half way between the traditional airline company market and that of private carriers.

Skyjet (bought out by Bombardier), a pioneer of the charter reservation sites, aims to improve the image of charter flights. With its SkyClub loyalty program, Skyjet is targeting three types of customer: companies in Fortune's top 500, individuals with a very high income and the big, first class customers (more than 100,000 miles per year). The idea is to offer advantages which largely compensate for the airline companies loyalty programs, e.g. price reductions and possibility of transfer from a small to a larger aircraft. Other on-line companies offer charter reservations on business aircraft, for example eJets.com targets the very high income bracket, companies and government organizations. This is also the case for ebizjets.com which offers reductions of about $200 per hour relative to normal charter companies against a minimum contract of $100,000.

There are also auction sites for reservations on business aircraft, such as legfind.com or myjets.com. The site receives the request for a journey from a broker and transmits it by mail to the appropriate operators, who then make offers.

Finally there are support sites which have developed such as aircharteronline.com or aircharterguide.com which offer updated databases on-line, listing different charter offers available from more than 200 operators.

3.3 Flexibility strategies

Alongside traditional strategies, "flexible strategies" are now being developed. For the company, this means adapting to strongly fluctuating circumstances. Among the price determining factors, sudden changes such as monetary fluctuations, new international regulations or the raising of protectionist measures, have been cited. The abruptness of these changes lead the company to modify their strategy:

In time: This is where the company adapts to a changing economic situation such as a growth recovery or a recession, as well as events sparked off by large competitors. To react to a monetary realignment, the company may eventually have to modify its initial strategy.

In place: Multinationals export their products to countries with very different situations in terms of competitor pressure, purchasing power and efficiency of distribution channels. Consequently it is usually impossible to adopt the same strategy in every country.

3.4 Yield management

For a long time, passengers have flown by class and at a given price. The tariffs, fixed according to the cost price of the particular journey, vary mainly according to season. Following the 1978 Deregulation Act in the United States, and the increase in competition, the airline companies have had to develop more efficient strategies.

8. Pricing Policy

In order to ensure their financial stability, the companies have followed strict cost reduction programs and searched for gains in productivity. Airline competitiveness has thus meant adopting yield management, ruled by supply and demand. Originally developed by the large airline companies for passenger transport, it has now spread to other service activities such as air freight, tourism, shows, car rentals, etc.

The objective of yield management is to improve profit per passenger transported, in other words maximize the income from a flight by varying the price as reservations progress. The airlines are thus favoring maximization of profit against traffic growth. This allows them to ensure greater long-term financial stability. The companies fill the planes with passengers who pay the maximum (Sinsou, 2000) and the consequences of this are that passengers from an identical segment, e.g. economy class, rarely pay the same price even though they have the same type of seat and the same service. The aircraft is divided into tariff blocks, if demand is strong, the company reduces the number of low tariff seats and vice versa. At the same time, if the objectives are not fulfilled in the weeks and days before the flight, then the distribution of seats can be modified. Actually, the seats remaining empty will never be recovered, because of the non-stockability of services[8].

The principal of yield management is therefore based on the optimization of global income from the sale of volatile services (Dubois and Frendo 1995; Lehu, 2000). This presupposes capacity management taking into account the specificities and profitability of the different products/services offered, which means using highly sophisticated models simultaneously taking into account a very large number of variables (sometimes several hundred). The companies have software using simulation algorithms for the demand, as well as a load factor database, because passengers must be placed by class of reservation. There are large teams of analysts working permanently in the airlines on yield management (Figure 8.14).

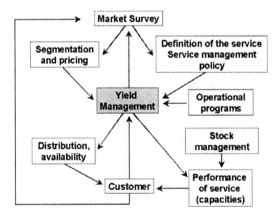

Figure 8-14. The yield management[9]

The concept of segmented pricing also allows the following two issues to be addressed:
- How to recuperate those customers who are unwilling to pay a sufficiently high sale price which covers the cost price and gives enough margin ?
- How not to lose turnover by under use of the service capacity ?

Carriers like American Airlines, or in Europe, SAS, have been among the first to use processing systems allowing them to improve the load factor on their aircraft, thus saving several tens of millions of dollars annually (McCartney, 1997). The improvement in yield is due to all of the following:
- higher prices,
- a decrease in the availability of reduced price seats,
- the setting up of complex procedures for over-booking,
- short term extremely reactive tariff changes,
- acceptance of a more restricted volume of traffic.

Yield management requires that the airline has a large market share (very large carriers, hub domination, etc.). Profit maximization also comes from the reduction in costs obtained thanks in particular to the hubs. The stopover platforms multiply the connections, allowing costs to be reduced by making the flow more dense (André, 2000). Setting up yield management implies the existence of a price sensitive demand which can be segmented on this criterion, very high fixed costs and a finite volume of resources (number of seats in the aircraft). It is essential to segment and identify the customers. Generally there is the price sensitive "leisure customer" and the "high contribution customer" who is above all influenced by the flight time, the service on offer, the frequency, etc. On airline markets with high penetration levels as in the United States, the demand comes essentially from leisure travelers sensitive to price variations. Simply offering greater discounts would increase the volume of traffic without increasing profits (opportunist demand). Yield management allows sales to be simplified by reducing the number of intermediaries and by ensuring greater transparency. Without yield management airlines would sell more reduced tariff seats and business travelers would have a job finding room. In this way it participates in improving service quality by supplying "the right product to the right customer at the right time and at the right price" (Cross, 1997).

The recent use of yield management in air freight is not based on seat capacity but rather the management of weights, volumes and compatibility of goods. In this way, Air France Cargo has used yield management to make gains of 4 to 5% of the turnover (André, 2000).

BOLD EASYJET: INTERNET, YIELD MANAGEMENT AND NON-CONFORMISM

While in the United States low cost airlines represent 25% of the market, they have only recently appeared in the Europe with the creation of the Irish company Ryanair (6 million

8. Pricing Policy 269

passengers), that of easyJet and Go (British Airways), plus the Anglo-Dutch company Buzz (KLM).

In 1995 the British company easyJet, was presented and popularized through the media by its President and Founder Stelios Haji-Ioanou, as the first company in the world to do without agents for its ticket sales. Today easyJet positions itself as a cheap and user-friendly company. They offer neither meals nor free newspapers on board, but regular flights 30% cheaper than anybody else. Its method of functioning is to search for the slightest extra charge and then generalize any economies made. For example, its fleet consists of 18 aircraft (with 32 on order), all Boeing 737's, in order to be better placed to negotiate purchase contracts and minimize maintenance costs and pilot training. Similarly, easyJet does not distribute tickets to passengers but only a recyclable plastic boarding card, thus avoiding the need for ticket machine rental at airports. In addition, it uses Internet to short circuit intermediaries (no commission paid to travel agents) and has an internal organization with a minimum number of hierarchical levels to gain in reactivity.

Finally it strives to optimize its resources using advanced yield management techniques. Prices are adjusted as a function of the demand to fill the aircraft to a maximum. As the date of departure approaches, the tariffs rise. If the plane is half empty the day before departure, discounts are made. The aim is to sell each seat at the best price for the company and this flexibility has allowed them to achieve a load factor of more than 80%. A freephone number and the website address adorn both sides of the company's planes (Figure 8.15). Reservations can be made using a call center, or directly on-line with a 4 euros discount. The company does not hesitate to launch anti-competitor campaigns displayed directly on the fuselage of its aircraft! The result of this willingness to shake up the sector is: more than 3.5 million passengers transported only four years after launch, and 60% of all tickets bought via Internet.

Figure 8-15. Example of advertising on the aircraft of easyJet airline[10]

3.5 The development of the "gray market"

Price strategies are particularly difficult to define for the multinationals, who must take into account the different national contexts and at the same time develop in a market which is more and more open. The name "gray market" is used to denote all those imports made directly by distributors looking to short circuit the national producer's structures, by getting their supplies directly through subsidiaries in the cheapest country. This obviously upsets decisions made by the multinationals whose pricing strategies take into account the local context and also the necessity to make a profit.

4. PRICE-ADJUSTMENT POLICY

4.1 Adjusting the conditions of sale

Often companies put together all their customer conditions of sale in a tariff charter. This tool responds to two objectives:
- it allows attractive prices to be set to gain access to a new customer who can then be offered a system of discounts and refunds in order to gain loyalty ;
- it avoids any sales discrimination and stays firmly within the context of the regulations.

The conditions of sale used must be chosen and implemented in conjunction with the communication policy and financial policy, and must be coordinated by these two departments.

The system of discounts and refunds is applied differently according to sector and above all buying practice. It is not adapted to markets attributed by tender, but is frequently used for sales to a multi-brand intermediary.

Quantity discounts

These discounts are based on a sliding percentage corresponding to the volume of orders. The greater the order, the greater the percentage discount and this is publicized to all customers in the same way. Generally the calculation uses the volume of the order in itself, however companies sometimes allow cumulative totals of different orders over a given time, to be used.

Discounts by quantity are made directly on the invoice, e.g. if there is a 15% discount the amount invoiced is reduced by 15%. This is applied systematically, in line with the discount table which the customer has been given. The various rates normally apply for a whole year, even in the case of deteriorating relations with the customer.

The thinking behind discounts is above all economic ; the customer groups up his orders to get a higher discount rate resulting in a lower cost price in logistics and production management for the company.

Rebates

This is a "reimbursement" that the company pays the customer when he fulfills pre-defined conditions. Unlike discounts, rebates are not immediate, but rather deferred. They are sometimes called "back rebates", and there are also "annual rebates". Calculated over a period of time, generally a year, they are based on an overall picture of the deliveries made to several offices, or subsidiaries within the same group. In this way the payment of the rebate is centralized, i.e. made in one payment to the customer's head office who can then decide to what account it will be allotted.

Unlike discounts, rebates are never systematic but rather conditional, i.e. the pre-determined conditions must have been fulfilled before the rebate can be paid, and these conditions can be quantitative or qualitative. The former are summarized on a rebate scale of fixed percentages corresponding to the band of the company's turnover. Qualitative conditions on the other hand remunerate a function carried out by the customer, e.g. logistically, if the customer, once supplied, sends out the products to his different offices himself.

The main objective of rebates is to gain customer loyalty, i.e. if, over the course of a year he is tempted by another, cheaper supplier, he runs the risk of not reaching the turnover band which entitles him to the 5% refund he budgeted on. For large companies, rebates from the different suppliers represent an important sum in cash management and these are usually paid at the beginning of the year, based on the previous 12 month's transactions.

Promotional discounts

Promotional discounts are more product based in that the object is to inject new life into product sales, as a function of the season, phase of life cycle, or state of the stock. At the same time the discounts give customer satisfaction through reduced purchasing costs.

Promotional discounts must not be continuous because they would then become reduced tariffs in disguise. Their one-off character makes them a flexible tool allowing sales to be directed as a function of the company's interests. Originally firmly rooted in industry, sales promotions are now being used in other sectors such as services.

Cash discount

This is usually a fixed percentage, between 1 and 3%, which can be deducted from the amount invoiced if payment is made within 30 days. In this way rapid payment is rewarded and cash-flow management improved. Unlike the other three systems – quantity discounts, rebates and promotions – the cash discount can be applied in almost any situation, including markets by tender.

4.2   The leasing

Leasing can be thought of as prolonging the pricing policy. Using this method, the firm offers a financial solution to the customer company, and this has a bearing on the price. Leasing has been widely developed in industrial marketing, especially in the office, computing and machine tool sectors, and also in airline companies.

For the customer, purchasing is replaced by a rental system with or without a buying option. Instead of gradually writing off the equipment, the company pays monthly bills which they can then deduct completely from their costs.

For the customer company's management, this method has the advantage in terms of accounting and is a better tax solution compared to normal writing off or interest payment. Leasing also avoids dilution of capital which would normally increase to finance purchases, and like a loan it makes progressive financing of large operations possible. Finally, leasing allows the customer company to guard against the risk of obsolescence. To obtain more up-to-date equipment all the customer has to do is sign a new leasing contract, replacing and extending the original one, which is why it is also a very effective means of getting customer loyalty. This 'continuity' solution avoids the internal discussions necessary for an investment, and is thus also a way of simplifying the acquisition process.

After all these advantages, one of the disadvantages of leasing is that it is usually more expensive than direct purchasing in terms of spending, not taking into account any deductions. For the seller, there are often two different prices according to whether the goods are sold or leased. Sometimes the leasing sale price is higher, but there are occasions when the opposite is true, where for example the company has created a subsidiary to finance customer leasing. Here the company has two sources of profit: the sale of the equipment and its financing.

In the aeronautics sector, the leasing companies ("lessors") supply business to the manufacturers and engine builders, especially from start-up airlines and small operators. In the United States the start-up airlines work only with rented aircraft: their job is not to buy the planes but to exploit them, fill them with passengers and freight. The main leasing companies are American, ILFC (International Lease Finance Corporation), GECAS (General Electric Capital Aviation Services). But there are other international players on the world market such as SALE (Singapore Aircraft Leasing Enterprise), CIT Aerospace, Kuwait Finance House, GATX / Flightlease. These companies have become major players, influencing the aircraft acquisitions market as a function of leased aircraft availability. "Lessors" have developed in Ireland and then Japan because of their favorable tax policies.

8. Pricing Policy

The leasing companies (*Aéroports Magazine*, 2000), because of their importance in terms of orders and their financial footing, can sometimes negotiate better deals with the manufacturers than the airlines (Table 8.16).

Aircraft possessed by...	1986	1996	2006
Airline companies	75 %	49 %	33 % ?
Banks	17 %	30 %	33 % ?
Leasing companies	8 %	21 %	33 % ?

Table 8-16. Evolution in the numbers of aircraft held by airline companies, leasing companies and banks[11]

With large cash-flow they can place big orders and negotiate reductions on the catalogue price for airframes, components and engines. The more planes the customer buys, the tighter the constructor's prices. The airlines can gain from this situation by profiting to some extent from the conditions obtained by leasing companies from the manufacturers and/or the banks. For the airline, renting planes by month is often less expensive (15-20%) than purchasing their own fleet. In the latter case, almost 10-12% of the company's income is devoted to the purchase of new aircraft (financial package, immobilization of capital, deposits, choice of equipment and engines, financial and tax write-offs frozen for 7-10 years, pay back of the principal and interests, etc.). However, the pay-off conditions are usually longer for the leasing company than for the carrier and thus an aircraft purchased for $50 million by a company can be completely paid for and written off in 12 years. If the company then wishes to sell it, this will be for the $25 million residual value (for an estimated life of 25-30 years). In the case of a leasing agreement, it is the leasing company which makes a gain even if the aircraft continues to be used by the same carrier.

Airlines sometimes lease when the economic situation suddenly changes or when there are problems in the fleet. These could be seasonal, such as the repatriation of expatriates for holiday leave or the traffic overload experienced by Maghreb airlines (Royal Air Maroc, Air Algérie, Tunis Air) during the Mecca pilgrimage. By using leasing companies, the airlines can benefit from newer aircraft doing a more efficient job.

For the engine builders like CFM International, IAE, GE, PW or RR, the leasing market generally behaves like the normal one in terms of engine choice. Nonetheless, it is a specific market in that the leasing company is not the end user and it is the leaseholder (the airline) who pays the maintenance costs. The cheaper the acquisition price, the lower the monthly rate for costs. In fact the costs per aircraft are calculated per month and include an additional sum which is a sort of contingency for the future replacement of the engines. For an A320, the monthly rate is around $200,000 and $120 per flight hour for the engine. It is these provisions as well as the resale of the aircraft, which constitute the main part of the leasing company's income.

4.3 The development of fractional ownership on the business aircraft market

Over the last few years the business jet market has undergone considerable growth[12] with new models being launched by the manufacturers such as the leader Bombardier and also Gulfstream (General Dynamics), Cessna (subsidiary, alongside Bell Helicopter, of Textron), Dassault and Raytheon. This market which oscillated between two and four billion dollars annual turnover, has almost tripled to $9 billion, and should continue to grow with the arrival of airliners, specifically designed for business use like the Airbus A319CJ and the Boeing 737BBJ.

The customer base has also grown thanks to the development of fractional ownership, most of whose customers are first time owners. But these customers are not an elite, insensitive to prices, and so there is tension. In the space of a only a few years, the fractional ownership companies have made an important place for themselves and now hold almost 10% of the world business jet market. And this share could increase in the next few years, reaching 25%. They have brought new customers into the business aviation market, some of whom will buy their own planes.

One of the rare independent fractional ownership is EJA, while certain important players (Travel Air, FlexJet) are financially linked to the manufacturers (Raytheon and Bombardier respectively) with whom they have exclusive contracts. But there are still several other players on the market, e.g. Flight Options, NetJets, etc.

Unlike traditional customers, the fractional ownership companies order a large number of planes in one go, and this market concentration gives them a negotiating power which leads to lower prices from the manufacturers. In line with this power, the fractional ownership companies demand waiver (even annulment) clauses from the manufacturers, based on market conditions.

THE NETJETS (FROM EXECUTIVE JET) FRACTIONAL OWNERSHIP PROGRAM

With its NetJets program, Executive Jets[13] has tried to make business aviation accessible to a greater number of customers. The principal is simple: the customer acquires a share of a plane which gives him the right to a certain number of flight hours per year. The size of the share is determined by the number of flight hours needed. In Europe for example, an eighth share gives 75 hours per year, a quarter, 150 hours. Thus by buying a share, the customer reduces his initial investment enormously.

The number of flight hours correspond to effective flying time. There are therefore no additional costs for waiting, convoying or lateness. The service covers everything from airport taxes, on-board catering, fuel, regular maintenance and possible repairs. For the price of a

8. Pricing Policy

monthly rent and an hourly flight rate, the customer can use the aircraft when he wishes (24 hours/day all year). A few hours notice is required to use the plane and the customer will not necessarily use his 'own', but sometimes an identical model (same layout) personalized to his tastes (usual newspapers, catering, films and music, etc.). Fractional ownership also gives the customer the possibility of using two planes simultaneously, unlike simple co-ownership. Finally, customers must take a share in the plane for a minimum time interval, beyond which they can either sell it directly to another client or, at the market rate to the fractional ownership company.

NOTES

1. See Chapter 10, Project Marketing
2. See Chapter 2, Individual and Organizational Purchase.
3. See Chapter 13, Brand Management.
4. See Chapter 6, Innovation and Product Management.
5. See Chapter 9, Selecting Distribution Channels and Sales Team Management.
6. Source: Kotler, P. and Dubois, B., (2000), *Marketing Management*, Paris, Publi-Union / Pearson ; Kotler, P., Armstrong, G., Saunders, J and Wong, V., (1999), *Principles of Marketing*, Prentice Hall Europe.
7. See infra 3.4 Yield management
8. See Chapter 7, Marketing of Services
9. Adapted from Daudel, S. and Vialle, G., (1994), *Yield Management*, Les Presses de l'Institut du Transport Aérien.
10. Illustrations taken from easyjet.com
11. Source: ATR
12. Source: *Air & Cosmos*, (2000), Les jets d'affaires, un marché qui devrait rester au sommet, 6th October, n°1765, p 6-9.
13. See *Air & Cosmos* (2000), Les Falcon font leur nid chez NetJets, 6th October, n°1765, p10-11.

REFERENCES

Aéroports Magazine, (2000), Leasing d'avions: les compagnies y gagnent-elles ?, September, n°311, p. 46-49.
André, M., (2000), Le yield management: une arme discrète mais efficace, *Aéroports Magazine*, n°310, July-August, p 24-25.
Cross, R., (1997), *Revenue Management*, Broadway Books.
Dubois, P.-L. and Frendo, M.-C., (1995), Yield management et marketing des services, *Décisions Marketing*, n°4, January-April, 47-54.
Kortge, G.D. and Okonkwo, P.A., (1993), Perceived Value Approach to Pricing, *Industrial Marketing Management*, p 133-140, May.
Kotler, P., Armstrong, G., Saunders, J. and Wong, V., (1999), *Principles of Marketing*, Prentice Hall Europe.
Lehu, J.-M., (2000), Internet comme outil du yield management dans le tourisme, *Décisions Marketing*, 19, 7-19.

McCartney, S., (1997), Airlines rely on technology to manipulate fare structure, *The Wall Street Journal*, 3rd November.

Sinsou, J.-P., (2000), *Yield & Revenue Management*, ITA ed.

Treacy, M. and Wiersema, F., (1995), *The Disciplin of Market Leaders*, Addison-Wesley Publishing Company, Reading, Mass.

Chapter 9

SELECTING DISTRIBUTION CHANNELS AND SALES TEAM MANAGEMENT

When aircraft builders and parts suppliers sell to businesses, distribution is often highly simplified. In general, there is no intermediary between the supplier and the customer and consequently, it is possible to focus on managing the sales team (direct channel). However, a product as complex as a plane can also be sold to a leasing company. In this case the latter plays the intermediary role. This phenomenon which began with civil aircraft is rapidly developing in the business jet sector. However, it is in the context of sales to individual consumers that indirect distribution is especially important. The airlines have sometimes adopted very distinct strategies in terms of ticket distribution. First we will look at the main functions of distribution from a logistics and commercial point of view, and then we will discuss selecting and managing a distribution channel. We will look at merchandising which allows travel agencies to best adapt supply to demand and the main specifications of services distribution and industrial trade.

1. LOGISTICS

All of the players who handle the activities which take a product from production to consumption make up a distribution channel. These intermediaries are used because they efficiently accomplish certain functions, the most important of which is logistics (Malaval, 2001).

A distribution channel is mainly characterized by its length, in other words, the number of levels crossed by the merchandise before getting to the end consumer (Figure 9.1).

278   *Aerospace Marketing Management*

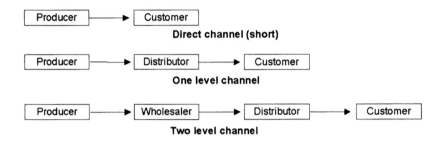

Figure 9-1. The different types of channels

The shortest channel does not have any intermediary: it goes directly from the producer to the customer. This is rarely used for industrially made consumer products. However, it is very common in business to business. The one level channel has an intermediary link between the producer and the customer: for example the specialized retailer.

Wholesalers or commercial agents are often the intermediaries in a two level channel, which sometimes are joined by semi-wholesalers in a three level channel. There are often complex, multi-level channels in certain industries, weakening the control of the producer over the whole channel.

In an analysis of the air cargo market carried out by MergeGlobal, the main distribution channels were listed. In fact, cargo activity depends on the flow of merchandise within the different sectors of activity. Thus upstream, the flow goes from suppliers to industrial companies who directly distribute their products to their industrial customers, wholesalers, retailers or internal distribution platforms (Figure 9.2). Upstream, shipments are made once or twice a week and generally are heavy loads (greater than 250 kilos).

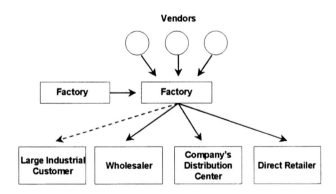

Figure 9-2. The flow of merchandise upstream in the supply chain

Downstream in the supply chain, shipments are lighter (less than 35 kilos) and more frequent (daily). The flow concerns those products distributed by:

9. Selecting Distribution Channels and Sales Team Management 279

- retailers directly to end customers,
- or from the distribution platform to other retailers, which then deal with end customers. In addition, the flow of merchandise can be between end customers (Figure 9.3).

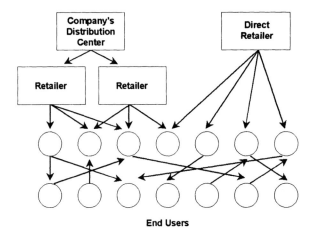

Figure 9-3. The flow of merchandise downstream in the supply chain

All of the distribution channels (Filser, 1989) through which merchandise circulates from where it is produced to where it is used allows the manufacturer to reach his market. Thus distribution consists in stocking, selling, and transporting the merchandise to the customer for it to be used. In particular, for widely distributed consumer and industrial goods, logistics are obviously very important: an intermediary distributor decreases the flow between producers and sales points (Figure 9.4).

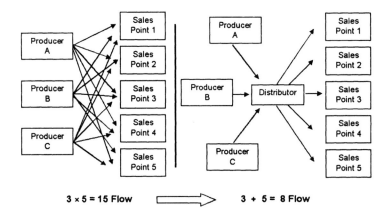

Figure 9-4. Rationalized logistics thanks to the intermediary distributor

The economy from a logistics point of view is clear: in the figure on the left, each sales point receives the manufacturer's products, delivered by three different vehicles. On the right, one single delivery vehicle is necessary, delivering products from the three manufacturers at the same time. This gain in efficiency concerns both spare parts sales as well as small equipment.

Besides these physical activities, one of the essential roles of distribution consists of adjusting the producers' supply to customer demand, whether in terms of quantity, variety, time or place. In the case of tickets sold in agencies, this is the aim of merchandising[1]. Beyond the logistics function, intermediaries play a role in taking orders and billing, negotiation, transmission of the producer's promotional operations and gathering of information or feedback. The same functions come into play whatever the type of merchandise distributed. However, distribution in the industrial sector has certain specific characteristics.

2. *CHOOSING A DISTRIBUTION SYSTEM*

The company has the choice between an internal solution (1) when it can count on its own sales team or an external solution (2) when it needs to use the sales team of players outside the company (Figure 9.5).

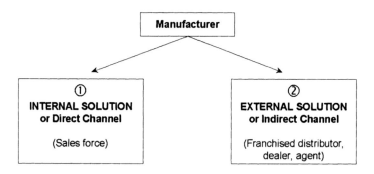

Figure 9-5. Choosing between a direct and indirect channel

The advantage of the first solution[2] mainly lies in better control and reactivity: the sales team has only the company's interest in mind. It is often well suited to complex, high added value, technological products (large civil aircraft, satellites, heavy equipment, etc.) which require an experienced, cohesive sales force capable of technically working with customers. However, this solution is less flexible and is less adaptable to sudden changes in activity. On the other hand, the external solution can be more easily adapted to changes in the market but it takes time to convince the distributor to carry and promote the product. For consumer products it is often indispensable to go through a distributor to cover a large enough territory.

2.1 Choosing an external solution

A manufacturer should work with a distributor in several cases:
- when a consumer product targeting a very large population is concerned. In this case, only a distributor can ensure wide enough diffusion of the product in an area;
- when the company does not have the expertise to deal with a new type of clientele (business aircraft, military equipment) or when it is breaking into a new geographical market;
- when the clientele is geographically dispersed;
- when the customer buying potential is too weak to pay for a salaried salesperson, whether for consumer goods or professional goods.

In all of the above cases a distribution channel is necessary (Benoun *and al.*, 1995). Once an external solution has been selected, the manufacturer must answer an essential question: is it better to distribute the product in multi-brand or single brand sales points (exclusivity) ? This question is important for the airlines but also for all travel service providers whether tour operators such as Preussag, Thomson Travel, Fram or Travel Asia, or vacation clubs such as Club Med.

2.2 Choosing a multi-brand or exclusive distributors

Choosing a multi-brand sales point network: The first advantage is that these sales points already exist. The objective for the company is for its product/service to be carried, which is less costly than trying to create a single brand network. The second advantage is that it is more attractive to have several brands or even several categories of products in the same sales point. A customer initially interested in another brand can ultimately purchase the company's product. In fact, the customer perhaps has chosen this sales point so as to feel more free to buy or not, considering that the sales people are not linked to one brand. The main disadvantage lies in the difficulty for the supplier to control the marketing mix. The multi-brand merchant is independent and usually wants to make its own marketing decisions. Thus the products offered are not complete: the distributor builds his range as a function of his different suppliers. The decisions he makes concerning prices, product presentation, sales arguments are based on his own interest, which does not necessarily correspond to those of the supplier. Image coherence is more difficult to obtain with multi-brand stores.

Choosing a single brand sales point: The product range, prices, communication and merchandising are all more coherent. The distributor is linked to the supplier company. He must aim to use all the possibilities available in accordance with the signed contract. The main disadvantage lies in the high cost of this type of solution and the longer delay necessary. For these reasons the geographic coverage is generally more limited than that obtained with a multi-brand sales point. In addition, the customer has more limited choices before making a decision.

Everyday products/services: In the case of simple ticket sales, the main goal is to obtain the best presence rate so as to increase the chances of the customer finding the ticket he desires. In this case, the airline chooses intensive and multi-brand distribution, with the aim of reaching the greatest number of sales points.

High end, status brands/products: The latter generally call for a network of exclusive, single brand stores. The distributor or dealer commits to selling only one brand and thus concentrates all of his energies on promoting it. In exchange, the manufacturer generally agrees not to intervene directly in a particular territory.

BREITLING'S SELECTIVE DISTRIBUTION

An intermediary solution called selective distribution is when the manufacturer chooses specific multi-brand distributors: a contract is established with reciprocal commitments from the two parties. While this solution does not allow as wide distribution from a quantitative point of view, it more easily preserves a high range image. This is why this form of distribution is especially used for brands having a higher positioning than the average.

This is the case for Swiss Breitling watches which are distributed through a selected network. The Breitling Emergency watch includes a micro-transmitter (a 121.5 MHz version, the frequency used in civil aviation and a 243 MHz version for military purposes). It was developed to help pilots in emergency situations. These watches were tested by elite European and American squadrons (Frecce Tricolori, Patrouille de France, Aeronautic Team Royal Air Force, Swiss Air Force Team, Blue Angels, Thunderbirds, Team 60...).

Choosing the distribution channel is an important marketing decision, given its consequences on other factors in the mix. Thus the company must make sure that its distribution policy is coherent with the other aspects of its marketing strategy:

Price: A low priced positioning could be accompanied by reduced service in terms of delivery. On the other hand, with a high price, the customer expects impeccable service and organization.

Product: A highly innovative and wide product range requires a partnership with distributors that are capable of managing complex stocks.

Promotion: If product availability is stressed, this must be backed up by a sufficiently dense distribution network to serve local customers.

2.3   *Selecting partners and managing the network*

The selection must take into consideration the distributor's technical know-how, experience, solvency and motivation for the product itself and for any complementary products. The geographic coverage must also be considered as well as the type of management, storage capacity, the capacity to provide after-sales service as well as customer financing. The size of the distributor is also important. In addition to its financial power, a large distributor also has many sales points and

9. Selecting Distribution Channels and Sales Team Management

sites for stocking, as well as a solid reputation in its market, which can attract customers.

The selection cannot always take into consideration all of the distributors in a market, as the most efficient in a given territory are not always available. Furthermore this choice is not fixed in time: the system evolves with changes in the market in terms of products and segments, and the distributors themselves. Of course the latter also have their own criteria for choosing producers with whom they wish to work. Distributors judge the reputation of the producer, his financial solidity, image and production capacity. They will study the risks of supply (ruptures in stock, defects, returns) and logistic infrastructures (warehouses, transport). They will also evaluate the producer's motivation and capacity to cooperate: training, after-sales service, technical assistance, support for communication and promotion. Finally, the accepted level of credit also largely determines which producers a distributor is willing to work with. Above all, the partnership between the manufacturer and the distributor depends on the potential profitability of working together.

Managing the distribution channel: Careful selection of distributors does not guarantee an efficient system. In fact, the enthusiasm of a distributor evolves over time and his motivation is based on mutual trust, which means respecting commitments and making sure that information is circulating in both directions. The supplier has several means of motivating distributors similar to those used for motivating his own sales team[3]. These could include discounts offered to distributors depending on the size of orders, their regularity, or on the contrary, their exceptional character, and depending on the management of stock and the achievement of annual objectives. These discounts can be combined with financial incentives for distributor's sales staff or more psychological types of stimulation such as training technical personnel in the latest products. Contests designed to reward distributors who privilege the company's products rather than those of competitors are also a means of motivation.

The organization of annual meetings between the supplier and his main distributors facilitates the exchange of information and strengthens the relationship. A support service for distribution can also be set up with documents on new products, a development plan, etc.

Evaluating the performance of the network: It is important to follow up on the distribution network as the market is always changing in terms of technology, customers, competitors, distribution, company strategies, etc. No matter how careful the supplier is, certain distributors can go out of business because of bankruptcy, death, a takeover, etc. Market evolutions, the appearance of new competitors, new products, new forms of distribution can very quickly make a distribution system obsolete. Finally the company's own evolution and development can also modify its distribution policy: external growth with buying out of distributors, cost cutting policies and re-centering of the activity, etc.

The company should never consider its distribution system as a final given. It should continually evaluate and put into question the system's effectiveness and whether it meets the company's marketing objectives.

THE DEVELOPMENT OF E-TICKETING [4]

Initially, the airlines turned to the Internet defensively to counter retailer sites. Then as the Internet developed, the airlines moved into cyberspace allowing them as well as their customers to optimize the use of information systems. Air transport portals, increasingly efficient search engines and new wireless systems (Wap, organizer, etc.) offer numerous ticket sale opportunities to the airlines. The Internet is interesting both for its sales and marketing potential: a website offers the possibility of containing a huge amount of information on customers (personalized data accessible to all staff in contact with customers). This information allows for better Customer Relationship Management and improves the company image as a service provider[5]. For example, if it is observed that a customer was dissatisfied with a previous flight and yet he still reserves a ticket for his next flight, he could be compensated by offering him a better seat or a reduction in the ticket price, among other things.

The e-ticket

E-tickets, which appeared in 1996, have allowed the airlines and the sales points to offer ticket sales from their sites. Initially, the sites offered the possibility of reserving a ticket on-line, then e-tickets allowed the airlines to sell without having to physically deliver the ticket: the customer only has to answer by e-mail with the flight information included. The e-ticket saves the airline money: thus according to Southwest Airlines, it costs 10 times less to produce an e-ticket than a traditional ticket distributed through a travel agency. Experts estimate that an e-ticket can lower distribution costs from $2-34 per ticket (Airline Business, 2000).

The e-customer can buy his ticket from home, comparing the services and prices on the market. The customer is no longer passive but rather has the freedom to find what he is looking for at the price he is willing to pay.

After having reserved his trip by phone, agency or on the Internet, the customer arrives at the airport and goes to the check-in counter with proof of identity and a file number: the ticket is no longer paper, rather it is electronic or virtual. The digital image is stored in the database accessible throughout the entire trip. This eliminates the risk of loss or forgetting the ticket. The reservation, payment and any modifications can be carried out with a simple phone call.

Kiosks situated in the airport facilitate the check-in of customers having reserved their e-tickets. On the Internet site, customers can if desired obtain a refund or exchange of the ticket on-line, check their frequent flyer account, down-load flight schedules either directly on the Internet or by telephone using Wap technology (offered by Icelandair for example) or via a personal digital assistant such as Palm Pilot. Swissair, in addition to its Internet sales has developed a system using a chip embedded card containing the customer profile so as to facilitate check-in, boarding and even identity controls. In fact, the goal is to get as close as possible to meeting customer expectations.

For the traveler, the presence of multiple offers on the Internet makes it easier to find good deals. Delta Airlines allows the customer to benefit from cost reductions obtained through direct reservation (without going through an agency). These are Internet Special Fares that are around 10% cheaper than the price offered by the travel agencies. However, certain passengers have been disappointed by the e-ticket in particular when their are problems such as lateness or cancelled flights: in the latter cases, the paper ticket seems more reliable as far as obtaining a transfer to another flight. This is why airlines such as Northwest Airlines and Continental Airlines have set up a worldwide e-ticket that allows passengers to change their itinerary or take an international or domestic flight on any other transporter without ever having handled a piece of paper.

Direct sales

Companies such as Southwest Airlines or Virgin Atlantic prefer to control their service offer on their own independent site. This is also true of the low cost company, easyJet (which directly displays its internet address on its planes). The Web represents one fourth of the ticket sales of this company. Internet can be a potentially important distribution channel. Following a promotional operation, the Irish company, Ryanair sold half of the operation's tickets on-line. Many American airlines estimate that over half of their ticket sales are through the Internet.

In an effort to differentiate and improve services, companies are increasingly focusing on providing information and complementary services such as on-line sales of games for children, down-loading of time schedules, airport plans, an exchange converter, weather information using a postal code, lists of hotels and restaurants and links with tourism offices. Recent trends include the alliance of airlines on one common site, with powerful search engines both on a quantitative and qualitative level, presenting the global world offer in terms of air transport but also associated services (hotel rooms, car rentals, etc.).

Several companies have created specialized sites for specific markets. This is the case of America West and its site awagolf.com (resold to an Internet provider of travel reservation systems) or Air France which in addition to its regular site has developed a site for low cost flights for young travelers. United Airlines was one of the first companies to understand the interest of specialized sites and partnerships with other companies on the Web. It has established links with star-alliance.com and collegetravelnetwork.com. It is also working on the creation of BuyTravel.com with Buy.com, while preparing for instance sites specialized in leisure activities. However, in spite of these efforts, the turnover generated by the Internet currently only represents 4% of the total turnover of United Airlines with the hope of reaching one fifth of turnover in a few years.

Sales via portals

Independent sites offer consumers everything for a complete trip from seat reservations to the hotel room and car rental upon arrival.

In light of the development of specialized sites, airlines have re-grouped to offer reservation and ticket sale portals on the Internet. This is the case of the nine leading European companies including British Airways, Air France, Lufthansa and KLM. In the United States several companies including American Airlines, United Airlines, US Airways have developed the site hotwire.com.

Other airlines offer ticket sales through powerful Internet players such as Amazon, Expedia, Travelocity, Priceline, which draw an impressive number of visitors. As the purchasing power is greater, the purchasing prices offered are lower. However, these companies reserve their best deals for loyal customers on their own sites.

Finally, certain airlines use travel portal websites to offer promotional prices as well as new products and services, becoming on-line travel agencies and provoking strong reactions from traditional agencies. In fact users often visit sites simply to find out about good deals rather than to actually reserve a flight at a particular date (congress, trade show, school holidays...) for which they often rely on a traditional sales point. The Dutch company, KLM, auctions tickets to the main European destinations via an agreement signed with QXL.com. The aim is to acquire Internet experience to optimize its on-line marketing (behavior regarding auctions, discounted prices...).

Auctions sites

Auction sites were developed to sell at the last minute still vacant seats on certain flights. Specialists in auction sales such as qxl.com, lastminute.com, ebookers.com, expedia.com, cheapflight.com have sold thousands of reduced price tickets this way. The ticket is put up for auction on-line and visitors make their bids until the fixed closing of the bidding at a specific time. Priceline.com offers reductions to travelers if they are willing to be flexible as to the time of the flight and the airline. Travelers set the price that they are willing to pay for their chosen destination.

The site then asks different airlines if they are willing to accept this price. The interest of this method is the traveler feels that he is in a strong position when setting his price. At the same time, this allows the airlines to fill their empty seats without having to directly use low prices, which can discredit the normal prices and dissuade passengers from reserving their tickets ahead of time.

3. *MANAGING THE SALES POINT: ADJUSTING SUPPLY TO DEMAND*

The distributor must first define his catchment area as a function of his priority target. Then he must define his product range and the services he intends to offer. Merchandising can be defined as optimizing sales space depending on demand. In fact, merchandising can be seen as marketing for the distributor.

3.1 The basis of the merchandising approach

Bernardo Trujillo, known as the "pope of modern distribution", was the first to write about the principles of merchandising in the fifties in the United States as applied uniquely to consumer goods (Minichiello, 1990). Today, in aeronautics, the principles of merchandising are in the same position with the sale of tickets or organized trips by travel agencies. Merchandising thus aims to best determine the right place for the sales point and any arrangements necessary as well as the best presentation of products and services.

The attractiveness of the sales point: Whatever the store and its image, a sales point rarely attracts enough customers on its own. This is why location is so important. For example, a travel agency could benefit from proximity to a shopping center, or a highly frequented shop, or from being in an airport. These "locomotives" multiply the number of people likely to pass by the sales point and notice the products/services displayed. The attractiveness of a sales point is thus measured by the percentage of people who come in to ask for information or make a purchase.

Within the sales point, the same logic applied to different positions reveals the "cold zones" (little frequented by customers) and the "hot zones" (very frequented). Using his knowledge of the different attractiveness rates, the manager of the sales point can use the most asked about products and services to introduce other lesser known services. For example, a travel agency can present a promotional offer for an exotic destination out of season at the ticket sales counter for more common everyday trips. It should be noted that the same logic applies to Internet sales. Loss leaders are always used to draw in potential customers with the lure of attractive prices. The sales point in this way hopes to increase the sales of products with a higher margins.

In merchandising, the probability of purchasing is linked to the probability of seeing the product. A product that is seen less than a competitor's product will be purchased less. This rule of visibility is also applicable to the sales point. This could be sales counters, ticket windows, payment terminals. Services are characterized by their intangibility, which makes visual information especially important: displays, posters, television ads, or brochures are all geared towards attracting interest and providing the customer with specific information.

In addition to visual signals, music is increasingly used by travel agencies to create a positive environment, evoking certain destinations: wood flutes for South America for example. A more recent trend is the use of pleasant odors such as coconut or aromatic plants like lavender or pine.

Two types of purchase: Premeditated purchases are those which have been thought about beforehand reflecting a conscious decision on the part of the customer who knows what he is looking for. Impulsive purchases however are provoked by the sight of the product itself or its picture.

TRAVEL AGENCY MERCHANDISING

To know their travel purchasing intentions, customers are interviewed before coming into the sales point. When their answers are too vague, the price level and service brand desired are asked for. Later, customers are again interviewed after leaving the agency. They are asked to state their final choice (Figure 9.6).

Analysis of purchasing intentions

Out of 100 purchasing intentions, 58 were exactly translated into action while 32 resulted in a modified buy, for example the choice of the service provider, while 10% made no purchase at all.

Analysis of purchases

Out of 100 purchases, 49 correspond exactly to intentions, 42% to intentions with modifications, and 9% are purchases that do not correspond to an initial intention.

⇒ The goal of merchandising is always the same: to minimize the −10%, in other words, to ensure that the maximum of initial intentions result in purchases. The second goal is to maximize the +9%, in other words to develop as much as possible spontaneous purchases.

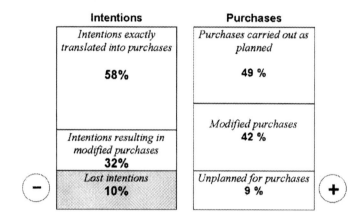

Figure 9-6. Analysis of intentions and purchases

Analysis of the 10% of lost intentions

The first reason for a non-purchase is the absence of the desired service: a destination that is not offered, in any case with the desired airline, or the wrong dates. The second reason is the price which exceeds the customer's budget.

For the desired destination...	Modify the planned on purchase	Do not make the planned on purchase
No seats available at the desired dates	84%	16%
No flight with the desired airline	76%	24%
Unsatisfactory price	48%	52%

Table 9-7. A too narrow offer results in loss of purchases

So as to not lose the purchasing intention, the agency has to offer all of the flights available in terms of prices, hours, dates and competing airlines. However, presenting the

9. Selecting Distribution Channels and Sales Team Management

complete range makes it hard to distinguish the different services offered: "too much information, kills information".

Analysis of the 9% of impulsive purchases

Table 9.8 shows customer explanations of impulsive purchases (several possible answers).

Desire	
- Upon seeing the product	27%
Reminder	
- Remembered upon seeing the product	34%
Special promotional offer	56%

Table 9-8. Clearly displayed services/products facilitate impulsive purchases

It is in seeing the products offered, whether promotional or not, that customers felt the desire to purchase. Consequently it makes sense to feature a particular targeted product. But the latter contradicts the preceding conclusion: therefore merchandising management of a sales point consists in finding the right balance between a wide enough choice of services and strong enough visibility to stimulate impulsive purchases.

3.2 Merchandising objectives

These must be considered from two points of view, from that of the distributor and that of the service provider.

For the distributor: Merchandising objectives can be quantitatively defined in terms of volume, turnover, and margins. For a travel agency, it is often the last criteria, that of profitability, which is the most important, even if it needs low margin, highly attractive loss leaders. In comparison, a group with many sales points can modify its objectives depending on the age of the sales point: a new sales point first looks to increase its volume of market share. An older sales point is more concerned with increasing the profit margin to finance the opening of new sales points.

In all cases, the merchandising logic of the distributor is global without necessarily favoring any one supplier, except when the agency is part of a group or when it has exclusive contracts with suppliers.

For the producer: The merchandising logic is very different ; the aim is to increase sales by increasing market share. This means very high visibility for his products. For the supplier's sales team, merchandising objectives can be broken down into three points:

- *Assortment.* The aim is to be present in this sales point with the services, destinations and prices the most asked for in the catchment area.

- *Visibility:* What display materials should be offered to ensure that the products are as visible as those of competitors ?
- *Sales price:* What is the final price that should not be exceeded, depending on the catchment area ?

The aims of the two partners, the distributor and the producer, are not necessarily the same. It is thus up to the sales department to find a win-win solution which will satisfy both parties.

A SPECIAL APPLICATION OF MERCHANDISING: MANAGING SPARE PARTS

Derived from the merchandising of consumer products, this involves assigning each product category a space corresponding to the needs of technicians.

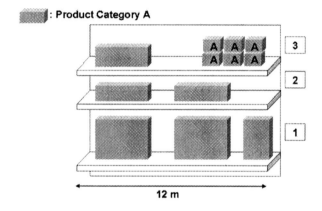

Figure 9-9. The main measurements on the spare parts racks

- Shelf-space on floor: This corresponds to the length on the ground of the furniture presenting the product category, 12 m in Figure 9.9.
- Developed or available shelf-space: This corresponds to the shelf-space on floor multiplied by the number of levels, 3 in the example above or a developed shelf space of 36m.
- Facing: The space devoted to each product is measured in centimeters, meters or units when the dimensions are important (sub-assembly).
- Shelf-space share of A: This corresponds to the space occupied by the products in category A on the whole shelf-space. If together they occupy 9m, the shelf-space is equal to 25% (9/36m).

PRODUCT CATEGORY	A	B
Shelf-space share	25	40
Use share	30	35

Table 9-10. Comparison of product categories A and B

Table 9.10 shows a comparison of the respective use of product categories A and B. A represents 30% of the use although it only takes up 25% of the shelf-space: A is therefore under-represented and B over-represented. Rationally the 5% of the space available for B should be transferred to A.

4. DIRECT CHANNEL: THE ROLE OF THE SALES REPRESENTATIVE

The sales force appoints the team of sales representative responsible for customer follow-up. This is an integral part of the company. Three main tasks are carried out by the sales representatives (Malaval, 2001): communication, sales and information feedback.

4.1 Communication

The relationship between the sales representative and the customer begins with communication. From the very first meeting, there is dual communication, inter-individual and formalized: beyond the relationship between the two speakers, there is an exchange between two companies taking place. The sales manager is in this way the first communication vector for the company. He is often backed up by communication campaigns (direct marketing, sales documents, professional press, etc.), which although not really pre-sales, do at least facilitate the first contact with the customer (brand renown, company image, etc.). First impressions are vital, and will influence the customer's impression of the company, and its products and services. The salesperson must be careful about how he presents himself. Once the relationship is established, the sales manager must ensure that it continues, by for example sending snippets of information about a past meeting or a deal made a few years back. This follow-up in human terms is highly important to develop the initial feelings of confidence. Obviously, the sales manager also talks about the company's services and goods when prospecting or selling, and in addition, he must keep customers informed about the company: financial performance, investments, present projects, etc. This sort of information proves to customers the continuity of efforts being made to satisfy him, and this is one of the objectives of the sales action plan[6], that of revealing the company's progress in terms of innovation, sales communication, etc.

But the sales manager's main task is sales. At the same time, sales are not immediate, they are the end of a relatively long road (Macquin, 1998).

4.2 Pre-sales: prospecting

The existing customer base is not normally big enough to ensure the achievement of growth objectives. In addition, certain customers will be lost to competitors, financial problems, etc., hence the necessity to win new customers and pave the way for the company's future growth. Before being an actual customer, the latter is a potential one, known as a prospect and these are often divided into "cold", "warm" or "hot".

In order make a "breakthrough" with customers already linked to other suppliers, the sales representative must visit non-customer companies where competitors are present, and where he has no internal contacts. This can be done in the first instance by taking small orders which are in fact little tests for building a relationship and to prove the serious nature and commitment of the company. The thankless and sometimes discouraging nature of the prospecting and its low returns (in the short term), often lead certain established companies to neglect it, by for example not taking any measures to encourage their sales staff to do it. The prospecting is often left to junior members, whereas experienced ones understand much better how to detect exploitable sales information (Bergadaà, 1997).

For efficient prospecting, it is essential to seek out the main sources of information available (guides, directories, internal files, internal data bank enriched by returns from marketing operations like mail shots, shows, etc.). The outcome of a meeting often depends upon information already on hand before the first contact (i.e. best choice of contact). Part of a sales representative's role is identifying companies, contacts and business potential. Over the course of meetings with the company being prospected, the sales person must gather information about the customer's outfit (suppliers, length of service, quantity, brands, price, etc.), why it is satisfied or otherwise (availability, after sales service, etc.) and on the way decisions are made (purchasing frequency, main people involved, their functions, their respective influence, etc.)[7]. The sales person can in this way fill out prospecting reports which are very useful for preparing future sales programs (arguments, choice of contact, time of visit, etc.). The aim of prospecting is not only to collect information, to prepare for future visits, to discover new potential, but also to create the beginnings of a favorable relationship.

To summarize, the prospecting steps are:
- detection and selection of target customers,
- identification of decision-making channels,
- analysis of customer needs,
- preparation of the deal.

4.3 Sales presentation and selling process

It is only after several contacts with the prospect that the latter may decide to work with the firm, having been finally convinced of its capacity to serve it efficiently and in a competitive fashion. Sales presentations dealing with the product or service offered as well as company know-how and capacity to adapt, have been covered over the course of visits. The sales representative supplies the prospective customer or customer with more specific information about the services or goods offered. He answers questions and solves any problems. Success depends on the capacity to understand customer needs and answer them. It is not a question of selling a standard proposal but a personalized offer, made up as a function of the customer's expectations, gift-wrapped with services such as financing, delivery delays, etc. The sales person must therefore be dynamic without being too much in a hurry: it is essential to really understand customer needs, know how to listen and ask the customer questions (Moulinier, 2000).

The selling process is based on exchange of information and communication. It is a question of evaluating what the customer is ready to do (his ideal and negotiating limits) and adopting an attitude which leads to consensus (Figure 9.11). In a win-win negotiating approach, neither of the two parties carry off the deal completely, rather, each of them obtains some partial gains and is satisfied with the agreement.

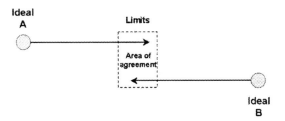

Figure 9-11. Negotiation

During the negotiation phase, sales techniques allow analysis of customer objections, by distinguishing real disagreements from astute arguments put forward by the customer in an attempt to lower the price. The prepared sales pitch must be presented, giving the advantages of the solution relative to competing offers. Product demonstrations also permit involvement of the customer.

On account of the importance of the contracts at stake, particularly in the aeronautics and space sector, the legal aspects must be prepared well before the actual negotiating phase. These legal aspects cover management of intellectual property (patents, licenses, franchises, copyright, know-how, software, brands, etc.). It is in the interests of the sales teams to involve the in-house legal services as far upstream in the sales process as possible, in order to benefit from advice and avoid obvious pitfalls (international regulations, new legal measures, etc.).

The main tasks during the selling process phase are:
- technical arguments,
- financial arguments,
- fine tuning of the offer,
- presenting a proposal,
- taking the order.

4.4 After-sales: the follow-up

Even if this stage comes after getting the contract, it is nonetheless an integral part of the global sales process since it ensures future orders. In fact, the existence of an after-sales service is increasingly written into the customer's specification. Regular contacts with the customer provide the long term future for exchanges and ensure loyalty. This follow-up phase is even more crucial when :
- the product is complex (after-sales service, technical service, etc.),
- the transaction is costly (guarantee),
- the offer is personalized.

The main functions of after-sales are:
- technical follow-up in the factory and with the customer (reception, setting up),
- insuring customer payment,
- information, training to use equipment,
- maintenance (spare parts),
- contact with users.

THE RECENT EVOLUTION OF THE SALES REPRESENTATIVES[8]

The sales representative's task has evolved. For many years he was the only one in charge of negotiation and covering the project's every aspect (technical, financial, legal, after-sales, etc.). On the strength of his technical ability he was the one and only contact for the different members of the customer's buying center. Progressively his role has been modified to become that of coordinator and supervisor of the many individuals involved in the project (preparation of the reply to the customer, negotiation, signing). He is at the heart of the "sales center", in symmetry with the customer's buying center, made up of internal managers (technical, legal, financial) or external ones (consultants, partners, sub-contractors, etc.). The idea is to create a response structure adapted to the customer. It is the sales representative's job to define the sort of response (alone, grouped, by sub-contracting or as prime contractor with sub-contractors). He must mobilize specific internal and external sources to carry out the project (choice of partners, task selection, level of involvement, etc.). In this way the sales

9. Selecting Distribution Channels and Sales Team Management

representative becomes someone driving business forwards, a polyvalent manager and integrator, capable of managing a multi-faceted team.

Since the end of the nineties there has been another evolution, in that it is not enough for the sales representative to just reply to the customer, in effect, he must anticipate, and even set up the demand upstream by creating the project. The sales representative thus becomes a business developer. Business deals are made over the long term and are based not only on technical arguments and the merits of the particular project, but also on the highly intuitive process mobilizing relationships, alliances and support. Specific arguments must be found to influence the different contacts and persuade them to join the project.

4.5   Information feedback

Along with communication and sales, information feedback is the third after-sales task. Meetings with customers give information as to reasons for satisfaction, dissatisfaction and about competitors. Sales representatives are thus essential intermediaries for the marketing information system, by giving them information about products (directions in which customers are developing, competitors) and on the companies (takeovers, subsidiaries, shutdowns, changes at the top, etc.). Companies can formally organize the work of sales staff by giving them dossiers to fill in and by asking them to write progress reports or pass on information.

Overall, the sales force is not only there to sell, but plays an important role as a source of potential information for marketing and as a communications vector. It must be:
- the company's communication support,
- the prospector, ensuring the company's future income,
- the confiding ear for the customer,
- the source of valuable market information,
- the expert advisor for purchasers, capable of putting together the best adapted solutions.

THE SALE OF AIRCRAFT

Air shows like Paris-Le Bourget or Farnborough are very often ideal shop windows to have a close look at aircraft and see them flying. They are often also the places where aircraft sales contracts are signed. These have actually been prepared a long time in advance, but are revealed at the show to have the greatest media impact (the trade and general press give big coverage to these events).

What is the sales cycle of a plane like ? The aircraft manufacturers such as Boeing or Airbus start by making market studies in order to measure the potential[9]. In this way, by taking an average growth in world traffic of 5% per year between now and 2020, they arrive at a global estimation of 13,600 new aircraft necessary, i.e. $1200 billion. Careful study of the market allows manufacturers to uncover opportunities and potential new markets.

In order to conquer their customers (airline companies or rental organizations such as ILFC or GECAS), the manufacturer's sales staff (airliners, regional aircraft) spend a large part of their time traveling around the world. The sales representatives must initiate, develop and finalize the internal and external arrangements necessary to sign a sales contract. Seasoned travelers by taste and necessity, it is not surprising that they are at the top of airline loyalty programs in terms of numbers of miles. For the sales staff, prospecting consists of identifying customer's exact needs, defining these with them and offering advice as to how and what to buy (type of plane for the particular line, market perspectives, possible ways of using the aircraft, help with financial evaluation, etc.). Systematic reporting gives an assessment of the visit and any actions to be taken (who, what, when). As soon as a need is identified, the way is open for competitors and the negotiation phase starts to obtain the contract. Companies often play one manufacturer off against the other to benefit from the best offer. According to the magazine Airline Monitor, reductions made by Boeing in the nineties were in the order of 15 to 17% of the catalogue price, and up to 22 to 24% for the B737-300, B757 or B767. But in general, manufacturers offer more than just the aircraft, and envelop their offers with special features which make the difference (increased engine thrust, compensation where maintenance costs are exceeded, pilot training, etc.). Airbus in particular uses the argument of commonality, i.e. common features between the aircraft giving gains in terms of training, spare parts management, maintenance, etc. Future economic circumstances must also be taken into account in the proposal, in order to avoid big differences between optimistic forecasts and sudden turnarounds. Non-repayable deposits are required as protection against canceling of whole orders. Another major element in the negotiation is the financial aspect with increasingly sophisticated financial packages. Certain manufacturers also deal in second hand sales: they take back the company's old aircraft to sell new ones.

Negotiations are often long, lasting from a few months to several years and require a high degree of patience from the sales staff, who alternate between periods of doubt and enthusiasm in the face of the frequent passivity of the purchasing organizations, particularly when the state is involved. Among qualities also required by sales representatives are a force of conviction, independence, active listening, a sense of communication and relationships, as well as a desire to excel.

5. *MANAGING THE SALES TEAM*

In order for the sales team to fulfill its different tasks correctly, it should be organized in such a way as to be adapted to the nature of the products being sold and the type of customers. This organization means fixing precise objectives for the sales team to achieve[10], and setting up a dedicated sales structure. There are various flow charts corresponding to specific activities. Generally the choice of organization is aimed towards a structure by type of product, by customer, by geographical zone or by a combination of these. The team of sales representatives must then be managed, starting with their recruitment, training and organization (Zeyl and Dayan, 1996). These individuals have a particular status in the company due to the fact that they work for most of the time outside, and so sometimes feel "cut off". To a large extent the responsibility for generating results is on their shoulders. Sales team managers therefore need to be well trained, well informed and well in touch with their team.

5.1 Defining objectives

The objectives consist of defining the sales team's roles in the light of the marketing strategy opted for by the company. These roles are multiple: from prospecting for new customers to actual selling, service provision such as advice, technical assistance, delivery, not forgetting communication, whose endpoint is the transmission and gathering of information. In view of the multitude of tasks, the time given to each one should be regulated as a function of present priorities and recent changes. Management objectives[11], are designed to prepare the detailed activities of the sales team, especially by fixing results to be obtained, sales to be made and thus the performance of the team and its different members.

The objectives must be:
- sufficiently high to mobilize the team,
- realistic enough to be attainable,
- equitable between the different team members, and especially taking into account, the different sector potentials, the types of customers belonging to each salesperson and the background (i.e. length of service in the sector or with this type of customer).

There are two types of objectives:
- quantitative, calculated in sales volumes and turnover which translate into financial margin objectives,
- qualitative, which are more difficult to check.

For the latter, it could be for example the sales staff's capacity to:
- explain to customers the choice of arguments put forward in the advertisements and the choice of any eventual supports,
- present new products,
- give advice in the use of products or services,
- prospect for new customers.

These qualitative objectives are not directly translated into sales results. They are nevertheless the guarantee for the future activity of the company. Their attainment is difficult to evaluate in that the salesperson alone, cannot be held responsible for the results obtained (positive or negative). They depend on the efficiency of the "back office". For all these reasons, qualitative objectives are more difficult to include in the remuneration program than quantitative ones.

5.2 Choosing the structure

There are different ways of structuring the sales team as a function of the strategic approach chosen, the type of company, the nature of the products and the type of customers.

Organization by sector: Each sales person is given a sector in which he sells the entire range of the company's products. This is the solution chosen when the range of products is homogenous and not too spread out. And this also allows the company and its activities to be easily identified by customers, and gives precise geographical zoning for each sales manager as well as rationalized traveling expenses.

Each salesperson knows and understands his sector and it is relatively easy for those in charge to check on his activity and results.

Organization into families of products: This is specializing the sales team by category of product. It is the solution chosen when a company offers technically complex, heterogeneous or very numerous products. This is a product-oriented organization. When different ranges of products are bought by the same customers, there is a risk of damaging the company image through lack of coherence and coordination between sales staff. The resulting additional expenses should be taken into account. In addition, this organization risks standardizing the sales approach too much.

Organization by market or type of customer: This arrangement is adopted when there are various customers with different needs and purchasing behavior. The customers can be classified according to different parameters such as geographical region, size, activity sector and sales volume. The sales approach is made more efficient because the staff are meeting customers from the same activity sector, with similar problems. At the same time, the sales staff cannot have in-depth knowledge of the whole range of products if the latter is very wide. As with the previous one, this structure offers perspectives for change and promotion for the best sales persons. It is a solution which is well-adapted to the knowledge and satisfaction of the customers, even if it is sometimes difficult to classify customers with diverse activities. This method of organization is found in particular in the satellites and aircraft sales sector.

Matrix Organization: In order to benefit from the advantages gained using the organizations covered, some companies combine them. When it is a large customer like an airline company for example, the sales team will put a specialist in charge backed up by product specialists.

5.3 The size of the sales team

Determining the optimum size for the sales team is a delicate business. In view of the cost of a qualified team and the company's profitability objectives, decisions linked to its recruitment are very important. A restricted number of sales staff will result in inefficiency: customers will not be visited or not in time, prospective ones will be forgotten, service will be neglected. On the other hand, too many sales staff will weigh down the company unnecessarily.

$$\frac{200 \text{ euros}}{5\%} = 4\,000 \text{ euros}$$

This means that most companies calculate the number of sales staff they need by analyzing the work load. They decide how many visits the company must make for each major customer and then for each type of customer[12]. The total needs in terms of number of visits can then be estimated. By combining these with the realistic number of visits that one sales person can make, an approximation of the number of staff needed can be made. This figure needs to be moved up or down according to things like the personal characteristics of staff or the market context. There are other ways of determining the size of the sales team, like using financial methods based on the average cost of a visit. This means starting from this average cost, e.g. 200 euros, including travel. This is then put into the context of the gross profit margin, in order to check that the additional turnover developed following the visit is enough to cover the extra cost. In this example, a 5% gross profit margin means a minimum turnover of 4000 euros to cover the cost of the visit.

Then the customer's annual potential turnover must be estimated. If this comes out at 20,000 euros, then it means that there can be a maximum of 5 visits per year to this particular customer. The same calculation is made for all customers, in order to obtain the total number of sales visits necessary. By dividing this number by the number of visits that one person can make in a year, the theoretical number of sales staff needed is obtained. Take the example of a company supplying spare parts in the aeronautics sector who need to make a total of 6,000 visits per year. A staff member can make on average 320 visits per year, i.e. 8 per week for 40 weeks work in the field. Therefore 19 people will be needed.

This method is interesting because it is relatively prudent in financial terms, but it ignores the company's sales policy which could sometimes consist of investing heavily in one customer, leaving another under-invested.

5.4 Recruiting sales representatives

The recruitment of sales staff must answer the estimated, quantitative needs but also the qualitative requirements (profile, personality, team spirit, etc.). The job, the tasks to be carried out, working conditions and company demands must be defined. Then there must be a sufficient number of candidates: these can come from small ads, contacts with teaching organizations, from specialized recruitment agencies, or by recommendation. Only the best individuals must be retained. The selection is based on a series of tests and interviews with the aim of evaluating each candidate's motivation, flexibility, dynamism, sociability, team spirit, sense of initiative, etc. A good salesperson is capable of negotiating and convincing, but should also have certain psychological qualities and intellectual aptitudes associated with a strong organizational ability. Finally he should have an excellent knowledge concerning the company's products, which may sometimes imply precise technical acumen. For the international markets and aeronautics in particular, the nationality of the customers is taken into account in order to recruit staff with the same origins. More in touch with the customers, they will be more effective.

5.5 Supervising the team

The role of the sales director is to run his team of sales staff, at the same time as advising and motivating them. It is not just a case of looking at their performance via their turnover. It means setting up key business indicators, adapted to the specifications of the company. This is creating an effective management tool including specified customer visit and prospecting norms.

It is for each person in charge to make up this list of key business indicators, using performance indicators and ratios deemed to be useful (Table 9.12): turnover for a period, number of orders per customer, number of visits per customer, payment delays, percentage prospects transformed into actual customers, costs per order and per visit, etc. These indicators are then compared, period by period and allow comparisons to be made between different salespersons. Evaluation of sales staff is based on analyzing the individual's evolution over time, comparing his performance with others as well as the satisfaction of his customers.

Turnover/period	Payment delays
Number of orders	Cost per order
Number of visits	Cost per visit
Turnover/visit	Sales charges / number of visits
Turnover/order	Effective sales time
Number of orders/number of visits	Number of kilometers
Number of visits	Number of time late
Number of orders	Number of errors/sales documents
Number of sales and turnover for priority products	Team spirit
	Punctuality and presentation
% small orders	Number and quality of information collected
% discounts on tariffs	

Table 9-12. Examples of quantitative and qualitative performance measurements for a sales team

The results are examined and discussed between the sales director and each salesperson during regular meetings, as much to let him improve his work as to encourage and advise him. Checking of the results obtained is mainly based on evidence supplied by the staff concerned, especially the progress reports. To encourage regular and precise monitoring, it is best to provide a certain number of documents such as prospecting and follow-up forms, road books, planning sheets, order form slips, various statistical tables, etc. A good report not only informs the company as to the work really carried out by the salesperson, the results obtained and the means used to obtain them, but also includes interesting information about the evolution of the market and the competition.

It is not enough to compare the work of one salesperson to another, in terms of turnover. The profitability derived from this must also be considered, taking into account the different working conditions between sales staff. The level of customer loyalty generally translates the level of satisfaction which must also be part of the sales person's evaluation.

The sales director must also make sure that the balance between time spent on customers and that on prospecting is respected. Personal contacts should remain dominant for sales and irreplaceable for customers.

5.6 *Remunerating sales representatives*

A sales representative's remuneration can be fixed, variable or a mixture of both. In addition, in real terms there are various fringe benefits to take into account.

The fixed remuneration system

This is above all simpler to manage: The first advantage of a fixed system for the company is that it is easy to manage and simpler to budget. For the detailed marketing plan, it is easy to calculate the payroll, independent of the unknown factors from sales. The other advantage for the company can be seen above all when the context is favorable: as the remuneration remains fixed for a high sales level, the company can better absorb the costs of its sales team. Any excess can only be positive for the sales director.

A "guaranteed income" but less stressed sales representatives: For the employee, the first advantage of a fixed income is the guarantee of a check at the end of the month, regardless of how many sales have been made. This guarantee has for a long time been seen as something negative, since it is associated with an "easy number". However, the fixed salary brings the employee a level of happiness which can only profit the company. His job has a certain thankless side to it due to the frequent solitude when he is with the customer or on the road. His dynamism would be open to question even more if he thought that his salary at the end of the month was going to decrease, which would be the case if he was paid by results. The doubts that he harbors as to his own efficiency are joined by doubts about that of the company. He has the feeling of being "dropped". Conversely, the employee receiving a fixed salary under the same conditions, can count on the company during the bad times. This calm and happiness is especially necessary during difficult negotiations.

It would be simplistic to make a global judgment about fixed remuneration. The idea of belonging to the company because of a finite salary level needs to be treated with caution. In addition, the fixed salary system remains a real motivation in companies which have an annual evaluation of sales staff, leading to updating of the salary. At Airbus or ATR for example, remuneration is mainly fixed (without bonuses for contracts signed). This method avoids tension and jealousy between the different departments working on the same contract such as marketing, studies, accounts and legal services. Within the American-Turkish joint venture Turkish Aerospace Industries, sales staff remuneration is also fixed. The company gives out bonuses, covers health and social security charges and participates in housing expenses and travel for the sales representative's.

Variable remuneration

A flexible and motivating system: The first advantage comes from the flexibility of the system and its adaptability to wide fluctuations in sales: low sales give correspondingly low remuneration. The second advantage is in its capacity to stimulate a team, since the remuneration is linked to achievements. The members of the most efficient team can serve as an example by their level of remuneration, etc.

A risk of pre-eminence in the short term: Under these conditions, a salesperson can in fact neglect to do things which don't have a direct influence on the turnover:
- presentation and explanation of the advertising campaign (demonstration with video cassette, Cd-Rom, etc.) ;
- information feedback, operations which necessitate patiently gaining the confidence of the potential customer ;
- prospecting for new customers.

A remuneration which is mainly variable, risks pushing the salesperson into becoming polarized on the products which are easiest to sell, the "loss leaders", and not necessarily on the priorities from the sales action plan.

A system which is often unfair: Another drawback of variable remuneration is that there is a risk of injustice: as a function of the evolution of a particular geographical zone or even some important customer, there can be a "guaranteed income" giving the salesperson a high remuneration which bears no relation to his work or colleagues' situation.

Commissions and bonuses: The variable part is made up of commissions and bonuses. The real difference is not based on the way it is calculated, since the bonuses, like the commissions can be worked out either as a forfeit, or on a pro rata of the turnover or margin made.

The commission cannot normally be questioned by the company. It is therefore an integral part of the salary. The bonus is much more flexible to use, since it can be offered for a given period. But there is no guarantee for the employee: the company is not obliged to renew the operation.

In US companies, non financial bonuses such as going to dinner in a small group with the CEO, are highly sought after. The financial bonuses can be of a fixed amount, agreed beforehand, or a one-off addition to the normal salary. For example, in a sales competition over six months, the first salesperson in the national team wins three months extra salary, the second, two months, the third, one month and the others get nothing extra. In this case, the overall amount cannot be budgeted for because it will depend on who wins. A bonus can be collective when for example a regional team with the best score wins a holiday trip.

Companies wishing to ensure that their sales team makes a continuous effort, usually opt for the mixed remuneration system with a very large fixed part. The latter corresponds to a payment for being "patient" which is necessary when dealing with an important customer, but also acts as a reward for prospecting and feeding back information to the marketing department, etc. Sikorsky International use a mixed remuneration in this way: the sales staff get a monthly salary as well as a percentage of the company's overall profit (this percentage is performance related and worked out by their manager).

In general, and contrary to popular belief, the main method used for the remuneration of sales staff in Europe and Japan is still a mixed one, with the fixed part accounting for 75-95%. On the other hand variable remuneration is very common in the United States where it is more in keeping with the North American culture.

Fringe benefits

The car: Originally instigated to reimburse the sales person's traveling expenses, this budget item has become an indirect remuneration method. For cars, there are two possible systems:

• Repayment of kilometers traveled is based on the official rates published in the press, with a fiscal horsepower limit according to responsibilities. This costly system is gradually being abandoned except for staff people, who travel very little;

• The company car is a vehicle which can be used by the salesperson only for professional purposes. Not considered to be push money, this solution, although more economical, is gradually being abandoned by companies. Sometimes, this can be used for personal purposes. In this case, the model has to be chosen from amongst an assortment of various makes, and as a function of the hierarchy. A sum is deducted from the salary, with the agreement of the tax authorities, so as to avoid having to declare it as a fringe benefit. This is the most common system.

The mobile-wap phone: This tool has gradually become indispensable for sales staff who have to travel around. Although only moderately expensive, it nonetheless gives an improvement in reactivity by allowing customers to get in touch with their sales contact (transfer of files, pictures, e-mails, etc.).

Other fringe benefits are offered by the company, for example they could participate in housing expenses. Faxes and telephones can also be partly or wholly covered.

The Ukrainian company Aerocharter based in Kiev, is a 50/50 passenger/cargo carrier. At the beginning of the nineties it had 10% profitability on its turnover and a workforce of 105, with salaries composed of 95% fixed and 5% variable. Because of the instability of the market, the company was forced to re-examine its organization, from the fleet, to its market approach, right down to the management of the sales team. At the end of the nineties, the workforce had been cut back to 65 people: 5% of these had a fixed salary, 30% a variable one (sales and marketing department), and 65% a mixed salary. At the same time, conditions had been improved, with paid holidays and housing expenses. The motivation of the sales team has been much improved, and the present profitability of almost 25% is much higher than that made with the previous arrangement.

5.7 Sales representatives: motivation, training and career management

To try to increase motivation by just using money, risks being costly and inadequate.

Motivational methods

There are many of these, starting with one-off events like a three month sales competition to select the best national or international salesperson. The rewards can be very varied as a function of the activity sector: from a holiday abroad to equipment of some kind. Such operations can stimulate the sales team but, being purely financial, they don't really reinforce the sense of belonging to the company.

In addition to these individual motivation exercises, there are some for teams, designed to develop solidarity between the members.

The achievement of objectives must be acknowledged by the company. Financial reward remains the most appreciated and the most used, hence the importance of the remuneration system. However, the salesperson also needs non-financial stimulation, and nominating someone for membership of an informal group, called for example " The 10 best performers club" or an invitation to an informal meal with the company CEO, are good examples. Invitations to more specialized shows is another way for the company to show its acknowledgment. In general, the best stimulation, and that which is the most eagerly awaited by sales

9. Selecting Distribution Channels and Sales Team Management

representatives, is the proposal of new responsibilities, a customer with greater potential, or a geographical zone known to be attractive to the person concerned.

Training

Training's main objectives are to enrich the human resources in order to make those in charge more competent in the tasks they are given. The time and investment necessary to train a sales person is longer the more complex the goods are that he is to sell. For the sale of aircraft or satellites, it takes about ten years for an engineer or administrator to become an independent sales manager.

There can be initial training, updating or technological adaptation. In terms of content, basically there are product (particularly new ones) oriented courses, courses on sales techniques covering customer financing, and those on managing a team. Most courses have an operational aim, in line with the actual or future job of the person concerned. But the company can also offer more general, prestigious and costly training courses. MBA's for example, offer a double investment for the company and the employee. The latter, conscious of his improved CV, is even more indebted to his employer.

Career management

From a manager's point of view, there is nothing more motivating than watching the progress of colleagues. Better than all the speeches, internal promotions demonstrate to employees that they can progress by staying in the company. This allows the employee to gain on two fronts: get new experience with new functions and thus avoid routine, while at the same time totting up points for experience and length of service.

Finally, the company's aim must remain the motivation of its workforce. The principal of motivation is simple: the more motivated the sales manager, the more he invests in his work and the better his results. It is a positive circle: motivation, acknowledgment, satisfaction and more investment in return (Figure 9.13).

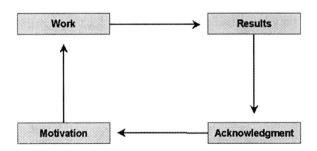

Figure 9-13. The positive motivation circle

The Sales Director's main task is to verify the coherence of the four components of the motivation policy, which are the system of remuneration, the various stimulation exercises, the training plan and the career management of the sales managers.

NOTES

1. See 3. Managing the sales point: adjusting supply to demand
2. See 4. Direct channel: the role of the sales representative
3. See 4. Direct channel: the role of the sales representative
4. Source: *Aviation Week & Space Technology*, (2000), March 27, 2000.
5. See Chapter 7, Marketing of Services.
6. See Chapter 5, Marketing and Sales Action Plan.
7. See Chapter 2, The Individual and Organizational Purchase
8. Source: Cova, B. and Salle, R. (1999), *Le marketing d'affaires*, Paris, Dunod.
9. See Chapter 3, Business Marketing Intelligence
10. See Chapter 5, Marketing and Sales Action Plan
11. See Chapter 5, Marketing and Sales Action Plan
12. See Chapter 5, Marketing and Sales Action Plan

REFERENCES

Airline Business, (2000), The big e, November, p 85-86.
Benoun, M., Héliès-Hassid, M.-L. and Alphadeve, M., (1995), *Distribution, acteurs et stratégies*, Paris, Economica.
Bergadaà, M., (1997), *Révolution vente*, Paris, Village Mondial.
Filser, M., (1989), *Canaux de distribution*, Paris, Vuibert.
Macquin, A., (1998), *Vendre*, Paris, Publi-Union.
Malaval, Ph., (2001), *Marketing business to business*, Paris, 2nd Ed., Pearson Education.
Minichiello, R.J., (1990), *Retail Merchandising and Control: Concepts and Problems*, Irwin, Burr Ridge.
Moulinier, R., (2000), *Le cœur de la vente*, Paris, Les Echos Editions.
Zeyl, A. and Dayan, A., (1996), *Force de vente: direction, organisation, gestion*, Paris, Les Éditions d'Organisation.

Chapter 10

PROJECT MARKETING

The design, manufacture and sale of complex equipment such as an aircraft, an airport site, or a satellite launcher require a specific marketing approach. Thanks to the work of Bernard Cova and Robert Salle as well as the Industrial Marketing and Purchasing Group (IMP), the project marketing approach has been defined (Cova and Salle, 1999).

1. THE SPECIFIC NATURE OF PROJECT MARKETING

Project marketing pertains to business to business since the customer is always a company or a public organization. Project marketing is mainly characterized by:
- high financial stakes,
- a generally one-off project,
- predetermined buying procedures,
- a discontinuous relationship with the customer.

1.1 High financial stakes

To understand project marketing, it is necessary to distinguish between "autonomous projects", such as the construction of an airport, and "prototype projects", which latter are subject to traditional industrial marketing. For example, a new commercial aircraft such as the Airbus 340-600 is a project during the design and industrialization phase. It corresponds to project marketing for all of the players involved in this phase.

By their very nature, projects involve a large budget and the use of cutting edge technologies whether in the space, aeronautics or building sector. For example, given the increase in air traffic, saturated airports have been forced to expand, building new runways and improving sites. A case in point is the Chek Lap Kok airport in Hong-Kong; built on artificial land to the north of the island of Lantau, it was opened at the end of the 90's and cost 18,3 billion euros. It will handle up to 89 million passengers a year. Designed by Mott Consortium (Mott Connell, Foster and Partners and BAA) this new terminal required the building of a 6 km long and 3.5 km wide island (visible from space) and over 197 million cubes of materials (parking lots, buildings, runways, bridges). A new city, Tung Chung was even created next to the airport with 200,000 inhabitants (Figure 10.1).

Figure 10-1. Exterior and interior views of Chek Lap Kok in Hong-Kong[1]

In Malaysia, the new elegant airport of Kuala-Lumpur, close to the tropical forest, will be able to handle 56 planes simultaneously. Kennedy airport in New-York has benefited from the renovation of its 5 terminals for a budget of 6 billion dollars with the construction of 3 parking lots, the creation of a train link between the airport and a new road network.

THE BEIJING CAPITAL INTERNATIONAL AIRPORT

In recent years, Beijing Capital International Airport has had considerable growth of air traffic (Tong Yong Fu, 1999). Considered to be the 2nd Chinese airport in 1988 with close to 5 million passengers a year, 10 years later it had become the biggest Chinese airport with over 17 million passengers. In light of the increased traffic, the airport had to expand its infrastructures and become an international air traffic platform. The extension of the Beijing airport was a huge financial undertaking with close to 1.1 billion dollars invested (Figure 10.2). The project was launched in October 1995 and was based on a traffic forecast for 2005 of 35 million passengers and 780,000 tons to be transited through BCIA for 190,000 flights. The extension program included 336,000 meters squared for the passenger terminal with 16 ground constructions, an area of 464,000 meters squared for maintenance and repairs as well as a multi-level parking lot and a cargo complex. The project was completed at the end of 1999. The new additions included 36 bridges, 51 elevators, 61 escalators and 24 moving

10. Project Marketing

walkways. The cargo complex as well as the parking lots and the road access ways have been finalized. This project involves numerous political and economic players.

Figure 10-2. The Beijing Capital International Airport[2]

Project marketing is also used by computer engineering consulting companies, as well as in organization and space engineering, among other fields. Such projects are generally long-term over many years. This is the case of the new Rolls-Royce Trent 700 engine project for the Airbus A380, which took six years from the launching of studies to the first delivery. The construction of a new airport in general takes 3 to 5 years from the drawing board to inauguration.

1.2 A "one-off" project

By its very nature, a project in a context of project marketing is unique; if an airport is "ordered" by Hong-Kong, there will not be a second one like it. Another airport ordered by another country will involve different specifications. Thus, the service is generally unique in terms of:

- Technical content: specifications required by the customer in terms of surface area, infrastructure, services, architecture and decoration,
- The human relations approach: first of all because the organization of two private or public companies is rarely the same, the supplier company cannot develop the same approach from one customer to the another
- The customers' decision-making method, which is different depending on national regulations, but also as a function of culture and sometimes individual personalities. The decision-making methods are thus often unique in project marketing in spite of the efforts of the national authorities to harmonize the managerial practices of their representatives and administrations. For example, if the mayors of major cities are analyzed, it is generally possible to distinguish between the "builders" and "managers". The builders are those who want to use their time in

office to positively transform their city or their region. More like entrepreneurs, they hope that their name will be later associated with the development and prosperity of their city. The mayors are mainly concerned with balancing budgets. Often conservative, they rarely defend innovative projects. As a result, they have less influence on technical or architectural decisions. Independent of the political system and regulatory constraints, the "builders" are much more involved in the decision-making process and in the technical or architectural choices.

- The way of financing differs from one project to another, even when two projects are comparable. For the replacement of an airline's old aircraft, financing is often a very important commercial asset. Thus, constructors try to facilitate negotiating with banks to enable financing and in turn the purchase. In the field of electric power plants, the Built Operated Transfer system is often used: the constructor will remain the owner of the plant for a certain number of years, allowing him to operate it at a predefined price with the customer country. He can thus make a profit before ownership of the station is transferred to the customer.

More generally speaking, financing varies according to the percentage of the amount which must be paid right from the order and the final payment schedule. For all the mentioned reasons, a project is never the same as an other.

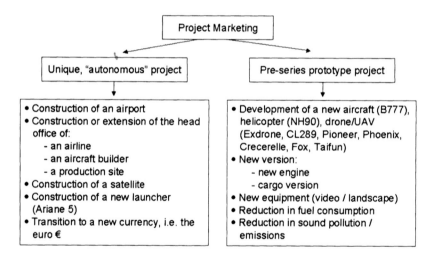

Figure 10-3. Project marketing: two types of project

1.3 Generally predefined buying procedures

Projects' customers can be private companies as well as public organizations. In many countries there is legislation for public markets to answer several objectives:
- obtain the best supplier's response to the customer's needs,
- obtain the best quality-price ratio possible,

- guarantee equitable conditions for competition between the different players,
- prevent corruption.

In terms of these objectives, different countries have defined procedures based on a certain budget. Private companies often follow some of the same procedures even if they are not required to. There are 3 types of procedures:

- *The negotiated contract or mutual agreement:* this is possible below a certain predefined financial amount. This procedure is the lightest to handle as a customer is free to negotiate directly without publishing the market to inform the different potential suppliers. If the customer is satisfied with a particular supplier's proposal, no competition is involved. This procedure is the most common in terms of number of purchases, but it is less important in terms of value, as it only concerns small transactions.

- *Formal procedure by allocation, principle of the lowest bidder:* In this case, the customer establishes extremely exact technical financial and administrative specifications. This procedure is required above a certain predefined amount. It is accompanied by an obligatory announcement in official publications such as the Official Journal of the European Community. The goal is to expand competition by informing potential suppliers from other countries who will thus have the possibility of submitting proposals. The response time is also fixed. Generally, a shortlist of proposals is made based on technical criteria and necessary references. Sometimes, this selection phase also involves administrative requirements. The winner is chosen from this shortlist. In this case, the company with the lowest price proposal wins the market. This procedure is the most objective making corruption difficult. However, it sometimes favors cost cutting to the detriment of long-term quality. This is why this procedure is less and less used in developed countries.

- *Formal procedure for invitation to tender, principle of the best tender:* this procedure is close to the preceding one. It is obligatory above a certain amount and also must be published in the same journals. However, the difference lies in the fact that it is the supplier with the most efficient service as opposed to lowest price that is chosen. The customer can accept a much higher price if it is accompanied by long-term guarantees or additional options. This procedure favors customer-supplier exchanges since it authorizes the candidate companies to ask questions, exchange information and make counter-proposals. It is wildly developed today.

There are of course variants about the bids described above in particular bids that are open to a wider number of competitors. In this case, the customer from the start is not able to define the best technical solution. Competitors can thus propose solutions using different technologies with different cost prices.

Project marketing involves all of these bidding procedures. Once the bid has been published it is a precise framework into which the marketing plan of the different supplier candidates must fit. This is why it is important to distinguish the different phases of the project.

1.4 A generally discontinuous supplier-customer relationship

The very nature of the project activity explains the discontinuity of the relationship between customers and suppliers. If the project involves a new head office or production sites, it is possible that the customer will not order anything further even if he was entirely satisfied with the supplier's service. The same is true of a prototype project. When the new helicopter or aircraft has been completed, the product generally has a fifteen year life and it is rare that such a big project is developed again in later years.

This discontinuity is a problem in terms of customer follow-up. Without enough contacts, the supplier is often not informed early enough of a new project opportunity. This is why the supplier must find a solution to stay in contact with the customer after completion of the project. A minor solution consists in advertising in the customer's in-house publications and making frequent visits. A heavier but more efficient solution is to establish a maintenance contract which will allow the supplier to remain in technical contact with the customer and therefore stay informed. There are three possible phases in the customer-supplier relationship: outside the project, upstream, and during the project.

- *Outside the project.* This is when there is no project on the horizon with the customer. All commercial investment must be seen from a long-term relationship perspective. Thus, greater intimacy can be developed with the customer so as to better know him and in particular his potential weaknesses or difficulties. This can involve advice and information exchange possibly leading to future demands.
- *Upstream.* In this case, the potential customer has already analyzed his weak points or the risks that he will have to deal with. By intervening upstream, the supplier company can enrich the debate with its expertise thus favorably orienting future specifications.
- *During the project.* In this case, the supplier company discovers the potential market through the published bid. Specifications have already been defined and it is probable that they have been established in part under the influence of a competitor, which places the potential supplier in an unfavorable situation. It is thus necessary to see how the potential supplier can participate in elaborating demand to optimize his chances of success.

Phases	Possible actions	Possible Influence
Outside the project	Construction of the offer	Strong
Upstream the project	Co-elaboration of the offer	Strong / Average
During the project (1)	Case 1: Nothing	None
During the project (2)	Case 2: Try to co-elaborate a new offer	Strong

Figure 10-4. Types of possible influence according to the phases of the project

10. Project Marketing

2. BUILDING DEMAND

It should be remembered that project marketing is fundamentally different from consumer marketing, which starts with the idea that demand exists and can be defined. The purchase of a car or clothing can be predefined by the customer who has enough references in mind. This is not the case in project marketing where the customer understands his requirements but does not have the best solution to satisfy them.

For example, the operator of an airport can require the construction of a 4,000 car underground parking lot. The goal is to remove 4,000 cars from ground level, but the management does not know if it's better to have two 2,000 place levels or four 1,000 place levels, nor what technology to use for smoke extraction in case of a fire or what payment system to adopt, etc. The difficulty for the customer to perfectly define his demand is an opportunity for the supplier. The main idea is to help to create demand rather than to submit to it. This process involves 4 phases: latent demand, expression of dissatisfaction, imagining a solution, creating a solution.

2.1 Identifying the customer's latent demand

First, it is necessary to recognize the customer's main requirements, in other words those that have an impact on the production system, the position of the customer in the value chain or on the customer's commercial efficiency. Latent demands fall into 2 main categories (Figure 10.5):

- Those that come from an existing risk and its consequences, such as the presence of asbestos in an old building or the risk of accident for personnel or customers. This type of expectation is usually one-off.
- Those related to the need to improve productivity or profitability which are on-going. The consulting groups tend to focus on meeting this demand.

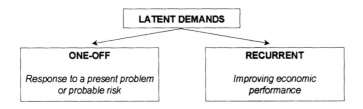

Figure 10-5. Two types of latent demands

2.2 Helping to formulate dissatisfaction

It is during the initial face-to-face contact that the supplier creates a climate of trust. At this point it is too early to address a group of managers. This first meeting is an opportunity to bring into the discussion other situations observed at other companies. The selection of these reference customers is very important. They must be as close as possible to the potential customer. Rather than giving average or anonymous examples, it is better to choose a small group of 4 or 5 named companies, even indicating what production site was involved. To avoid any problem of confidentiality, average weighted performances should be presented.

This presentation of results allows the customer to compare his own situation and it shows him that he can improve his own results, on condition that the sample has been carefully chosen. This phase of questioning the customer lasts over several consecutive appointments. Analytical tools are presented until an original interpretation of the customer's own objective is reached.

The supplier's objective is to convince the customer to recognize the problem. To do this, the customer must see that it will benefit him within his company. This "empowerment strategy" consists in giving power to the representative of the customer company thanks to the expertise of the supplier, who has already dealt with similar problems or situations in the past. This contact thus becomes an "ambassador" for the supplier as it is in his own self interest and that of the company.

Progressively, other players within the customer company will take an interest in the problem, which will require the supplier's "sales center" to work closely in coordination with the customer's "buying center".

2.3 Developing a solution

The different heads of the departments concerned in the customer company are now aware of the problem and looking for the best solution. The supplier must collaborate with them to develop a solution, while at the same time highlighting the required expertise to carry it out.

Together the supplier and the customer must establish a solution which corresponds to the needs created by the dissatisfaction. This vision should not be imposed by the supplier. It is essential that the customer progressively makes it his own. Given the sophistication of many projects, expertise from different companies is often required. The main supplier, the leader of the group, can propose several alternatives to the customer, depending on the technological choices.

AN EXAMPLE OF CONSTRUCTING DEMAND: THE BUILDING INDUSTRY AND AIRPORTS

Building and public works companies have seen the industry progressively change, even as far as airports are concerned. In the past considered as simple transit areas, airports have increasingly become places where one works, eats, makes purchases, and even sleeps. With the increase in air traffic, airports represent a very large project potential. For the building

10. Project Marketing

industry, this trend is an opportunity: cold, austere, soulless airports are going to become a latent risk for their operators. In fact, airports can be very important commercial centers, such as London Heathrow or Seoul Incheon with their hundreds of shops (Figure 10.6).

It is up to the building industry to suggest new commercial perspectives to airports and the potential risk of revenue loss if action is not taken. This is why increasingly the architectural design of airports is entrusted to internationally renowned designers such as Renzo Piano and Paul Andreu (the architect of Roissy) who were asked to design the undulating structure of the Kansai airport in Osaka, Japan. The British designer, Sir Norman Foster, imagined the light filled spaces for the recent airport Chep Lap Kok in Hong-Kong and the Japanese designer Kisho Kurokawa designed the airport of Kuala Lumpur in Malaysia.

Figure 10-6. The Incheon airport in Seoul (Korea) [3]

The idea is to provide a pleasant reception area with an increasingly commercial outlook: private rooms, game areas, religious facilities, malls, shopping centers. In the United States, passengers stop off in Pittsburg, sometimes just to take advantage of the duty free shops there. In Chicago, in the international airport a giant aquarium was built to relax passengers; in addition, works of art are also displayed. In Singapore, passengers can sunbathe in a cactus garden next to a swimming pool. Other possibilities can be imagined such as the setting up of movie theaters. For the renovation of Kennedy Airport, a 18,000m2 duty free shopping center has been planned. When approaching airport customers, some suppliers have even gone so far as to provide post-project training courses for the staff in charge of directing passengers in 30 languages.

2.4 Drafting a solution

The product or service proposed by the supplier is based on four different dimensions (Figure 10.7):

The technical-functional content: This involves ensuring the technological heart of the project. It is the first condition to win the market. Beyond the intrinsic characteristics of the equipment proposed, this includes the personnel training necessary and product support.

The legal-financial content: This means providing a reassuring and rigorous legal framework, then building the financial backing with reliable partners (which often involves intermediary financial organizations). The conditions of the financial package are crucial as well as the rapidity with which the backing is organized.

Developing relationships with political leaders: The idea is to develop strong political support in the customer's milieu. For this, it is important to understand the relationships between the directors of the customer company and the political leaders in their area, as well as their background. The supplier thus must invest in creating a strong relationship based network in the customer company.

In the aeronautics sector, political considerations often play a crucial role in the signing of contracts. Thus Boeing has been able to develop very strong relationships with Israel and the national company El Al. Public figures such as a Secretary of State or spokesman of the American State Department don't hesitate to support Boeing in Israel. Thus a Secretary of State was heard to say: "It would be difficult to make Congress see the importance of aid to Israel if Boeing were turned down". Thus although a 350 million dollar agreement was signed between El Al and Airbus for the purchase of aircraft, the company later changed its mind under pressure from the United States government (Rouach, 2000).

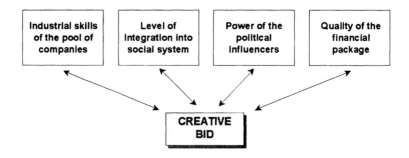

Figure 10-7. The four dimensions of the "creative bid strategy"

Building relationships with diverse social groups: Particular attention must be paid to social groups who could take sides for or against a project such as local inhabitants associations near an airport, staff unions, consumer groups, minority groups and communities. The supplier must highlight improvements which can be easily identified by these various external players.

The creation of a pool of different companies is an opportunity to integrate local companies, especially when the leading supplier comes from another geographical area.

3. CUSTOMER INTIMACY

Customer intimacy is based on the ability of the supplier to become accepted and known as the regular partner. Customer intimacy creates a virtuous circle: the better the supplier knows the customer company with its objectives and difficulties, the better able he is to provide an optimal solution. The more adapted the supplier's product or service is, the happier the customer will be, and the stronger the "intimacy" between the two parties.

To develop customer intimacy, the pre-established product or service proposal must be avoided, because it will be perceived as a solution which comes from outside. Project marketing requires solutions that are co-defined with the customer (Azimont *and al.*, 1999). Customer intimacy is made possible by the interaction between the supplier company and the customer company from the co-designing stage right up to the use and maintenance of the projected equipment. This interaction can be broken up into two dimensions:
- the depth, which corresponds to the level of involvement of the supplier but also the degree of openness of the customer, i.e. the degree to which he accepts working in common ;
- the width, which measures the number and extent of the different contacts involved between the customer and the supplier.

3.1 The depth of the interaction

Four depth levels can be analyzed.

Level 1: Simple Supplier This basic level is really just a passive form of acceptance, answering the call for tender and staying exactly within the specifications. It is the minimum degree of interaction, since the supplier is passive, only reacting to the customer's appeals. By staying at this first level, the supplier company has only one possible way of interesting the customer and standing out: by making a proposal which is much less expensive than competitors. It should be noted that small companies are often in this category, and as sub-contractors, they are dependent on the prime contractor.

Level 2: Solution Provider Here, the supplier company is capable of proposing an alternative to the specifications, in order to resolve the problem. The supplier now places himself as the one offering a solution, i.e. he makes an improved counter-proposal. He thus begins to improve his status from simple supplier to something approaching a partner. Here, the contacts chosen by the supplier are more respected and their opinion has more weight.

Level 3: Expert The supplier entirely restates the customer's problem and sets out different methods of analysis plus recommendations. The supplier in this way puts himself in the position of expert vis-à-vis the customer's problems. Such a position can only be envisaged after having carried out a certain number of projects in the same activity sector, thanks to which the supplier company has developed valuable expertise. From this level onwards, the supplier has succeeded in re-positioning himself. He can be thought of more as a partner than a supplier. This pre-supposes that the supplier will give priority to a long-term relationship with the customer company.

Level 4: Actor of change This fourth level is used to describe a supplier who is capable of helping the customer in daily operations. He thus elicits change, providing information about technological innovations which could improve the organization or productivity, and thus the customer's economic performance. This fourth status level is mainly found in large, often multi-sector companies with international bases. By having a better mix of customers, such a supplier can make the sample of companies on which he has built up his expertise, more representative. The downside of this partnership is a potential sense of interference on the part of the customer company. Management is often very conscious of being in the supplier's grip, with a host of bilateral relationships with his different departments. This negative feeling, based on the fear that the supplier will set higher prices and his staff will become arrogant, can have negative consequences. The important thing is to make sure that the advantages of such a collaboration with the supplier far outweigh any disadvantages. However it is illusory to think that perfect unanimity can exist between the various people involved.

In conclusion, it is in the supplier's interest to successfully move up the various levels ultimately becoming an actor of change. This is the only solution for helping the customer to put prices into perspective; he will see the latter as being compensated for by the different members of the buying center. Such a strategy is only realistic in the case of a long-term relationship with the customer company.

3.2 *The extent of the interaction*

There are three possible degrees of interaction, i.e. the extent of the relationship between customer and supplier.

First degree: individual relationships This is where a supplier has only set up one relationship between a salesperson and a purchaser from the customer company. This is obviously the most fragile situation for the supplier because the sales relationship depends on that of two individuals, with all the risks that this entails. The salesperson cannot count on the support of his technical colleagues within the

10. Project Marketing

319

customer's organization. Negotiations risk coming down to a discussion about prices.

Second degree: multilateral relationships The supplier company has been able to develop multiple relationships between opposing parties: a production specialist with the head of production, a logistics specialist with the head of logistics, an expert in quality with the head of quality control, an engineering advisor with the service methods team, etc. This type of relationship has already been described in Chapter 2 within the Airbus Case. An empowerment approach has been set up by members of the supplier company in order to obtain the maximum number of allies within the customer company. In this way the customer-supplier sales relationship is much stronger, since it no longer depends on an individual relationship. In addition, although it is unrealistic to look for unanimity within the customer's organization, those contacts who are the most supportive will defend the supplier in the face of competitors.

Third degree: "selling center" / "buying center" relationships Here the supplier must coordinate what his different technical and sales managers are doing in an effort to optimize coherence. This degree is reached by most of the large companies in the aeronautics sector. Certain groups succeed in establishing communication between their team and the extended buying center, i.e. by integrating members from the customer's network. This could be external influencers such as specialized journalists, engineering advisory bodies, management consultants, architects, and also politicians, those in charge of associations in contact with the company, and even in some cases, influential shareholders of the customer company. This situation maximizes the chances of developing a sustainable sales relationship, and it is obviously in this case that the supplier can obtain the greatest level of customer intimacy. Figure 10.8 below summarizes the different interactions possible to obtain a more intimate relationship with the customer company.

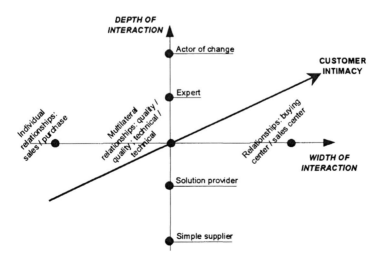

Figure 10-8. Possible positioning for the supplier company[4]

4. INFLUENCING SPECIFICATIONS

From the moment that the customer has chosen a pre-defined procedure for informing and selecting his suppliers, the main thrust of marketing will be to work on the contents of the proposed products or services, i.e. the tender. Thanks to the customer intimacy which has been established, it is possible for the supplier to work upstream of the deal well before it has been put out to tender. This means helping the customer to formulate his demand. There are two possible cases depending on whether the deal has already been launched or not.

4.1 *Intervening upstream of the deal*

Here, a partnership must be set up with the customer company as soon as the specifications are being drawn up. This consists of trying to look at things from the customer's perspective while he is drawing up the specifications. This is possible at any time that the supplier company has greater expertise than the customer. Depending on how much confidence has been built up over previous transactions, the following influence is possible.

The specifications contents: For an airport extension project, take the example of a telecom services supplier who shows, with studies to back it up, that passengers on the ground expect Internet connections. By persuading the customer that this is true, he indirectly obtains modification within the specifications concerning the connecting plugs and necessary cables. Obviously he will make sure that the technical specifications are as near as possible to his company's products, and under these conditions, it will be almost impossible for competitors to conform as closely to the specifications.

Choosing the consultation procedure: This is influencing the customer on the composition of the different parts or lots of the bid. There are two basic cases:

- When the supplier judges himself to be highly capable in one category, but less so in another, it is in his interest to persuade the customer to separate these offerings into two distinct lots. In this way, in one lot will be the service where he is uncompetitive, and in this case, he could decide not to bid, or make one just for appearances. However, the other lot consists of services where he is competitive and where he will have every chance of winning against a more general competitor, capable of making a global offer for the whole market. This separating out of lots is in the interest of specialized companies. Take the example of an electrical engineering supplier specialized in low voltage supplies who advises the customer to separate low from high voltage by using the following arguments: by breaking up the purchase into more restricted lots, the customer will be able to negotiate the best price for each lot, whereas in the opposite case there could be a more "global" supplier who lowers his prices on one service, but gets the shortfall back on another one, which is part of the same lot.

- When the supplier feels that he can make an offer which is more extensive than that of his competitors, this means persuading the customer to group up the lots into one package. The object is to eliminate the competitor who is incapable of making a global offer. This will also eliminate those who lack a solid financial base to cover a large project, and who have weak coordination between their various sites. This process of grouping up lots is in the interest of general companies. The example of building fighter aircraft can be used. A global group will propose that the security electronics and counter-measure systems should both be in the same lot, and the customer is told that this will simplify the procedure. There will be one supplier, which means rationalizing tasks for the purchasing department, and one group responsible for future maintenance.

Another solution is to create a dedicated consortium, grouping together the different companies with specific and complementary know-how, in order to answer the tender globally, while at the same time offering adapted solutions.

INDUSTRIAL PARTNERSHIPS TO WIN BIDS: THE TENDER CONCERNING MISSILES FOR THE BRITISH EUROFIGHTER (Neu, 1999)

The restructuring of European defense industries with, in particular the creation of an important 'missiles' section, Matra BAe Dynamics (MBD), opens new perspectives for collaboration with other industrial groups. In this way, Boeing has signed a strategic partnership within Meteor, with Alenia Marconi Systems, Casa, Saab Dynamics, LFK (Dasa 70% and MBD 30%) and the leader MBD, which is composed of Aerospatiale-Matra, BAe Marconi and Alenia Difesa (subsidiary of Finmeccanica). Meteor is a consortium created initially to allow the Europeans to answer a British call for tender aimed at equipping its future Eurofighters with new generation, long range, air to air missiles. There is a lot at stake in this tender, in that the missile chosen by the British will be likely to equip all the new generation combat aircraft of the European forces (Eurofighter, Rafale, Saab Gripen). Without an American presence, the Meteor has absolutely no chance of equipping the American forces. Boeing thus formed an alliance with these industrial groups to open up the American market for them, and at the same time position itself on its home market. In fact the production of missiles for the American Air Force has, for a long time been the preserve of Raytheon for the F16 (the highest selling aircraft in the world, with several thousand sold), the F15, Phantom and F18. With this alliance, the partners are aiming to take Raytheon's place. The latter has offered to associate itself with Great Britain to develop an improved version of its AMRAAM missile. Great Britain will handle half of the project, including that concerning the American Forces.

The list of companies consulted: Where there is very strong intimacy, a supplier can try to eliminate one of its competitors from the list of companies being consulted by citing disappointing services at a technical level or, more traditionally, very high prices. It should be noted that this sort of case is much rarer and more delicate than the two others. The opposite also occurs when a supplier gets other competitors added to the pre-selection, with the aim of better positioning the company's offer. For example, if there is fear that the proposal stays at a high price, it is always possible to add one or two more competitors who will be even higher, allowing the customer to put things into perspective.

To conclude, it is by intervening upstream of the specifications that the company has the greatest latitude for bringing in concrete additional material to the customer's demand.

4.2 Intervening in the deal

This is much more delicate because the technical specification has already been defined and sometimes published, perhaps influenced by a competitor. The ideal would therefore be to somehow reconstruct this demand using a "risky approach".

Instead of submitting to an unfavorable specifications, it would be better to try to question it. This is obviously very delicate since the customer, alone or with other companies, has worked long and hard to set up this technical platform. Avoiding any attitude which could be construed as arrogant, this requires studying the risks which have been identified by the customer; in fact, these risks are generally considered to be one of the main driving forces behind buying behavior (Chapter 2).

It is therefore necessary to look at the customer's perception of risks. This mainly depends on:

- What is at stake in the project in terms of cost, delays, but also technical characteristics of equipment and their eventual impact on the customer's activity ;
- The profile and motivation of the members of the buying center according to whether they are looking more for safety, or rather extra performance. In other words, to succeed at this phase implies that the members of the buying center have been identified well beforehand.

The method to follow therefore consists of analyzing the risks identified by the customer and those he runs according to the supplier, whereupon the two can be compared to see whether there are any openings.

Analysis of perceived risks: Here the detailed elements of specifications must be examined in order to determine which perceived risks they are based on. It is better to validate the conclusions by exchanging points of view with the people concerned in the customer company.

Analysis of incurred risks: This is where the supplier looks objectively at the real risks being run by the customer. These are risks which he can identify thanks to his wide experience on previous projects carried out for other customers.

Comparative analysis between perceived risks and incurred risks: It may turn out that there are no significant differences between the two types of risk, which

10. Project Marketing

means that there are no opportunities for the supplier. But apart from this extreme case, the supplier can act in three directions:

• Reduce the risks identified by the customer, especially by using the modifications proposed in the brief. In this way, regardless of which candidate is chosen, they will have to conform to stricter specifications.

• Launch a fresh estimation of the different risks identified by the customer. This could be a question of decreasing these risks thanks to special points contained in the supplier's offer. In the technical specification for example which states that parts are welded together, the supplier could make a counter offer where the piece is machined from a single billet, avoiding welds and their associated risks. Or on the contrary, it could be a case of highlighting risks linked to a competitor's offer. The risks that the vibrations during flight will have on the different types of weld.

• Open the customer's eyes to risks which up to now he has not identified. This cannot be direct risks on equipment being purchased, rather, those which are indirect and more difficult to perceive. With back-up studies as proof, this could be for example the risk of a component deteriorating after 'n' thousand flight hours. But it could also be that this new risk does not come from the technical-functional contents (cf. point 2 in this chapter) but rather from political or social issues such as the emergence of an opposition movement of local inhabitants or unions, or a movement in defense of a national supplier, etc.

In conclusion, this method is very delicate to set up because the customer must adopt the new risk and build it into his thought plan. In the same way, the customer structure must be given time to accept the criticisms of the previous specifications, rather than jumping in with an immediate ready-made solution. The latter will have in this case every chance of being rejected by the customer's technical departments. In the end, there could be a new specification perimeter from the customer, by way of an offer which is original because it has been constructed with the customer and with his active participation (Figure 10.9).

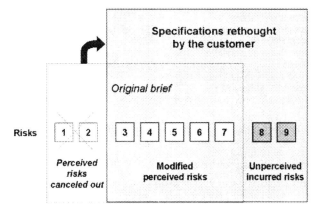

Figure 10-9. Strategy for redefining the specifications

If the supplier has been able in this way to get the demand redefined, he will stand a much better chance of carrying off the market thanks to better coherence between this demand and his proposal. But, even in the case where he doesn't succeed, the supplier has spent time and money on this customer which is an investment for another deal.

SERVICES AT THE VERY HEART OF THE CREATIVE BID STRATEGY: APPLICATION TO THE MILITARY MARKET[5]

The Syldavian armed forces[6] decided to buy "smart weapons" in order to renew their military equipment. A call for tender was put out to three world suppliers whose propositions were not made public. One of these was from the North American multinational STX company, leader in numerous defense sector activities. The competing offers, (and in particular that from the German supplier DDX) were carefully prepared, in that these suppliers had better anticipated the Syldavy project than STX.

The offer presented by STX was entitled "Syldavy and STX... facing the challenges of the new century together". It was made up of three parts: what is at stake for the customer, the STX solutions, and the benefits for the customer. In the introduction to its offer, STX recalled that the Syldavian army was looking for a multi-function arm. In addition, the Syldavian government wanted there to be an industrial cooperation program with economic spin-offs for the country. For example, pollution reduction is one major problem in the capital Klow, criss-crossed by over 9,000 buses a day. In addition, one of the main priorities of the government is improving the level of education in the country and this is why the STX offer will be based on weapons systems and technologies adapted to the Syldavian military and socio-economic requirements. The analysis of Syldavy's expectations by STX, brought out four major concerns coming from the four main players (Table 10.10).

Expectations	Players concerned
• Equip the armed forces • Improve the quality of the environment and reduce air pollution in Klow • Improve the level of education in the country • Bring growth and diversity to economic development of the country	• The Syldavian Army • Klow town hall • The Ministry of Education and its ENFICLES project • The Ministry of Finance and Foreign Trade

Figure 10-10. Taking into account the expectations of the enlarged purchase center

The STX offer is split into three parts:
- An operational part containing the technical and functional aspects answering the expectations of the military forces. This concerns the supply of military aircraft adapted to the needs of the Syldavian Army and full support services linked to the US Air Force. The overall

10. Project Marketing 325

cost for use of the aircraft proposed is lower than that for the competitor aircraft (Gripen, Mirage, etc.).

- A "para-operational part" made up of the industrial return in aeronautical terms. Thanks to the global experience of STX, the NSAC (National Syldavian Aeronautics Company) and the STSC (Syldavian Technology and Systems Company) are offered industrial cooperation. These direct compensations from STX to the NSAC mean obtaining technology transfer and help with improving productivity. For STSC as well as the NSAC, there will be technical help to improve their maintenance capacity and marketing support will also be provided to facilitate sales promotion.

- An additional non-aeronautic part grouping together the proposals destined for the other players identified. This in particular would mean the transfer of STX technologies using innovative systems for hybrid diesel electronic control applied to buses in the capital. STX also offers to transfer expertise gained in distance learning when acting as service provider to the US Department of Defense. In addition, a partnership between STX and HW is also set up to create a Syldavian Transport and Production Center on a continental scale. The creation of this inter-modal hub will allow development of the Syldavian production and its industries in order to cover Eastern Europe and the USA, stimulating the overall national economy.

According to STX, this offer gives Syldavy sustainable advantages:

- better defense, associated with the development and growth of the national military industry, at a cost adapted to the resources of the country ;
- a cleaner environment thanks to the reduction in pollution of the city's buses;
- an improved education system and access to education for all at a reduced cost ;
- rejuvenated economic growth thanks to valid marketing initiatives.

The offer proposed by STX answers the different expectations of the members of the extended buying center. There is a network of players who are more or less implicated in the decision to buy arms. The offer from STX does not limit itself to a precise answer to the needs of Syldavy for weapons, it is enveloped in services which give sustainable value for the customer. It allows the cost of using the product to be reduced and favors the customer's development by giving him a competitive advantage. These services are of two types according to whether they are directly linked to the product (technical support, etc.), or indirectly associated with the product (customer training for optimal use of product, etc.). The Syldavian government can thus use the additional services offered by STX (in terms of education, fight against pollution, etc.) to justify their choice of supplier in the face of public opinion. In addition, these services also concern Klow City Hall, the Ministry of Education, the Ministry of Tourism, etc., who can exert a favorable influence on the purchase decision (Figure 10.11).

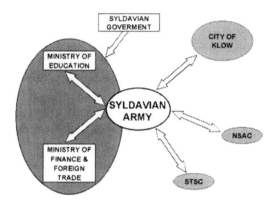

Figure 10-11. The purchase decision network for the Syldavian arms systems project[7]

NOTES

1. Source: fosterandpartners.com/news
2. Source: Photo n°0032954, Beechener, J., (2000), "Beijing rides market downturn", *Jane's Airport Review*, June, p 8.
3. Source: Incheon - *Airline Business*, (2000), September, p. 84-85.
4. Source: adapted from Cova, B. and Salle, R. (1999), *Le marketing d'affaires*, Paris, Dunod.
5. Adapted from Cova, B., Dontenwill, E. and Salle, R., (2000), "The Product/Service Couple in Business to Business Marketing: From Service Supporting the Client's Actions to Service Supporting Client's Network", *Séminaire International Eric Langeard de Recherche en Management des Activités de Service*, La Londe Les Maures, 6-9 June.
6. For reasons of confidentiality, the identity of the protagonists is replaced by imaginary names.
7. Source: Adapted from Cova, B., Dontenwill, E. and Salle, R., (2000), Op. Cit.

REFERENCES

Azimont, F., Cova, B. and Salle, R., (1999), "Vente de solutions et marketing de projets: une même recherche d'intimité client pour une construction conjointe de l'offre et de la demande", *Revue Française de Marketing*, n°173/174, p 131-140.

Cova, B. and Salle, R. (1999), *Le marketing d'affaires*, Paris, Dunod.

Neu, J.-P., (1999), " Boeing signe un partenariat stratégique avec les missiliers européens", *Les Echos*, 20 October.

Rouach, D., (2000), "Airbus-Boeing en Israël: un combat symbolique", *Les Echos*, 11 May.

Tong Yong Fu (1999), "Beijing's New Terminal", *Air Transport World*, vol n°36, n°6, p 101-105, June.

Chapter 11

COMMUNICATION POLICY

In the aeronautics and space sector, communication policy is based above all on individual relationships, as well as on specialized media such as trade shows and the trade press; however first of all there must be a coherent communication policy. In line with company strategy, the communication plan defines targets, messages to be developed, supports, and resources required.

1. DIFFERENT TYPES OF COMMUNICATION

The sales team, backed up by a multi-support communication policy, is the vector for personalized communication in business to business (Chapter 9).

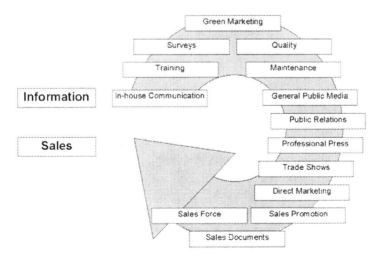

Figure 11-1. The sales and information function of communication tools

328   *Aerospace Marketing Management*

Communication actions as a whole must be part of a homogenous and coherent policy, reinforcing their impact and benefiting from a synergy effect. In this way communication can be a vector for information and a sales aid. Communication tools always have to varying degrees both an informative and a sales function (Figure 11.1). In this way, market surveys or organized training sessions can be considered as part of the communication policy, even if this is not their primary function.

1.1 The objectives of communication

The act of a customer buying something is the result of a chain of events starting with motivation and going from perception, attitude and opinion to behavior which is transformed into a trial or a purchase (Chapter 2). In total complementarity, advertising and sales promotion can intervene in this process. Advertising is based on the " pull " effect ; its main objective is to draw the customer towards the product or service. It also acts on the potential customer's motivation, which explains why it only works on a medium to long term basis (Figure 11.2). Sales promotion (Chapter 12) is based on a " push " effect ; its main objective is to push the product towards the customer. Sales promotion is designed to influence a customer's behavior and it has a short term effect. Advertising and promotion are increasingly run together or very close to each other in order to improve the effectiveness of the whole operation.

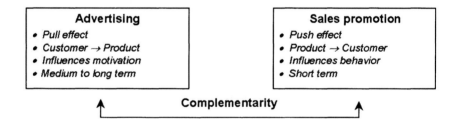

Figure 11-2. The complementarity of advertising and promotion

A company can have many communication objectives when they are defined relative to the sales and global strategy. These objectives could involve transmitting to the different partners, information concerning:
- the launch of new products and services,
- participation in a major professional event (trade shows, etc.),
- new industrial or commercial layout,
- the acquisition of production or distribution centers,
- the setting up of new processes and registration of patents,
- the launch of a communication campaign,
- the presence of an event covered by the media (cultural, sporting, etc.),

11. Communication Policy

- the presentation of the company's financial results,
- the recruiting of new staff or distributors.

Beyond transmission of information there are other objectives to communication which favor sales:
- increasing the renown of the company, its brands and its products,
- providing information about products and services by presenting them attractively,
- creating preferences (highlighting the distinctive characteristics),
- inciting and benefiting from derived demand,
- breaking down psychological barriers to purchasing and reassuring prospective and actual customers thanks to references and image,
- supporting the sales team's actions by furthering the arguments developed,
- reaching unknown or inaccessible people who influence the purchase,
- developing customer loyalty.

Communication tools should allow the sales team to be more efficient (Figure 11.3), facilitating entry into the companies prospected. It is the sales staff's job to win the first market and develop other business. Then, the customer company must be made loyal to the point where it can serve as a reference for new, prospective companies.

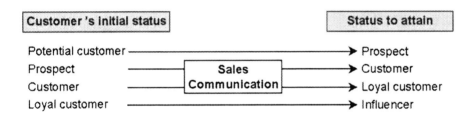

Figure 11-3. The role of sales communication: from potential customer to influencer

Taking into account the specificities of industrial purchasing (Chapter 2), communication actions should target key accounts and buying center members (Figure 11.4). The heterogeneous nature of the target populations require very different communication policies. Communication must by necessity aim at several targets to catalyze the act of buying. These include influencers, decision makers, purchasers, users and opinion leaders, that is five types of targets with different motivations and interests. This means keeping to a specific schedule and multiplying the messages and selective media in order to best reach the groups of prospective customers. And this is why the term "ricochet media policy" can be used, since the messages reach the target via several, successive channels.

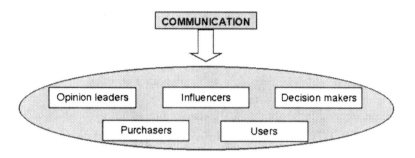

Figure 11-4. The targets of B to B communication

1.2 The four main types of communication

The company needs to monitor the coherence of the messages it channels through the four types of communication: in-house, corporate, brand/product and collective communication.

In-house communication

In-house communication is aimed at targets within the company. The objectives are to inform, motivate and increase the personnel's sense of belonging by using interactive information and investing in building personal relationships. Actually, in-house communication extends beyond the personnel (direct network) to partners (indirect network), distributors and even external targets. This form of communication usually is based on (Table 11.5) in-house publications (magazines, reviews, letters, etc.).

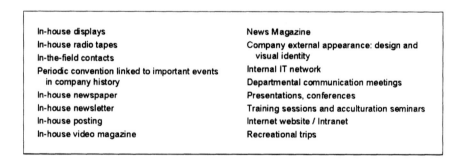

Table 11-5. Some tools for in house communication

Although really designed for in-house personnel (especially the sales team), the company magazine is a good medium for prospective and actual customers, distributors and influencers depending on its circulation and the confidentiality of the information. It presents new products and projects (new markets, new communication campaigns, events, shows). The objective is to make the in-house enthusiasm contagious to the exterior. In-house communication based on togetherness, makes the personnel aware of the challenges facing the company. The idea is to build up team spirit, reinforce cohesion and inspire confidence. It has a federating and mobilizing role especially for companies with sites scattered all over the world. With this in mind, Spot Image (world leader in analysis, processing and the marketing of data from earth observation satellites) publishes Spot Magazine, for the personnel at its different sites. Inside there are the latest important events, current projects, ideas developed by different teams, market information (Figure 11.6), etc.

Figure 11-6. Example of the Spot Magazine[1]

Corporate communication

Corporate communication is the logical outcome of a strategy seeking to build company awareness and this in turn reinforces its technical credibility. The objective is to construct or modify the company's image in the long term through public events. Thus Lucas Aerospace uses corporate communication to vaunt its local services for airline companies (Figure 11.7).

Figure 11-7. A Lucas Aerospace corporate advertisement[2]

A corporate campaign must therefore be a true reflection of the company: description of activities, leadership areas, values (technological excellence, environmentally friendly, citizenship, etc.).

Some corporate communication policies aim at federating the different brand products of the group. Thus United Technologies have put an advert in the British, French and German general press presenting the members of their group made up of Pratt & Whitney, Carrier, Otis, Sikorsky, International Fuel Cells and Hamilton Sundstrand. The corporate campaign also serves to heighten company awareness and the motivation of the sales team, based on the principle that a known and recognized company reassures customers and partners. These communication actions strengthen the identity of the company vis-à-vis different populations, from shareholders to financial partners (Figure 11.8).

11. *Communication Policy*

Figure 11-8. CFM International advert with a humorous, environmental approach " To help mother nature, we're fathering cleaner engines "[3]

Similarly, Rolls Royce, another engine builder uses the trade press to highlight its efforts to minimize the environmental impact of its Trent engines (Figure 11.9).

Figure 11-9. Rolls-Royce reducing environmental impact[4]

334   *Aerospace Marketing Management*

Corporate operations are often taken up by the general media. They include public and press relations, patronage, sponsorship, lobbying, direct marketing, prestige publications, exhibitions and factory visits. For example, Aerospatiale-Matra Airbus, via a specialized service provider Taxiway, offers throughout the year visits to its Toulouse complex, on Europe's largest aeronautics site.

Brand/product communication

This type of product or service-focused communication is used to support and promote sales especially for the launch of a new product or to help mature products. Aimed just as much at buyers, users, technicians and design offices as distributors and journalists, it can be mainly sales oriented (persuasive communication) or more information based (informative communication). The latter could announce a change of price, the existence of a new product/service, suggest new uses or set out the services on offer, etc. Persuasive communication creates brand preference by modifying market perception of the product's attributes. The objective of brand/product communication is therefore to demonstrate know-how and develop brand awareness on one or a range of products (Figure 11.10).

Figure 11-10. Example of brand/product communication: the Boeing Apache helicopter[5]

11. Communication Policy

All types of communication can be used, but those with a strong commercial emphasis such as technical documentation (Figure 11.11 and Figure 11.12), show rooms, trade shows, direct marketing and promotional sales are particularly effective.

Figure 11-11. Technical data sheet for the Vulcain 2 engine used in the Ariane 5 launcher[6]

Figure 11-12. Technical information on Bell Helicopter's, Bell 206B-3 Jet Ranger III

336   *Aerospace Marketing Management*

In business to business, sales documents play a very important role as they are the first point of contact with prospective customers. Industrial suppliers are using more and more technical documentation and catalogues about their products or services. These give setting up details, technical advice, assembly diagrams, explanations concerning norms and even help with estimates, providing a real customer tool.

Collective communication

The goal here is to institutionalize the product among professional influencers, users and the general public using mass media, the horizontal and vertical press, public relations and press relations. Collective advertising, by promoting the whole industry, activity, generic product or all of the companies in a given sector, contributes to regulating competition. It can for example, increase interest for a material such as aluminum, steel, wood, wool or glass, etc. Financing for such a campaign is generally handled by a trade organization, federation or chamber of commerce which collects funds in proportion to turnover (Figure 11.13).

Figure 11-13. A Government Electronics & Information Technology Association advert[7]

2. THE COMMUNICATION PLAN

Even if the main way to communicate remains the sales team, it is in the firm's interest to back up its actions by defining a communication policy which is coherent with its commercial strategy. Setting up communication is demanding ; it must fulfill certain objectives, target certain defined goals and meet budgetary and control imperatives. Any communication action must be preceded by an in-depth analysis of the situation, problem identification and definition of objectives as a function of new opportunities (Figure 11.14).

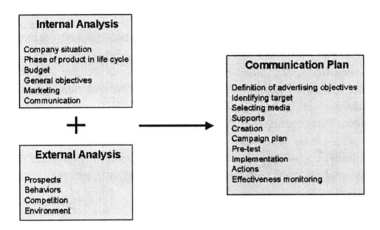

Figure 11-14. Setting up the communication plan

The communication objectives must therefore take account of the market, the general company objectives and both in-house (products, strengths and weaknesses of the offer, image, culture, etc.) and external (industrial customers, distributors, competitors, influencers, etc.) environments.

Therefore marketing objectives govern the communication objectives and actions in terms of advertising, sales promotion, public relations and the sales team. Obviously the targets and the resources need to be predefined.

2.1 Determination of targets and budgets

The different targets

By precisely defining the communication objectives, the company determines the targets. Creating 'increased awareness' as an objective means very little unless it is backed up by a detailed appraisal, i.e. increasing general company awareness by

5%, or by 2% on product X among the distributors, or by 3% among the airline company players.

The question is: who is the audience ? The influencers ? The actual or potential customers ? The buyer ? The decision makers or the users ? Defining a target in the aeronautics and space sector is fraught with difficulty. The choice of support depends partly on this target definition.

The customer base: Heterogeneous, made up of a mix of old, new and potential customers this is the main target for a commercial boost. The distinction must be made between the 'never have been' customers, the former customers and those with a current account. Targeting objectives must be frequently re-centered as a function of the priority members of the buying center.

Administrative authorities: This is mainly the national and international air regulations authorities such as the Federal Aviation Administration (USA), the Civil Aviation Authority (GB), the Direction Générale de l'Aviation Civile (F), the International Aviation Transport Authority (IATA) or the European Joint Aviation Authority (JAA).

Suppliers: These partners can contribute to the company's reputation. The firm has close relationships with the suppliers, especially during the buying decision, sending estimates, consultations, information requests, etc. It is therefore important to adapt the communication to this target. In addition, for purchase marketing, the supplier target is the number one (Chapter 2).

The personnel: Using in-house communication to present company projects mobilizes and brings people together, and this energy can be exported.

The influencers: Journalists, buyers, decision makers and advisors must be pampered because their opinions carry much more weight than any advertising campaign. The influence strategies which are set up must be perfectly coherent with the company's general communication.

The general environment: This is an amalgam of shareholders, public authorities, bankers, local administrative authorities and even the general public. The latter each represent a special type of influence: they wield an enormous amount of power over the company itself and the market in which it evolves. Communication policy in this case is lobbying, public relations, etc.

A wide array of communication means are used as a function of the targets. Prospective and actual customers are usually approached via Internet sites, letters, trade demonstrations and sales supports ; in-house letters and newsletters are used for the personnel ; annual reports are carefully worded to put a twinkle in the eyes of shareholders and financial institutions ; open days, press articles and public relations attempt to win over the general public.

To attain the objectives and targets which have been defined, the communication plan must also allot the necessary resources, and this implies a suitable communication budget.

11. Communication Policy

The budget

The communication budget depends on the objectives (in terms of awareness, image, etc.), and the initial conditions (first try, success or failure of previous actions, life cycle phase of the product being promoted, strength of the competition, etc.). Even if in overall terms, that part of the turnover devoted to communication in B to B is less than that for the consumer goods sector (1% as against 3% respectively), there is nonetheless a wide range of spending depending on the size of the company and its willingness to showcase itself to the general pubic.

There are large differences depending on the activity sector. Luxury products have the most advertising and budgetary back up, followed by general consumer goods (Table 11.15). Industrial products are a long way behind, which can be explained by the restricted target area and the use of less costly media (trade press and shows).

	% spending on advertising relative to turnover
Luxury goods	8 – 20
Consumer goods	2 – 4
Industrial goods or services	0,5 – 1,5

Table 11-15. Communication budget estimates by activity sector

Share of voice: In order to compare the advertising pressure exerted by the different players in an activity sector, the SOV is calculated. This is the company's share in the product or service category's advertising investment. It is obtained by dividing the brand's budget by the estimated advertising budget for the whole sector (Table 11.16). The SOV taken over a minimum of 3 years is compared to the market share.

	Brand A	Brand B
% Market Share	15	25
% Share of Voice	21	20

Table 11-16. Comparison of market share and share of voice

It looks as if brand A has over invested in advertising: with a 15% market share, it represents 21% of the spending on advertising for the product category, which is an investment for the future. However, if the gap persists, the effectiveness of the advertising could be open to question.

The bigger the company, the more they invest in advertising. Smaller companies spend proportionately more on sales promotion.

It is essential to effectively combine all the various channels of communication in the best possible way. However, the breakdown of the communication budget into advertising, direct marketing, public relations and sales promotion is not always clear due to overlaps between them, and this must be taken into account when evaluating the effectiveness of any single measure.

The amount to be invested in advertising can be determined by several methods.

As a function of competitor's budget: This is mirroring a particular competitor's budget in the hope of countering his actions. This takes into consideration neither the respective positions of the companies on the market, nor their strategies, nor the effectiveness threshold of the communication activities. By mimicking the competitor, the company is losing touch with its own objectives: it won't be able to anticipate large or one-off showcase actions, it won't incorporate the different modifications resulting from market circumstances and characteristics, different customer segments targeted or the number and pressure of advertising actions.

As a function of the company market share: This method is risky because there is no direct, proportional relationship between market share and share of voice.

As a function of the company's previous budgets and resources available: The company decides on the amount to spend based on previous actions: presence at shows almost a sine qua non, and catalogues must be published. This simplified calculation using a percentage of the turnover gives a conservative overall investment level and is based on an estimate. It risks being blind to changes in the competition.

As a function of an analysis of a database: Here the budget is determined using computer simulations such as Advisor which take into account the known parameters (life cycle phase, buy frequency, market share, profit models, advertising experiences, etc.) but is not particularly flexible.

As a function of goals and means available: This is the most rigorous method and also the longest and most difficult. It is based on a great deal of thought as to the company's marketing and trade objectives and in particular the definition of:

- the target (quantitatively and qualitatively),
- the style of communication (intensive or extensive),
- means necessary (supports and medias),
- how much these means will cost,
- marketing and advertising objectives in terms of audience,
- means of communication (spread, frequency),
- the best media to attain these objectives,
- expected results.

For each communication objective, the company establishes a certain number of messages per target. The idea is to invest, as long as the unit dollar sunk in promotion gives an equivalent return to the same dollar invested in another domain, such as product quality or distribution, etc.

2.2 Setting up the communication plan

Once the objectives, targets and budgets have been determined, the communication plan will define in detail:
- arguments,
- main lines and the themes,
- means selected,
- the schedule of operations,
- the type of control and evaluation of the effectiveness of the actions carried out as a function of the objectives.

It is essential to ensure unity and coherence between the communication plan, the marketing strategy and the company's general objectives. In their search for an identity, companies try to stress the uniqueness of the advertising message, its distinctive character and emotional power. Once the choice has been made, this image must be transmitted using all the various communication tools available while at the same time respecting the company's visual identity code (visible on buildings, vehicles, agencies, furniture, materials, work clothes, sales materials, technical documents, etc.).

The copy strategy: This involves the actual creation of advertising. It defines in one document (or creative work plan) the theme of the message which usually sets out the competitive advantage, the promise or benefit accruing to the customer as well as the main guidelines for the advert (tone of the message, etc.). This definition is based on pre-tests and on information gathered while prospecting (shows, sales force, mailing, etc.) or from studies made concerning the company image and customer satisfaction. The pre-test entails studying different aspects of the advertising message :
- source identification (brand, firm),
- the catch phrase (message attention value),
- information provided,
- comprehensibility (intelligibility value, capacity to transmit the advertiser's idea, symbols, cultural references, choice of language),
- interest aroused (read, skimmed, listened to, memorized, etc.),
- acceptability (message credibility),
- the power to incite a reaction (the capacity to incite, to convince, to induce desired attitudes, opinions or behavior).

The copy strategy consists of 4 points :
- the promise : the main benefit highlighted in the message,
- the proof : the demonstration that lends credibility to the promise from the customer's point of view,
- the benefit : the advantage that the customer stands to gain,
- the tone : the overall feeling of the advert (professional, entertaining, official, elitist, etc.).

The advert below illustrates the four elements of the copy strategy (Figure 11.17 and Table 11.18).

Figure 11-17. Figure 11.17 The Eros advert for its MC-10 Mask[8]

The promise	Confidence: "trust in the air"
The proof	The mask chosen for the pilots of the American President's plane
The benefits	Guaranteed protection
The tone	The veiled message: leak proof material, "no leaks" (which is important especially when there are presidential secrets around !)

Table 11-18. The advert copy strategy

To reinforce the credibility of their message, suppliers often highlight their collaboration with important customers. Where there is joint action or co-advertising, the advert could make a direct link between the name of the company and that of the customer (Figure 11.19). In this way the supplier company benefits from customer prestige (aeronautics program, new product, etc.).

Figure 11-19. Co-advertising linking Deloitte Consulting to its customer Rockwell Collins[9]

The copy strategy provides a creative and coherent platform which takes into account the communication objectives. The characteristics of the message are then fine-tuned as a function of the type of media (video, radio, the press, etc.) used.

There needs to be a strong, simple catch phrase to attract attention and make the message easy to remember. In B to B, advertising has to be very rational providing tangible information (solidity, reliability, etc.), while as the same time attracting attention.

The message can be diffused through various communication channels, each with its own specificities in terms of 'source effect' (in other words the credibility of the source : expertise, confidence, popularity). Advertising is transmitted through the (press, posters, radio, television, direct marketing, shows, etc.), the objective being to find the most suitable vehicles. For example, the aeronautics press covers publications like Aviation Week & Space Technology, Flight, Air Transport World, Air & Cosmos, Airlines Business, Jane's Airport, Journal of Electronic Defense, etc.

Media-planning: The diversity of media means that close attention must be paid to choosing the most effective form for the message and as much in line as possible with the communication policy, company strategy and above all the targets aimed at. Media-planning means selecting suitable media and defining how to use them : it establishes how much of the advertising budget should be allotted to each type of media. Choosing the media means analyzing them attentively as to global audience, diffusion, target audience, minimum publicity impact threshold, target audience calculations (Table 11.20).

Media audience: Overall number of people in contact with the media.

Global audience:
- press: number of readers who actually read the publication
- radio: number of listeners for the broadcast
- television: number of viewers who have seen the program
- billboards: number of people passing in front of the display.

Target audience: Number of readers, listeners, viewers belonging to the target population.

Unduplicated audience: Target audience from A added to the target audience from B which together contain the audience common to both A and B

Contact: Opportunity to Hear the message (OTH) ; Opportunity to see the message (OTS)

Table 11-20. A review of some definitions

The aim is to obtain the best exposure to attain the best trial rate or the greatest awareness. When choosing the most suitable media, the product, the message, the cost and the target population need to be considered. The number of the media selected can vary from one sector to another. Business to business for example rarely uses the mass media (television, radio, cinema). Special events or the adoption of a particular brand visibility strategy sometimes induces some large companies to aim their communication campaigns at a larger audience rather than just the professional players. These actions target the general public and can be used for product campaigns such as for the Airbus A340, Boeing 777 or corporate exercises concerning the environment, presentations of innovations, changes within the group, etc.

Once the media have been chosen, it is essential to program the timing of the campaign, and whether it will be intensive or extensive (constant pressure, one-off events, accompanying operations, etc.). Communication campaigns usually back up parallel operations whether these concern the sales team or the company in general (participation in shows, international conferences, etc.).

Finally, the effectiveness and impact of the actions must be monitored by comparing the results obtained to those found initially in image and awareness studies, etc. The " historical " approach, in other words using company experiences from the past (amounts spent, results obtained), and the " experimental " method (consisting of testing different markets with different advertising pressures), make it possible to accurately determine the campaign's effectiveness. The analyses and evaluations generally concern :
- the effectiveness in terms of sales or potential use,
- the identification of the company and its image,
- the memorization (the number of interviewees who recall it),
- the recognition (the number of interviewees who recognize the advert),
- the clarity (comprehension of the text and visuals),
- the influence on changes in attitude or opinion,
- the relationship between the share of the voice – share of company advertising spending on its market – and the market share.

This data allows the communication policy to be streamlined, increasing the effectiveness of any future actions. In business to business it is easier to evaluate the effectiveness of a communication campaign because the targets are more restricted and because of the nature of the media used (professional press, shows, direct marketing, etc.). In this way the effectiveness of participation in a show (Chapter 12) for example can be easily monitored by counting the number of persons who have visited the stand and who are really interested, and the percentage of these contacts which have eventually produced an order, etc.

The analysis of the results must however take into account the snail's pace and complexity of buying behavior. The impact can well be spread out over time and should therefore be measured several months after the event.

NOTES

1. Source: Spot Image
2. Source: Lucas Aerospace - Aviation Week & Space Technology, (1999), 8 February, p 14
3. Source: CFM International - Flight International, (2000), 23-30 October, p 48-49
4. Source: Rolls-Royce - Aviation Week & Space Technology, (1999), 25 October, p 33
5. Source: Boeing - Aviation Week & Space Technology, (2000), 16 October, p 1
6. Source: Snecma
7. Source: GEIA - Aviation Week & Space Technology, (2000), 13 September, p 93
8. Source: Eros - Aviation Week & Space Technology, (2000), 2 October, p 41
9. Source: Deloitte Consulting / Rockwell Collins – AW & ST, (2000), 2 October, p 4

Chapter 12

SELECTING MEDIA

The choice of media depends first of all on whether the clientele is professional or consumer. In B to B, the main media used are trade shows, the trade press and direct marketing. In consumer marketing, television dominates followed by the press, direct marketing and to a lesser degree billboard campaigns and the radio. Public relations, sales promotions, lobbying and Internet sites are used for both markets.

1.   TRADE SHOWS

1.1   *The specific nature of trade shows*

Trade shows are the most important vector of communication in B to B, first of all in terms of spending, but also because they in part determine the marketing schedule of companies and the use of other media (trade press, direct marketing). In the aeronautics sector there are many different shows[1] beginning with the most well known such as the Paris Air Show, Farnborough, Dubai Air Show, and Asian aerospace (Figure 12.1).

Figure 12-1. Logo for the Asian Aerospace in Singapore

Aeronautics shows are held all over the world (Table 12.2) and in particular in key countries whether customers or builders of aeronautics and space products.

Show	City / Country	Period	Central themes
Aero India	Bangalore (India)	December	Aeronautics
Aerospace Africa	Pretoria (South Africa)	September	Aeronautics
Air Cargo Forum	Hong-Kong (China)	September	Aeronautics cargo
Air Transport	Subang Jaya (Malaysia)	October	Air transport
Airex	Istanbul (Turkey)	May	Civil aeronautics
Airport Facilities & Equipment	Shanghai (China)	September	Equipment and technologies for airport
Airport Pan-Arab/African Exhibition of Airport	Cairo (Egypt)	February	Ground equipment and supplies
Airshow Canada	Vancouver (Canada)	August	Aviation
Airshow China	Zhuhai (China)	November	Aeronautics
AMD - Aviation Maritime Defense	Manila (Philippines)	April	Sea, ground, air technologies
Asian Aerospace	Singapore	March	Aeronautics
Astex	Ryad (Saudi Arabia)	February	Military
Australian International Airshow	Melbourne (Australia)	February	Aeronautics, defense
Aviation & Airport Exhibition	Hanoi (Vietnam)	September	Equipment and technologies for airport
Cidex - China International Defense Electronics Exhibition	Peking (China)	June	Defense and electronics
Defense Asia	Manila (Philippines)	September	Defense
DSA - Defense Services Asia	Kuala Lumpur (Malaysia)	March/April	Defense
Dubai Airshow	Dubai	November	Aeronautics
Expo China	Peking (China)	October	Aeronautics and equipment for airport
Fidae - Feria Internacional del Aire y Del Espacio	Santiago (Chili)	March/April	Aeronautics and space
Heli Asia	Kuala Lumpur (Malaysia)	October	Helicopters
Helitech Latin	Sao Paulo (Brazil)	December	Helicopters
IAA-INFRAERO	Rio de Janeiro (Brazil)	November	Aeronautics and space
Idef - International Defense Industry Aerospace & Maritime Fair	Ankara (Turkey)	September	Defense, aerospace, maritime
Idex - International Defence Exhibition & Conference	Abu Dhabi (United Arab Emirates)	March	Defense
Inter Airport	Singapore	April	Airport
Korea Aerospace & Defense Exhibition	Seoul (Korea)	October	Aeronautics and defense
LAD - Latin America Defentech	Rio de Janeiro (Brazil)	April	Defense
Lima - Langkawi International Maritime & Aerospace Exhibition	Langkawi (Malaysia)	November	Aeronautics and maritime
Maks - Mos aeroshow	Moscow (Russia)	August	Aeronautics and space
Security Israel - Middle East Security and Defense Exhibition	Tel Aviv (Israel)	July	Military aeronautics
Tate - Taïpei Aerospace Technology Exhibition	Taipei (Taiwan)	August	Civil and military aeronautics
Tokyo Aerospace	Tokyo (Japan)	March	Aeronautics
World Airline Entertainment	Brisbane (Australia)	September	Aeronautics and space

Table 12-2. The main aeronautics shows in the Middle East, South America, Asian Pacific region and Africa

12. Selecting Media

Numerous international shows are held in Europe (Table 12.6) and in the United States (Table 12.3), the main aeronautics and space market. In general, in the United States these events have a more national dimension.

Show	City	Period	Central themes
AeroSense	Orlando	April	Aerospace and military
Aerospace North America	Seattle	September	Aeronautics and space
Aerospace Technology Exhibition	Washington	September	Aerospace and military
AFCEA	Augusta, San Diego	December / January	Military
Aircon	Edison	September	Aerospace, aeronautics
ASME Turbo Expo, Land Sea & Air International Gas Turbine and Aeroengine Congress and Exhibition	New Orleans	June	Engines
GATC (General Aviation Technology Conference and Exposition)	Wichita	April	Aviation
Heli-Expo	Dallas, Los Angeles, Las Vegas, Orlando, Anaheim	January / February	Helicopters
Inter Airport	Atlanta, College Park	September	Aeronautics and space
NBAA	New Orleans	September	Business jets
Pama	Fort Worth	May	Aeronautics maintenance
Sea-Air-Space Systems and Technology Exposition	Washington	April	Technologies, military
World Space Congress	Washington	September	Satellites, launchers

Table 12-3. The main aeronautics shows in the United States

Usually, shows in the aeronautics and space sector (civil and military) take place every 2 years. However, there are a few annual shows such as Sea-Air-Space (Washington) or Security Israel (Tel-Aviv). Attendance varies considerably from one show to another, from several thousands for regional shows (12,000 for Heli-Expo) to almost 250,000 (including 120,000 professionals) for the Paris Air Show.

In addition to the general shows, there are more specialized ones such as:
- Heli-Expo (Dallas, United States) and Helitech (Red Hill, United Kingdom) for helicopters,
- Ila (Berlin, Germany) for satellites,
- Idex (Abu Dhabi) for defense equipment,
- Euronaval (Paris, France) for naval armaments,
- Air Cargo Forum (in Hong-Kong in 2002),
- NBAA convention (US National Business Aviation Association) and the EBACE (European Business Aviation Convention & Exhibition) for business jets (Figure 12.4).

Figure 12-4. The NBAA Business Jet Show at the Lakefront airport in New-Orleans[2]

There are two main groups of professional visitors at trade shows:
- *The users*: the airlines, Ministry of civil protection, Ministry of defense, private cargo companies, national space agencies, telecommunications companies and leasing companies. These organizations are represented by the pilots, the operating departments, the maintenance departments but also general management and the main shareholders.
- *The influencers*: governmental organizations such as the FAA, DGAC, CAA, journalists, pilot unions, airport management, the main leasing companies and financial organizations. They often have to advise customer companies on the latest technological market trends.

Visitors attend shows to learn about the market and the competition, discover new products, look at equipment on the ground, attend air demonstrations and inform themselves before making a purchase. Visitors not only make contacts with suppliers but also with customers and potential customers.

12. Selecting Media

Aeronautics shows are also important for commercial deals, prospecting and exchanging information. During shows important contracts (which have been prepared months ahead of time) are signed and spectacular announcements are often made (an alliance, creation of a subsidiary, buyout, new program...) (Figure 12.5). In fact, the presence of the leading players in the sector and the extensive media coverage given to such events offer excellent advertising opportunities for companies.

Figure 12-5. An ad for the British show at Farnborough[3]

Besides the major shows, there are other professional events that allow people within the sector to meet such as conferences and seminars. The latter give experts the opportunity to express themselves and discuss different technologies and the latest developments and trends. Companies in the sector are attracted by the quality of the information presented. These events are very important especially for professionals who have few specialized shows in their specific activity. This is true for companies in the satellite and space sectors, which can attend important conferences such as those run by Satel Conseil in Paris, Satellite 99 in the United States, the AIAA (Aeronautic Advents Astrological Association) in the satellite sector or the Joint Propulsion Industry in the space sector. There are many such conferences in the United States, which sometimes offer the possibility of presenting companies and displaying products. Conferences can be organized by customer companies, at which suppliers are encouraged to speak as experts.

Show	City / Country	Period	Central themes
Aerofair	North Weald (United Kingdom)	September	Aviation
Aerosalon	Prague (Chek Republic)	September	Civil and military aeronautics
Baltic Defence International	Gydnia (Poland)	September	Aeronautics and space
Central European Defence Equipment & Aviation	Budapest (Hungary)	November	Civil and military aeronautics
Defendory	Athena (Greece)	October	Defense
Ebace (European Business Aviation Convention & Exhibition)	Geneva (Switzerland)	April	Business jets
Euronaval	Paris (France)	October	Military armaments
Eurosatory	Paris – Le Bourget (France)	June	Defense
Farnborough International	Hampshire (United Kingdom)	September	Aeronautics
Helitech	Red Hill (United Kingdom)	September	Military and civil helicopters
Hemus	Plovdiv (Bulgaria)	May / June	Defense
Idet	Brno (Slovak Republic)	May	Defense
ILA (International Aerospace Exhibition)	Berlin (Germany)	May / June	Aeronautics, space, civil and military
Inter Airport	Frankfurt / Munich (Germany)	September / October	Airport technology and services
Le Bourget / Paris Airshow	Paris – Le Bourget (France)	June	Aeronautics and space
MSPO	Poland	September	Military, Defense
Passenger Terminal Expo	Cannes (France)	February / March	Airport design, technologies, services
Unmanned Underwater Vehicles Showcase	Southampton (United Kingdom)	September	Aeronautics and space

Table 12-6. The main aeronautics shows in Europe

1.2 Exhibiting at a show

The main constructors and contractors participate in most of the big shows. Parts suppliers generally participate in the leading shows in their geographical region. Both parts suppliers and constructors have many objectives when they exhibit at shows (Table 12.7).

12. Selecting Media

> Buying goods and services
> Developing a commercial relationship and facilitating sales (preliminary talks to facilitate the future sales at low cost)
> Developing corporate image (presence of the trade press...)
> Evaluating market prices
> Identifying the competition's strategy, gathering market information, new trends and innovations
> Introducing new products, testing new concepts
> Motivating the sales force
> Developing or strengthening contacts with prospective customers, customers, influencers, suppliers, distributors and partners
> Breaking into or prospecting a new market (in particular an international one)
> Giving visitors a chance to discover solutions offered
> Presenting and developing sales arguments for a rather wide, interested public
> Presenting very large usually untransportable equipment
> Meeting professional organizations, official or public institutes (public relations)
> Selling goods and services

Table 12-7. The main reasons for participating at a show

In general there are four main factors which lead companies to participate in a show:

- *As a tool for corporate communication policy:* In terms of image and awareness, a show allows the company to affirm its presence on the market, communicate its main messages and increase awareness by the different targets (organizations, public authorities, shareholders...) Generally, the top management in different companies in a sector as well as many political figures attend shows thus heightening the importance of the event. As a form of direct communication media, shows are a key opportunity to diffuse messages and gather information (customers, competition, technology).

- *Shows are also an excellent showcase for a new product or innovation.* Press conferences allow companies to present their latest developments. Aeronautics shows are also used for product demonstration with often spectacular air exercises designed to highlight the performance of the equipment presented. They are an opportunity to present scale one models of projects underway (Figure 12.8). Thus thanks to the models displayed, constructors allow visitors to see the cabin, test the seats or define the position of the kitchen or toilets.

- *Negotiation:* within the framework of negotiation, shows often serve as a deadline for carrying out the transaction. In addition, shows often initiate new commercial negotiations.

- *As part of the sales action plan shows are extremely important in stimulating sales and improving customer relationships.* Thus, certification flights are programmed to be carried out before or during shows.

Figure 12-8. At the Embraer stand, Paris Air Show: possibility of visiting the cabin of the Brazilian constructor's latest project, the regional EMB-190[4]

THE PARIS AIR SHOW - LE BOURGET: THE LEADING INTERNATIONAL AERONAUTICS AND SPACE SHOW

History

In 1908 the first aeronautics show was held in Paris at the Grand Palais as part of the second automobile show. The following year, the first show entirely devoted to aviation was held and from 1924 onwards German and English constructors exhibited in the show which took place every two years. 1946 marked the first flight demonstrations that were performed at the show which five years later would be held in Le Bourget and would rapidly become one of the most well known aeronautics platform in the world. In 1953, for the twentieth show, Le Bourget revealed a global configuration and an organization close to that of today's. This show became more and more international with the presentation of numerous prototypes and products from different countries.

The show today

With over 1,880 exhibitors from 45 countries and over 250,000 visitors (of which half are professional), the Paris Air Show is the number one aeronautics event in the world. The exhibit space is 50,000 m² for the inside stands and 30,000 m² for outside allowing the display of 200 aircraft. There are 450 châlets in which exhibitors can receive customers who can watch flight demonstrations while getting something to eat. The French president inaugurates the show on the first day and the Prime Minister closes it on the last.

12. Selecting Media

Figure 12-9. The Paris Air Show[5]

The show often marks the signing of contracts which have been negotiated months ahead of time. For example, there were over 18.3 billion euros of contracts signed during the first 4 days of the 43rd show. The announcement of these contracts benefits from the powerful media attention given to the show. As the show is a worldwide event, both the trade and general press give it extensive coverage. This is why constructors also use the show to announce commercial agreements with other constructors and parts suppliers, the launching of new programs, major innovations and new generations of aircraft or launchers. During the show, both professional customers and the general public can attend spectacular flight demonstrations (70 daily presentations) in particular of military aircraft (Table 12.10).

13 h 03	Su 26 M	14 h 29	A 340
13 h 07	CAP 232	14 h 35	Joint presentation DASSAULT
13 h 13	A 119 KOALA	14 h 48	K-8 KARAKORUM
13 h 19	Autogiro MAGNI	14 h 53	TYPHOON (Proto DA 7)
13 h 23	YAK 52W	15 h 00	DO 328
13 h 29	NH90	15 h 05	F 16
13 h 33	Joint presentation EUROCOPTER	15 h 11	C-130J HERCULES II
13 h 43	PROTEUS	15 h 16	GRIPEN
13 h 48	SF 260E	15 h 22	Joint presentation EMBRAER
13 h 53	MB 339 CX	15 h 30	B 717 BOEING
13 h 58	AEW 8C ERIEYE	15 h 36	MIG-21bis LANCER III
14 h 03	Joint presentation AERO VODOCHODY	15 h 44	PC-9M SWIFT
		15 h 53	APM 20-1 LIONCEAU
14 h 11	C 295	15 h 58	AT-802 AF
14 h 16	MIG AT	16 h 03	SWIFT S1A
14 h 21	Joint presentation SOCATA	16 h 10	SCUB (airplane)
		16 h 16	PARAMOTEUR BIPLACE

Table 12-10. A schedule of flight demonstrations for a day at the Paris Air Show

1.3 The different stages of participating in a show

A show is an event which must be prepared, skillfully managed and followed up. First the show's opportunities and constraints in terms of coherence, costs, possible feedback and budget among other things must be evaluated. A company cannot exhibit in all shows all over the world and must select those best suited to its interests (market penetration, consolidation of a position...). It must also evaluate the risk of not participating in a show. If a show is a must on the market, the absence of a company can be interpreted as a sign of financial problems, a non renewed product portfolio or a lack of customer relations.

Of course companies need to verify the costs of participating. Exhibit costs depend mainly on geographic location and the cost of a stand (variable depending upon the prestige of the show). In general, the biggest expense is renting the stand and providing food services followed by designing the stand, travel cost and sales promotion (Figure 12.11).

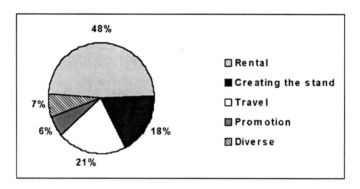

Figure 12-11. The budget breakdown for participating in a show

After having decided to exhibit and choosing the most suitable shows, the type of representation needs to be determined: direct or indirect exhibit, sharing a space with a customer, a supplier, an organization, etc. Then the main thrust of the company's message needs to be defined and the show has to be prepared on an operational level: position of the stand, selection of people in charge, documents, creation of the model of the stand, defining the necessary resources, the in-house aspect of communication, planning of events etc. To optimize participation, it is essential to ensure enough promotion to create traffic at the stand (press ads, a preview of products reserved for the specialized press, customer and prospective customer mailing...). In this way, combined with public relations, direct marketing and press relations, the show is profitable and efficient, benefiting from communication synergy (Figure 12.12).

12. Selecting Media

Figure 12-12. Sticker announcing Dassault's participation in Paris Air Show[6]

During the show, the exhibiting company must always keep in mind that the customer has made an effort to visit his stand and must receive an excellent welcome. Sales representatives must be present to inform customers about the technical aspects of products. Special events (contests, a draw, awards for innovation or quality, VIP dedications, conferences/debates, cocktails) attract visitors. It is also important to make sure that throughout the show the press department is supplied with enough company documentation.

Cost of the contact	$\dfrac{\text{Total cost}}{\text{Number of potential customer visitors}}$
Profitability	$\dfrac{\text{Amount of sales made (show + 12 months)}}{\text{Total cost}}$
Attraction of the stand	$\dfrac{\text{Number of visitors to stand}}{\text{Number of visitors to show}}$
Personnel yield	$\dfrac{\text{Number of potential customer visitors}}{\text{Number of salespeople} \times \text{number of days of the show}}$
Yield in terms of new prospective customers	$\dfrac{\text{Number of new prospective customer visitors}}{\text{Number of salespeople} \times \text{number of days of the show}}$

Figure 12-13. Evaluation of costs and results of participating in a show

After the show; it is important to analyze results and exploit new contacts. Using information gathered at the stand, it is essential to thank visitors and as quickly as possible to send them the documents required. It is possible to use awards obtained during the show for quality, innovation or design in a press campaign or for in-house purposes. It is difficult to analyze the results of a show in terms of profitability

358   *Aerospace Marketing Management*

especially given the fact that contacts only begin to pay off several months later. The show should be assessed qualitatively and quantitatively in relation to the determined objectives. Ratios such as profitability to contact help in evaluating the show's results (Table 12.13).

2. THE TRADE PRESS

2.1 Main characteristics

Read by professionals, the trade press includes all those publications in a specialized area. For its readers, it is a useful and practical information tool. Publications can include magazines, newspapers, reviews, specialized directories or confidential letters. Companies use this means of communication mainly to place ads for product campaigns. But corporate campaigns similar to those which appear in the economic and general press are also sometimes carried out. Trade press advertising is generally affordable. The average price of an advertising page can vary depending on the sector from 2,000 to 25,000 euros. This depends mainly on the number of publications available per sector, their audience, real distribution rate (number of issues sold) and the circulation rate (average number of readers per issue). The trade press generally has a good quality to price ratio given its targeted advertising (activity sector, determined function).

Most aeronautics and space publications come in paper format and are updated via the Internet. The aeronautics and space press reflects the international character (exchanges and partnership), size and complexity of this industrial sector (number of actors). Most of the weekly or monthly, generic and highly specialized aeronautics and space publications come from the United States. The American press provides very rich, up-to-date information, read the world over by professionals. Generally publications such as Aviation Week, Defense News, Space News, Jane, etc. work with journalists all over the world. In France and in French speaking countries, the main magazine is the weekly Air & Cosmos. In the United Kingdom, the main publication is the weekly Flight International which was published for the first time in 1908. The magazine Jane founded in 1898 and devoted to aeronautics since 1908 is also considered to be a reference in the aerospace sector in particular in the defense field.

AVIATION WEEK & SPACE TECHNOLOGY

In the aeronautics and space sector, Aviation Week & Space Technology is the most widely read publication in the world. Founded in 1916 and belonging to the McGrawHill group, Aviation Week has many offices all over the world (Washington, Los Angeles, Cape

12. Selecting Media

Canaveral, London, Paris, Moscow, Ankara, Yokohama and New Delhi). Published in English, the magazine has an editorial staff made up of people who have worked several years in aeronautics before becoming journalists (former pilots, engineers...). The review covers aeronautics, space and military news (Figure 12.14).

The publication has no limitation on pages, and content varies depending on the news. The review is longer during important shows such as Le Bourget or Farnborough. In the beginning of the year, a special 500-page issue is published to provide a complete presentation of products launched in the world and a panorama of the sector. Special issues are published to handle a specific theme such as space launchers.

Figure 12-14. Aviation Week & Space Technology

Aviation Week also publishes magazines, newsletters and directories and it offers a diversified range of media (Figure 12.15):

- Magazines: in addition to the horizontal publication, Aviation Week, there are vertical publications such as Business & Commercial Aviation and A/C Flyer devoted to business aviation, or Overhaul & Maintenance for the maintenance manager. Magazines in specific languages are also published such as Air Transport Observer in Russian or International Aviation in Chinese,
- Newsletters: Aviation Daily for air transport, Aerospace Daily for the military and space sectors, along with Airports, ATC Market Report, and Weekly of Business Aviation for vertical sectors of the market.

- Directories: World Aviation Directory published twice a year, the Cd-Rom Direct Select and the Aerospace Source Book.

- A complete media range: The publication produces specialized issues specifically for customers and prospective customers of a particular company. This communication can extend to radio, television or internet advertising. It is possible to sponsor programs produced by Aviation Week and shown on networks such as the Discovery Channel or Arts & Entertainment. Aviation Week participates in and organizes conferences and events ideal for meeting customers (Aerospace Expo, MRO, Space Business & Technology, Global Competitiveness Summit). Internet sites complete their range such as aviationnow.com or wadaviation.com.

Figure 12-15. Other publications in the group: Overhaul & Maintenance and World Aviation Directory

---

In general, there are two types of trade publications.

Horizontal publications: These include the general technical press read by individuals in different activity sectors such as Plant Engineering, which is of interest to people in the chemicals sector as well as aeronautics, the automotive industry, the mechanics industry etc. Companies that sell products to numerous firms in very different sectors advertise in these publications which are a good way of reaching a wide range of targets. They are widely used by airlines as an easy way to reach executives and by suppliers to reach the different departments of their customers.

Vertical publications: These include the specialized trade press and are of interest only to people within one sector (Journal of Electronic Defense, Jane's Airport, Rotor and Wing, Space & Communication, Air & Space Europe and Air Cargo News). Vertical publications are selective and efficient in terms of cost. They are a means of addressing specific targets depending on the products and messages promoted.

12. Selecting Media

Figure 12-16. Example of vertical trade publications: Space & Communication and Air Cargo News

There are other sector-based specialized publications providing both general and highly technical information (publications devoted to military activities, civil aviation, helicopters, telecommunications and electronics). It also exists very high technical publications or newsletters such as Advent Composite, Performance Material or Electronic Defense. While the aeronautics sector has a large number of publications, the military sector has fewer owing to the highly confidential nature of the information. The space sector falls between these two extremes in terms of number of specialized reviews (Table 12.17).

Aeronautics	Space	Military	Helicopters
Flight Aviation Week & Space Technology Air & Cosmos	Air & Space Europe Aerospace Daily	Journal of Electronic Defense Jane's	Rotor & Wing
Air Transport World Air Cargo News Aviation International News Airline Business Aircraft Economics Airlines Avionics Aviation Daily Aero International Aéronautique Business Air International Business Air News World Aircraft Sales World Aircraft	Space News Satellite Finance Space Technology Space & Communication International Space Industry Report	DGA Armada Armement Defense News Military Technology Defense Helicopter Armed Force Journal	Helicopter World Defense Helicopter Helicopter International Helicopter News Helidata

Table 12-17. Aeronautics and space trade publications: aeronautics, space, military, helicopters

Among other specialized reviews are the American monthly, Air Transport World, which covers international airport and aeronautics news or the American weekly, Aviation International News. Airline Business, published monthly by Reed Business Publications, provides information on the airlines and the cargo market.

The general press: It is used by companies to complement their professional campaign or by travel agencies to directly reach consumers. This includes the daily press (short lived advertising) as well as monthlies, periodicals and weeklies (which are saved and re-consulted longer). The regional and national daily press is fairly flexible to use (very short reservation deadlines) and provides very wide market penetration. It is mainly used for informative advertising (Table 12.18).

Advantages	Disadvantages
Wide audience for general dailies	Weak targeting according to social-demographic characteristics
"An official media form" providing legitimacy	Fairly limited geographic range
Close relationship between a journal and its readers: local information, practical life	Power of the media limited by different linguistic zones
Short reservation deadlines and highly reactive for urgent advertising	Short lived nature of a daily which is not re-read often
Suitable for promotional operations	Reproduction quality limited by the quality of the paper or printing constraints

Figure 12-18. Main characteristics of the daily press

Advertisements often present a new activity, or product, an exceptional result/profit or a particular event. For an industrial company this could involve affirming an ecological position or a positive role in the community. This is true of Boeing which used a national daily press campaign in Europe to show the important role played by its local economic partners in the building of its aircraft, and in turn the volume of work and wealth created. For their corporate campaigns, aeronautics companies favor general international publications such as the International Herald Tribune, Financial Times, Asahi Shimbun, and the Wall Street Journal. They also use international magazines such as Fortune, The Economist, Time, Newsweek, L'Expansion, etc. Magazines target a selective audience, offer credibility and a certain prestige but also a longer life time and better printing quality (Table 12.19). However, the cost price is higher and reservation deadlines are tighter.

Airlines use the general press extensively as well as economic magazines, which reach numerous executives. Generally they focus their campaigns on the product range (prices, in-flight services) displaying their positioning.

12. Selecting Media

Advantages	Disadvantages
Better targeting	More limited audience than dailies
Important credibility through specialized interests (tourism, leisure...)	Higher price for consumers
Stronger reader-magazine relationship	Longer reservation and technical deadlines
Longer lifetime, magazines are re-read, kept, lent and the readership rate is very high compared to the distribution one	
High level of visual quality: the possibility of more effectively arguing product advantages. Allows more complex advertising	
Good media for promotional operations (coupons...)	

Table 12-19. Main characteristics of the magazine press

Thus the campaign made by EuroRSCG Babinet Erra Tong Cuong for Air France affirmed the key points of the company's positioning: serenity, pleasure, proximity, accessibility and modernity. Awarded the Lion d'Or at the International Advertising Festival in Cannes, the slogan of the campaign was "Making the sky the most beautiful place on earth"; it placed the passenger at the heart of the company's concerns (Figure 12.20).

Figure 12-20. Air France's corporate campaign: "Making the sky the most beautiful place on earth"

It should be noted that the economic press (Fortune, Business Week, Financial Times) is often included in the trade press, even if its readership is much wider. The economic and financial press is used for corporate campaigns, prestige operations for shareholders (acquisitions, mergers, profits) or public authorities (ecological actions, contracts). The tone is often much more general, less targeted and therefore less efficient (higher cost price).

2.2 Resources and tools

The company can be in contact with the press through the purchase of advertising space, the organization of interviews and press conferences, "editorial-style advertising" as well as press releases. Communicating with the press is essential and is not to be improvised. The company must build up a file of journalists and newspapers, organize press conferences and press releases. It must manage its press relations and have different tools to project its messages.

The traditional insertion: A half page (or a page, two pages...) in the journal, review or magazine transmits the message delivered by the company. This type of paid insertion is widely used.

A full page insert: It is a less costly solution than the preceding one. It is widely used to announce major events such as a show.

Editorial-style advertising: It is half way between an article in its form and advertising in its content. In fact, the text resembles any press article but the content is totally controlled by the company. It buys the space for its advertising report as it would for a traditional ad. The company produces an "ideal article" while benefiting from the credibility of the publication. Advertising brochures are entirely dedicated to the company and created by it, published by reviews and magazines under their name. This is true for example for Airbus which created a special dossier on the environment published by Time. The review Avionics Magazine also offers special editions (offprints) which are an efficient marketing tool to position a company with a product. A good means of communication, they can be given to influencers and customers, lending credibility to the company.

Technical articles: This can be articles written by the journalist and based on information provided by a company. The feedback depends mainly on the original and innovative character of the content of the article. For very specialized subjects, the article can be written by a confirmed expert in the sector (research worker, pilot). Such articles generally present an analysis of current products, technologies and processes with their advantages and disadvantages. This technical credibility can then be used as a bridge to the communication of the chosen product: the highlighted quality of the new product will echo the article. Technical articles are informative and aimed at influencers via the specialized press. They provide information on new product applications, ongoing research, patents etc. The presence of these articles is left to the discretion of the editor-in-chief and requires great skill. Generally the article improves the company's technical image and pre-sells its products, reaching the people who influence the purchase. They are a source of important information for customers and are published in magazines read by industrial clients.

Professional directories: It is essential to appear in the diverse professional directories (classified by technique, branch...). This official presence is a reference in the eyes of other professionals. The World Aviation Directory is an example. With the development of the Internet, a company must also be listed in electronic directories and its site must be listed in the main search engines.

Company news magazine: This type of news magazine is created by companies themselves. It is often sent free of charge to customers and/or partners (suppliers, shareholders, influencers, the media...). It is designed like the traditional general magazines but it is published, prepared and written by the company itself. It generally deals with themes linked to the company's sector of activity (new developments, markets gained, customer successes) but also "general public subjects" (information collected during shows...). A vector of communication and information on the group, this news magazine can include external advertising. It aims to provide objective information while convincing readers of the validity of its choices, actions or products. This "customer review" should not be perceived as only advertising for the company that produces it but also as a real tool providing useful information. This form of trade press is very common in business to business, it is both a tool for communication, promotion and information.

Very targeted, these "customer reviews" are produced in magazine form or as an information letter (paper, Cd-rom, Internet). For example, Airbus published several publications such as the bi-yearly newsletter, The Link, focusing on services provided to customers (training, flight operations, maintenance training...), the semi-annual Above and Beyond for its North-American customers or the Bi-annual Fast Magazine (Flight Airworthiness, support and technology) for pilots[7]. Dassault publishes Falconer, which presents its business jets and is aimed at partners (Figure 12.21).

Figure 12-21. The Dassault Falconer

In the space sector, Alcatel publishes the quarterly review, Alcatel NewsLink (Figure 12.22), which features current news, markets and technologies and has columns for its customers (special dossiers). The quarterly review, Alcatel Space, is more technical with scientific articles. Spot Image produces Spot Flash, a quarterly information letter and Spot Magazine, a twice yearly publication to inform readers about new products, services and applications. Company internet sites are increasingly used with "info flash" or media centers to diffuse information.

Figure 12-22. NewsLink from Alcatel and Snecma Infos

Companies in the same sector can join together to publish a magazine such as Orient Aviation. The latter is produced by a group of Asian Pacific airlines including Air New Zealand, Air Niugini, All Nippon Airways, Ansett Australia, Asiana Airlines, Cathay Pacific Airways, China Airlines, Dragonair, EVA Air, Garuda Indonesia, Japan Airlines, Korean Air, Malaysia Airlines, Philippines Airlines, Qantas Airways, Royal Brunei Airlines, Singapore Airlines, Thai Airways International and Vietnam Airlines.

3. THE INTERNET

The main characteristic of the Internet lies in the technical integration of media which until now have been very distinct: the press, video, sound, the telephone, mail, directories, pictures... This makes it possible to communicate in very diverse forms: from simple texts accompanied by pictures to the showing of video clips, direct interviews or the sending of e-mails with attached files.

The other specificity of this media is the active role played by the user (questions asked on the site, recommendations to other Internet users, expression of opinions in discussion forms, public actions with the creation of protest sites or pages...). On the Internet, there is a very strong interactivity between the sender and the receiver.

However, this fast-developing media does not yet have as large an audience as that of the traditional media (Table 12.23).

Advantages	Disadvantages
Interactivity	Limited audience
A complete media (sound, pictures, video, text): allowing complicated messages	The general public is still not fully equipped
Practicality: permanent access, ability to quickly and freely download information	The Internet user profile is still specific and limited (socio-professional category, age, nationality...)
Specific targeting of professionals	The development of difficult to control non-official sites
Low cost of occasional communication operations (purchasing of space, banners...)	The high cost of e-based communication: dedicated organization (database and processing/updating of the site, answering messages)
A short reservation and technical deadline	
Used by the professional target	

Table 12-23. Main characteristics of the Internet

Internet communication is thus characterized by the strengthening of links within a community (customers, subscribers, employees...) and by a collaborative approach. It is part of the development of customized products or services and customer relationship management. Instead of the traditional communication based on a monologue (graphic or TV-advertising), interactive communication is developing based on the dialogue between the company and the customer (as long as a forum or an e-mail address exist). This interactivity and personalization have given rise to a unique culture with its norms and linguistic and symbolic representations (Nicovitch and Cornwell, 1999), thus net-etiquette sets the rules for discussion forms with their own Internet jargon (BBL for "be back later", ☺ smileys...) and a very direct linguistic style.

In the aeronautics industry (Herrera, 2000), Internet was first used for Intranets and later for corporate communication. The airlines have progressively used Internet for the on-line sale of tickets. Now they can find builders and suppliers on the market place for the purchase and sale of plane parts and services. They have also joined together to create their own marketplaces which allow on-line reservations and they offer e-commerce services (virtual travel agencies...)[8]. This is the case of the announcement on the Internet of the alliance of British Airways, Air France, Lufthansa, Alitalia, KLM, Iberia, SAS, Aer Lingus, Austrian Group, British Midland and Finnair. The latter will form a virtual travel agency to decrease the cost of ticket sales and distribution.

While the Internet can be a means to sell[9], it is mainly a communication tool. It is a worldwide interactive media allowing both mass and personalized communication. There are several communication modes on the Internet, depending upon whether it is one-to-many or many-to-many. There are two major types of company sites (Figure 12.24): presentation sites and information sites.

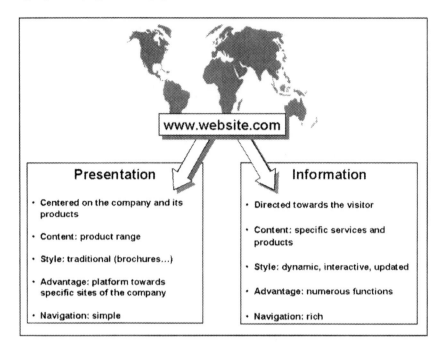

Figure 12-24. The two types of company sites

Presentation or functional websites: These are equivalent to general brochures presenting the company with its products and services (location, turnover, number of employees, history, financial information, products, press releases). Owing to its light content, this type of site is generally simple and quick to access. It often regroups the links of all the company's subsidiaries and is a useful navigation platform called a portal. Functional sites are not meant to be the main communication vector for the company. They are built around a central service ; it can be a thematic search engine for media sites, a form to fill in for reservations (airlines) or to obtain specific information (for example, tracking).

Information websites: These also provide information on the company and its products and services. They offer more technical information and aim to meet the demand of users via a search engine or high added value press kits. They analyze the actions of the user and determine his profile. These sites offer the user different services, some of which may be far from the company's major activity. For example, the cargo company Atlas Air offers a range of consumer products such as models of its planes stamped with its brand, t-shirts, caps, office or travel supplies (pens, watches, mugs…).

BOEING AND THE INTERNET

The aircraft builder Boeing has one of the richest Websites in the sector (Figure 12.25). Access from the home page is by clicking on the corresponding photo: commercial aircraft, business jets, military aircraft, space systems, electronics and information systems, missiles and tactical weapons, rotorcraft, satellite systems. Pride of place goes to the day's news item and in addition, the visitor can access other parts of the site by clicking on the headings: News, Exostar, Employment, associated products, the Boeing store, Investor relations, Corporate secretary, Capital Corp., Realty Corp., Search.

At the bottom of the home page there is the stock price (with a 20 minute delay) which allows shareholders to gain time when monitoring their shares. The business visitor can click on Customer Logon to open a specific page using an access code.

Figure 12-25. Boeing's home page

The visitor can also get more information and access another page where there are details of Boeing and its history. There is a product photo gallery, an education page (relations with schools and training organizations), a business page (supplier relations, a citizens page (teaching, environment, art and culture, voluntary work, contributions to charity, etc.), a general information page (boeingmedia.com, aircraft orders and delivery, employment, management biographies, an on-line revue: Aero Magazine, recent speeches, etc.). Finally there is a special page just for children: the Boeing Kids Page. Here there are games (mazes, crosswords, drawings to be made) or pictures with an aeronautical theme, to cut out or color (Figure 12.26).

Figure 12-26. The Boeing Kids page

Basically, in addition to its official site, the company can be accessed on the Internet:
- by being keyed on a maximum number of search engines,
- using links from suppliers, customers, partners or subject sites (specialized trade press, aeronautics shows organizers, etc.),
- via events which it organizes: chat, web conferencing, etc.
- by visibly supporting organizations or associations present on the web (pilot's, trade, top graduate schools and university associations, etc.),
- by buying advertising space on high visit, target sites (aeronautics and space information sites, marketplaces, portals, etc.) using pop-up adverts, banners, etc.

Wherever it appears the company must make sure that the messages have an overall coherence about them so as to derive the maximum benefit (image and awareness) from the advertising. There must also be a specific organization for this communication channel to keep the site functioning: updates, presentation modifications, adding services. One of the key elements is to pamper the Internet user ; the latter is often very demanding and wants the site to offer him something in exchange for the time taken to find and consult it. This means that any demand via the Internet must at the very least be recorded and taken into consideration. A reception acknowledgment is the baseline for the user and actually what he really wants is a personalized and intelligent rather than a standard reply to his demand. The company must show him that he is not just an IP number but in fact someone who they care about. This means for example keeping him up to date by sending him a mail or letter or giving him a call some time after his visit, providing information relevant to his original question. Respect is at the heart of the relationship with the user and it is measured in terms of the amount of attention paid. Powerful Internet user profiling tools like Datamining enable Customer Relationship Management.

4. DIRECT MARKETING

Direct marketing can be defined as a direct and interactive communication tool. It allows direct, targeted and customized relationships to be established with customer companies. Direct marketing facilitate a personal on-going dialogue allowing customer companies to react immediately (reply cards, prepaid cards, freephone numbers, etc.).

This direct, personal approach is particularly well adapted to business to business since one of the latter's characteristics is a restricted number of potential targets. Industrial customers with varied and complex needs can thus be alerted efficiently and benefit from precise and well-adapted information.

4.1 The objectives of direct marketing

Often presented as a sales tool, direct marketing is more than this. It is a flexible and rapidly deployed communication tool for strengthening image and awareness and stimulating buying intentions. It can be used to announce the launch of a product or the company's participation in a particular show. It is also often used to create traffic (invitations, making appointments, promotional operations, etc.). In this way direct marketing does not take the place of the sales team, rather it helps to convert potential customers into prospective ones and then the latter into actual customers. In view of the cost of sales visits, direct marketing actually is a relatively simple and effective method of sales prospecting (Figure 12.27).

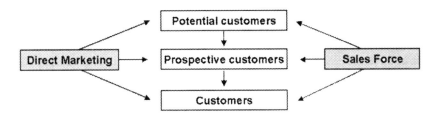

Figure 12-27. Direct marketing as a support for the sales team

Direct marketing tools require the use of a qualified database (customers, suppliers, prospective customers, influencers, etc.) in order to establish a relevant list of contacts. This requires an efficient information management system which uses general and specialized business to business files from professional bodies, specialized bureaus or professional publications. Indeed, direct marketing operations allow the databases used, to be updated (complementary information, segmentation, analysis of buying behavior, etc.).

4.2   *The different tools*

The medium used for direct marketing must encourage direct relations between the company and its customers or prospects. These operations leave customers the time to compare offers but also incite their active participation (telephone, read, cut out, etc.). Direct marketing can be in the form of mailing, telemarketing, e-buying, etc.

Couponing: This technique is not used in the aeronautics and space sector except in airline companies dealing with the general public (miles for frequent flyers, private cards, etc.). The use of coupons (inserted into adverts, in the press, in a prospectus distributed during a show) basically allow new contacts to be assessed and named. They can also promote a service or product directly to a target, by generating information feedback (offers of catalogues, brochures, sales visits). Replies from customers and prospective customers allow a file to be built up rapidly and cheaply and permit statistical analyses (profiles, etc.) to be made.

Mailing: Like couponing, mailing is primarily used with the general public, e.g. airline users. This type of posted document, usually has an envelope, a letter, a leaflet, an order form, an envelope for the answer and various sales supports (CD-Rom, DVD, etc.). It gives the company the possibility of personalizing the message and selecting the target. It has many objectives: it can just as easily be used to make tests, to build up a file, to obtain a direct order, as to qualify prospective customers. In business to business, it is often used in conjunction with the trade press and shows, to announce the launch of a product or participation in a professional event (product or material trial offer, coupled with participation in a competitive game). In order to reduce the cost of useful contacts, and to improve its yield, mailing must rely on good files, with good, up-to-date information as well as on a professional presentation so as not to detract from the company's credibility.

Bus-mailing: This is a grouping together of several products or services aimed at target prospects. The bus usually consists of 15 to 25 complementary, non competitive offers in the same pack. It offers good selectivity for a low unit cost, and the possibility of making tests and setting up a file by encouraging requests for information. Its easy and speedy formulation, the attractive aspect of the packet where the prospective customer can run through the cards, gives a spontaneous reaction and a rapid return. In aeronautics, bus-mailing is used to promote services such as the transfer by helicopter between the airport and downtown, and also for air freight. Just as for mailing, its profitability can be calculated by comparing the cost with the number of coupons returned.

Videos: videotapes, CD-Roms & DVD-Roms: Using a visual medium makes the message easier to recall. Therefore videocassettes are good for 3D demonstrations to show the technical development of the aircraft, or to broadcast testimonials from professionals in different countries. CD-Roms and DVD-Roms are easier to use and can hold more useful technical information (description of products, technical data sheets, diagrams, etc.) while at the same time giving a lively presentation of the products (audio, video, demonstrations, etc.). CD-Roms are mainly used as electronic catalogues presenting all the references in the company's range, with added commentaries or technical tools (Figure 12.28).

Figure 12-28. The Sea Launch CD-Rom

5. *TELEVISION, BILLBOARDS AND RADIO*

In general the three mass media, TV, radio and billboards are very little used in the aeronautics and space sector. They are more appropriate for airlines, which are looking to reach a larger target beyond professionals: the passengers.

5.1 *Television*

General TV has a triple role: to inform, to instruct and to entertain. The diversity of programs makes television the most appreciated medium by the different socio-professional categories.

Advantages	Disadvantages
Instantly powerful	Weak selectivity for socio-demographic characteristics
Wide audience but targeting possible thanks to theme programs	High costs: - buying time - shooting a film
Audiovisual gives strong demonstration value: moving pictures, colors and sound	Decreasing efficiency because of development of theme channels
Cautionary role	
Media which can stand alone or be used with magazines, billboards or cinema	

Table 12-29. Main characteristics of TV

Like the press, TV can be used in four ways:
- Buying time and broadcasting commercials especially in prime time to reach a wide target. An example is American companies like United Airlines or Delta Airlines which communicate in Europe using the message " the benchmark company for visiting North America ".
- Use of off-peak hours (paid) to broadcast a 30 to 40 minute program destined for shareholders and/or employees. This is broadcast several times to allow the greatest number of people to see it. This type of approach is used for important company events: launch of a new program, alliance with another company, etc.
- Sponsoring general public programs, whether the getaway sort or simply about nature. For example the top leisure tourism companies sponsor a program: a competition during the program has different holidays as a prize. In exchange for this participation, the company is highlighted in the presentation.
- Media spin-offs during TV news. Just as in the press, journalists will only include the company, when information is original and fresh.

Another original way to get exposure time for the company, its brand or product within a TV program or cinema film is the " tie in" or "product placement". This is when production of the film is helped by the supply of material or financial means in return for a worthwhile presence of the brand in the film. A good example was the presence of a Eurocopter Tiger helicopter used by the heroes in the James Bond, Goldeneye film.

AN ORIGINAL OPERATION: AIRBUS BELUGA "DELACROIX"

Airbus was associated in an original way with the Delacroix exhibition organized in Tokyo, in March 1999 by the Louvre Museum, under the auspices of " French year " in Japan. Airbus offered to transport the paintings (there and back) using a Beluga (A300-608 Super Transporter) fitted with a special compartment to avoid any vibrations and pressure differences which could harm the works of art. In order that the service from Airbus could serve as a support for a communication action on the event, it was agreed to reproduce the Delacroix painting, " La liberté guidant le peuple (1831) " on the fuselage (Figure 12.30).

On the outward journey, media cover was reinforced thanks to the reaction of the national Emirates authorities during a technical stopover, a case of " hide that bosom which I shall not see " (Molière, 1664) etc. Consequently a prudish veil was used to hide the nudity of liberty from the eyes of the airport technicians. On the return journey, the Beluga stopped over at Macao, Calcutta, Bahrein and Heraklion.

Figure 12-30. Photo of the "Delacroix" Beluga[10]

5.2 Radio

Radio in general is listened to a lot by individuals (as a complement to the press and TV), in the workplace or while traveling. For companies, it is an interesting medium especially for advertising, partnerships or coverage. Staff from the companies can for example participate in debates or reports/programs which allow their key messages to be favorably presented. Certain radios such as Radio Classic in France have been able to develop a clearly targeted audience mainly composed of managers. Which is why the airline company Cathay Pacific, has chosen it and why it sponsors certain programs. This airline reminds listeners in the initial announcement how it is the inescapable element for getting to the Far East: " The company which takes you to Hong-Kong and beyond ".

Advantages	Disadvantages
Radios: - possible to target wide geographical zones - age and socio-professional targeting possible using times and program information	Less credibility than the other media
Message repetition easily obtained	Excessive repetition of messages can cause listeners to get fed up and decrease effectiveness
Tactical media, reacts quickly to events	No visual element: demonstrations difficult
Campaigns quickly run	Messages are ephemeral
Cheaper than other media	
Short reservation time: usually one week	
Can be used for promotions: partnerships with advertisers and distributors easily obtained (airlines)	

Table 12-31. The main characteristics of radio

5.3 Billboards

This is the oldest media. Billboards can be fixed (walls or street) or mobile on normal vehicles (buses, trains, etc.) and specialized ones (delivery trucks, etc.). Aircraft can also serve as a support for the duration of a campaign, e.g. Star Alliance displayed on some of its aircraft the same visual as that used in its press campaign, repeating the logos of the other airline partners[11]. Similarly, easyJet uses its aircraft as an advertising support, with slogans or service offers (telephone number for reservations, Internet address, etc.)[12]. Billboards can either be made singly or in a network (when they are offered to advertisers). For network use, paper, plastic and synthetic materials are usually used (Figure 12.32) rather than " long life " billboards which are painted directly.

Figure 12-32. Example of outdoor advertising for the Chanel brand at London Heathrow airport[13]

12. Selecting Media

Spectacular, one-off displays are sometimes set up for an important event, using for example the space above a busy area (beaches, sports stadiums, etc.) with:
- an airship in the company colors, produced by Air Atmosphere (air-atmosphere-dirigeable.fr), Sky Power (skypower.co.uk) or Giant Advertising (giantad.com) ;
- an airplane towing an advertising banner, e.g. Gasser Banners (gasserbanner.com) or Imagesaloft (imagesaloft.com) ;
- an aircraft sky writing the company name or drawing the logo. These services are offered by specialized companies such as Airsign (airsign.com) or Skytypers (skytypers.com).

The main advantage of billboards is precise geographical targeting (Table 12.33) and the possibility of creating surprise by using original visuals or techniques (relief, lighting, animation, etc.).

Advantages	Disadvantages
Good cover of urban zones	Impossible to target demographically (age, sex, etc.) or socio-professionally
Very good targeting by geographical sector (districts, streets, etc.)	Difficult for demonstrations and sales pitch
Good for surprising and attracting attention: effective for launching products	Long reservation delays for periods required (e.g. end of year shows, etc.)
Increases renown	Insufficient when used alone
Well adapted for coordinated use with tour operators	High cost for downtown use and urban structures
Complements TV and the press very well	

Table 12-33. The main characteristics of billboards

6. *LOBBYING*

Getting the right decisions from public authorities can be facilitated using lobbying (Le Picard *and al.*, 2000): this means contacting very precise targets, capable of influencing the functioning of the market (public authorities, government, parliament, ministers, international organizations). Lobbying can be direct: face to face talks, interviews, publication of "white papers" or official reports, inviting politicians to visit the factory, sending out useful information such as company newspapers, technical specifications, etc., or congresses and symposiums can be organized with target personalities invited. Indirect actions can also be taken using the press, third party accounts, results from opinion polls, etc.

Lobbying is based on organizing activities which allow pressure to be brought on a single decision maker, a group of economic and/or political decision makers and/or an administration (Lehu, 1996). It is used when the economic interests of the lobby do not coincide with new regulations or legal measures being prepared. Action must therefore be taken to find out more and put pressure on those concerned.

LOBBYING THE REGULATORY ORGANIZATIONS

Lobbying is often used to influence projects aimed at modifying norms and regulations. Thus in 1999, there was a proposal from the American FAA which would have required all foreign airline companies flying into the United States to apply the same anti-terrorist measures as those in power for American companies[14]. The different aeronautical associations, the Asia-Pacific airlines and the Asian and European airports immediately reacted against this proposal (Table 12.34).

Governments	United Kingdom, Japan, Australia
Pressure groups	International Air Transport Association (IATA), International Civil Aviation Organization (ICAO), Association of European Airlines (AEA), Airports Council International (ACI), Association of Asia Pacific Airlines (AAPA)
Airline companies	Japan Airlines, All Nippon Airways, Air Canada, British Airways, Swissair, Virgin Atlantic
Airports	Germany Airports Association, Hong-Kong International Airport

Figure 12-34. The main opponents to the FAA proposal

For them, such a measure would have been very expensive (security equipment, checking procedures, fitting out security areas) and delay flights (checking individuals, luggage). The ICAO (Civil Aviation Organization) also considered that it would go against the spirit of the Chicago Convention (which is at the very heart of international air regulations), because the American measure would affect national regulations in other countries. In the light of these reactions, the project has been abandoned.

Another example illustrates the use of lobbying by aircraft manufacturers (Aubert, 1999). Boeing and Airbus confront each other in business, but also as far as regulations are concerned. Several, often technical debates, have occurred between the two groups whether about the amount of their respective government's grants or concerning sales at a loss. There is so much at stake commercially, that decisions on regulations can have enormous consequences on the sales claims for the aircraft offered. Because of this, the manufacturers inform the organizations in charge of regulations, as much as possible in an effort to influence their decisions.

A case in point is the three hour rule. This security measure requires twin-engine aircraft to always be within three hours flying time of an emergency airport that it can divert to in the case of problems with one of its two engines. Following a demand by the four American airlines, Delta, Continental, American and United, the FAA has re-examined this rule in order to allow twin-engine aircraft to cross the Pacific using a more direct route which takes them further away from emergency airports. The companies in question use the B777 twin which comes under this ruling, and thus use airports in the Aleutian Islands and other sites on the extreme east of Russia as emergency points. However in view of the often difficult weather conditions in this area, and in order to respect the three hour rule, the aircraft are often forced to fly even longer routes, thus extending overall flight time. A normal 11 hour flight from San Francisco to Tokyo takes twelve and a half hours, thus increasing running costs (fuel, etc.). And this is why the four companies have asked the FAA to modify security restrictions and extend the time to 207 minutes rather than the accepted 180 minutes. In this way they would benefit from a faster route over the Pacific regardless of weather conditions. However, this change would be a major disadvantage for Airbus in that they offer a rival to the B777, the four-engine A340 which, by definition is not constrained by the 3 hour rule and allows airlines to choose their own route. Obviously this is a very strong commercial argument which has often been decisive in making airlines choose the A340 rather than the B777. Boeing obviously backed its four customer companies to the hilt, presenting studies showing that in terms of safety, two, three and four engines are equivalent. In addition, the American manufacturer was supported by the main American pilots association who in return, demanded reinforcement of other safety measures. These debates concerning the regulatory bodies demonstrate just how much is at stake commercially and illustrate the pressures brought to bear, and the amount of power and influence at work between the manufacturers, the airlines and the pilots.

Although it is normally aimed at specific objectives, lobbying as a communications vector and by its mode of action can also resemble public relations. The latter usually uses different means to publicize the company and to create or maintain a favorable image.

LOBBYING FOR THE AIRBUS MILITARY COMPANY A400M PROJECT

Founded in 1999, the Airbus Military Company (AMC) groups together EADS as well as BAE Systems (United Kingdom), Alenia (Italy), Flabel (Belgium) and Turkish Aircraft Industries (Turkey). In order to answer the precise, joint needs of the 7 European countries, AMC has developed the 4 turboprop, A400M military transport aircraft, with high speed propellers (Figure 12.35). Its launch depends on firm commitments signed by the different member states.

However, the decision process of the national authorities concerning the procurement of military equipment, is particularly long and complex. Numerous socio-economic and political factors come into the decision. Each country has to take into account not only its military needs, but also the national industrial, economic and political context, its political relationships with potential suppliers and their country of origin, etc. Military orders like this involve a large number of actors, from the highest national authorities, the Defense Ministry teams, to the military authorities, the national political institutions (Congress, Senate, etc.), public opinion and industrialists. For a supplier, it is a question of rapidly obtaining favorable backing for their offer from members of the buying center. And it is within this context that AMC started lobbying right from 1999, in order to send the key messages to the members of the buying center.

Figure 12-35. The A400M military transport aircraft[15]

This lobbying campaign was aimed at accelerating the decision making process in favor of the proposed offer, and came upstream in the launch plan (Figure 12.36).

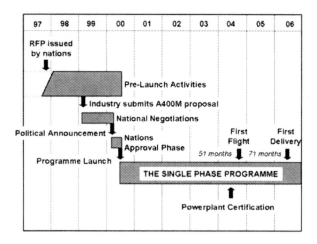

Figure 12-36. The A400M launch plan[16]

12. Selecting Media

In the wake of the Request for Proposal (RFP) at the end of 1997, AMC prepared the launch of the program by submitting the A400M proposal to the Defense Ministries of the 7 countries. At this point there was a clarification phase, during which there were intense negotiations between AMC and the military authorities. Then, with all the necessary elements at hand, the authorities evaluated the proposal with their different services (without there being any formal negotiations with AMC).

During this phase, AMC began its lobbying campaign in order to back up its proposal and convince the other parties within the project (Figure 12.37), with the objective of an official announcement in 2000. Only after the validation phase by each country could the actual program be launched.

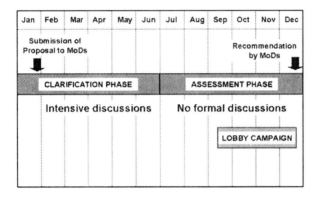

Figure 12-37. The lobbying campaign plan of action[17]

The lobbying campaign

The lobbying campaign had to deliver a certain number of key messages about the A400M. These were:

- Operational: the A400M is the only aircraft meeting the precise requirements of the armed forces in Germany, Belgium, Spain, France; Italy, United Kingdom and Turkey.
- Industrial: the A400M will allow the European aerospace industry to come back into line and it is based on existing industrial investments.
- Technical: The project will reinforce European technological expertise.
- Economical: it is the cheapest solution to produce and use.
- Political: This project will create (or at least maintain) employment in the high tech sector and contribute to economic activity.

Several advertising approaches to boost the A400M project were considered. One of them was based on strongly worded texts vaunting the advantages of the aircraft and comparing it to some of the competitors such as the Hercules. In some European countries, comparative advertising is not allowed and in others, this aggressive style would have sent the wrong message. So this option was not followed up since it could not have been used in an identical manner in all the target countries. Another possible, less aggressive campaign was based on a comparison between the plane and certain birds. Successive drawings showed for example the head of an eagle and the nose of the aircraft. However this natural parallel between the capacities of the aircraft and the assets of birds, did not give any technical information about the A400M. Finally, it was a campaign called " Road signs " which was chosen. All the European countries use the same road signs, well-known and easily understood by the inhabitants, and AMC used these universal symbols accompanied by a short text for its A400M campaign.

These visuals allowed the specificities of the plane to be explained: to understand each of the symbols given, the text needed to be read. This gives details about the basic characteristics of the plane (height, weight, speed, autonomy, technology, etc.), and so the A400M is presented as a multiple solution with adapted characteristics. For example, under a visual showing a maximum 3.85m height sign, a short text explains that thanks to a higher and wider hold, the A400M can take larger loads than the aircraft it is replacing, e.g. it can transport a Patriot missile system without having to dismantle it. Similarly, under a speed restriction sign, a text reminds the reader that the speed of the A400M is higher than the previous aircraft and is well adapted to urgent missions (humanitarian or military). However, whatever the support, the same strapline comes across to anchor the product's promise firmly: " The European Solution to European Needs ".

Three main communication targets were chosen: the decision makers, the influencers, the players involved in the project and public opinion. For each of these targets a specific approach and appropriate tools were used (Figure 12.38).

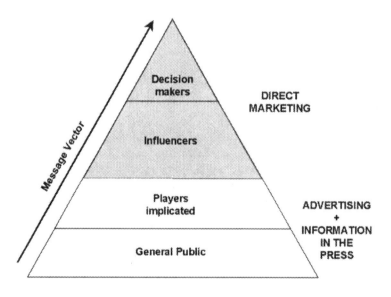

Figure 12-38. The tools used for the lobbying campaign[18]

The decision makers

The key decision makers were mailed: ministers, politicians, top civil servants, the military authorities and their advisors, supplier's top management.

12. Selecting Media

The package sent contained:
- a letter giving important and useful, previously unpublished information,
- a special " Executive Lobby " brochure and a 6-8 minute video presenting the A400M and recalling the key messages.

All this material was written in the language of each of the countries concerned, and due to the excellent quality of the way it is presented, was likely to be kept by the recipients. Owing to the cost of the package, it had to be given to the decision makers at the right moment for maximum effect.

The main messages were presented in a clear and concise way such that they could be understood and repeated with no difficulty. Although the messages were concise, they nonetheless incorporated the key ideas such as the rationality of the purchase, its operational advantages and the political and economic aspects. For example it was spelt out that the companies in each of the target countries would participate in the making of the project (Figure 12.39).

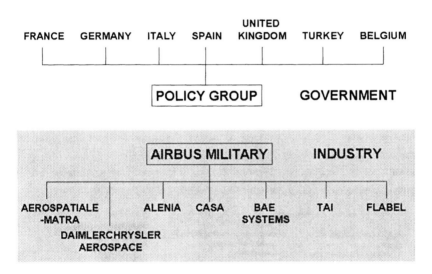

Figure 12-39. Airbus Military Company's response to the customer nations[19]

Each partner's contribution to the aircraft is shown in a diagram (Figure 12.40), and companies from each of the countries collaborate on the project. This European collaboration results in a growth in economic activity, employment and know how. It also symbolically demonstrates and opens the way for the building of a European defense force.

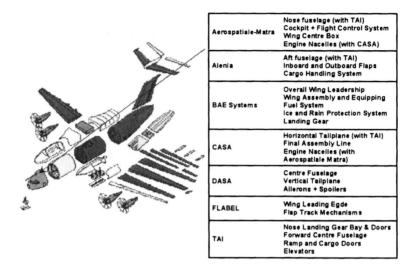

Figure 12-40. Share out of industrial work[20]

The influencers

Even more than the decision makers, influencers are looking for precise information about the characteristics of the aircraft compared to the requirements which have been defined. They are sensitive to a program's advantages whether they be economic, industrial, technical or military. These players are in general politicians who are specialized in the defense sector, other ministers, the press and media and the suppliers. The technical documents, written in English ("A400M Technical Review") were attractive and more complete, allowing specialists to increase their in-depth knowledge on the aircraft.

Figure 12-41. The main characteristics of the A400M[21]

12. Selecting Media

There was also a CD-Rom which reinforces the technological positioning of the aircraft and includes more specialized and complementary information to that found on the website (clips, etc.). It was designed around a standard format so that it could be read by all the different IT systems likely to be encountered in government offices.

In parallel, BAE Systems used a road show to promote the aircraft across the whole of the United Kingdom (London, Manchester, Edinburgh, etc.). This high tech truck presented the advantages of the A400M plus the opportunities it offered to industry, aeronautical suppliers, sub-contractors and politicians. Using a highly sophisticated graphics system and a comfortable environment, this road show presented the aircraft's concept and logistical capacities to a target population (Figure 12.42).

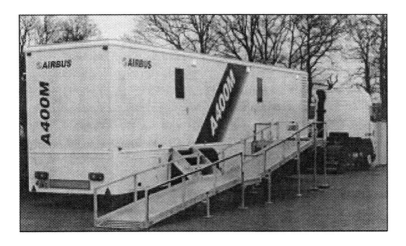

Figure 12-42. The A400M mobile road-show[22]

AMC's approach also aimed at individual influencers, i.e. the pilots. Up to then the latter used planes which did not have all the modern technological advantages found on board civil airliners. Conversely, flying the A400M, military pilots would use the same technological standards as civil aircraft, and there was a strong resemblance between the two. This approach was an attempt to interest the pilots, not only would they be able to fly a more rewarding aircraft, but they would also be able to transfer more easily to civil aviation once their military career was over.

AMC press conferences were also held with journalists, to build up interest for the project and improve knowledge about the A400M program. Details were given, and the advancement of the project updated. They were especially used at air shows where AMC was present, e.g. ILA-Berlin Air Show and Farnborough 2000 (Figure 12.43). A 1/20 scale model of the A400M was presented and wide wall video screens showed the main characteristics and capacities of the aircraft. Good media cover was ensured thanks to the presence of the specialized and national press.

Figure 12-43. Airbus Military at the Farnborough 2000 show[23]

Public opinion and the other players

To reach all the other players (personnel from organizations, employees from industry, humanitarian agencies and volunteers and public opinion, e.g. the electorate) ad campaigns and short articles in the press were used. Actions were different according to the country and their national concerns (industrial aspects, military, humanitarian, etc.). The publicity needed to catch the public's attention with visual aids emphasizing the advantages of the offer, plus simple texts with strong messages. As in all the various methods, the " European Solution to European Needs " strapline had to be present. The visual aids, pictures etc. were published in the national, international and specialized press plus the dailies. Press conferences with general journalists from the press, radio and TV were also planned. In addition, video clips were made (the aircraft landing, dropping material, etc.) and shown on the BBC in the United Kingdom.

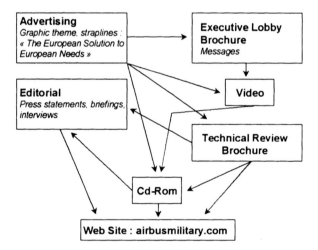

Figure 12-44. The complementarity of media used in the lobbying campaign[24]

12. Selecting Media 387

A specific website (airbusmilitary.com) was developed in order to present the aircraft's characteristics and the latest information about the project. The interior of the plane (cockpit and inside the fuselage) could be visited virtually with 360° vision.

The various methods used throughout the lobbying campaign reinforced the messages via different channels of communication (Figure 12.44). When these were relayed between the different players, (media, decision makers, influencers, public opinion), the information gradually built up to give an illuminating and convincing " beam " in which the key messages could not be ignored.

The results

The AMC lobbying campaign was a success in that the first, official positive feedback in its favor, came shortly afterwards. In May 2000, the British government decided to equip the Royal Air Force with a new, long range, air to air missile and new tactical transport aircraft to replace its aging C-130 Hercules fleet (Patri, 2000). It chose the European offer with on the one hand the Matra Bae Dynamics (MBD), Meteor statoreactor missile, and on the other, 25 A400M aircraft from the Airbus Military Company (EADS and BAE Systems together). By way of compensation for the Americans, the RAF announced a rental agreement for four C17 Globemasters and a provisional missiles order with Raytheon. The Meteor is a Beyond Visual Range Air to Air Missile (BVRAAM) and will be used to equip European fighter aircraft (Eurofighter, Saab Gripen, Rafale, Mirage 2000, etc.), and thanks to an agreement with Boeing, it will doubtless also be used on the F15, F18 and even the F22 and JSF. The A400M offer is based on the key concept of inter-operability within the European forces.

This result has not been without problems. At the end of 1999 for example, Germany, for political reasons, announced its preference for a solution involving the purchase of the Ukrainian, Antonov 70 aircraft. At the same time, the British government's choice has initiated a trend with the other governments interested by the A400M program. French members of parliament (MP's) have, in an article in one of the large daily papers, defended the A400M solution as the most realistic for their Air Force. Using arguments from the recent conflict in Kosovo, the MP's have demonstrated that their present fleet of Transall and Hercules C130's is no longer adapted to the needs of a modern and high performance protection force. They have explained in full the reasons for choosing the A400M, particularly in terms of employment in Europe (35,000) and its coherence with the European Union project (the euro, etc.).

A short time after the British decision, the French and the Germans announced a similar choice at the European summit in Mayence. In July 2000, during the Farnborough 2000 air show, the defense ministries from the German, Belgian, British, Spanish, French, Italian and Turkish governments, officially announced their intention to acquire 225 examples of the A400M military transport aircraft in preference to its American competitors (C130-J, C17) and Russian-Ukrainian (An-70). Germany agreed to buy 75 aircraft, Belgium 8 (of which one, in cooperation with Luxembourg), the United Kingdom 25, Spain 27, France 50, Italy 16 and Turkey 26, all preludes to signing a memorandum of understanding (MoU). For the different

European ministers, this program represented one more step towards setting up a European defense system and a coherent engagement from an economic point of view.

7. PUBLIC RELATIONS AND SPONSORING

7.1 Public relations

Public relations, groups together the company's communication actions in the form of a dialogue destined for the different sectors of the general public. It is a case of creating a link with society, i.e. integrating the company's way of thinking and sharing this with the public as a whole or in small sectors. The point of public relations is to establish confidence and understanding with the target, to give objective information about what the company has done, and anything else concerning their activities, without propaganda or advertising, following the principal of free use of the media. The aim is to enhance the image of the company and its products building public support rather than necessarily selling. There are various targets for public relations operations, corresponding to partners who could damage or enhance the correct functioning of the company:
- public authorities and institutions,
- socio-economic or socio-cultural groups such as unions,
- the stock exchange, the financial market, actual and potential shareholders,
- prospective and actual customers,
- professional associations,
- actual and potential suppliers, competitors and distributors,
- general public and professional media at home and abroad, press agencies, etc.

The objectives of public relations can be varied (Figure 12.45) including awareness, explaining the actions of the company, making others understand and absorb these actions, obtaining backing, helping with a distribution operation or a partnership.

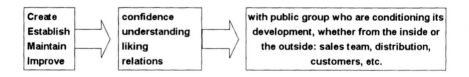

Figure 12-45. The different functions of public relations

12. Selecting Media

389

The aim is to elicit favorable reactions, obtain the collaboration of the particular groups targeted, whether it be the staff, the shareholders, the customers, the suppliers, the influencers or the partners. It could be a case of facilitating the delivery of authorizations, or public regulations, the financial help from shareholders or banks, obtaining larger coverage. Overall, public relations must awaken interest, understanding and support from the public such that they contribute to developing the reputation of the company and strengthen its image. For a relatively low cost, public relations are an excellent way of promoting and communicating with a long term impact.

In the interests of greater synergy, public relations actions are often linked to other communications operations, such as direct marketing or sponsorship. There are a very large number of ways of contacting the public, such as organizing conferences, round tables, seminars, company visits, participation in exhibitions, specialized shows, congresses. The main methods are:

- the publishing of information by the company for an important occasion or event (contract, takeover bid, anniversary, launch of a new product, acquisition of an important new market, etc.) ;
- the participation of the company and one of its top managers in events or operations (quality control, press conference, donations, company presents, support for other actions, speeches, congresses, seminars, interviews, etc.) ;
- the events themselves during which documents are handed out, question and answer sessions organized (visiting new factories, open days, sports competitions, trophies, challenges, exhibitions, hosting operations, non profit making but general interest activities, foundations, etc.) ;
- publications from brochures and catalogues (video cassettes, CD, etc.) to prestige editions, annual reports, company newspapers, letters and magazines. It should be noted that when there is a coherent identity given to all the media (company charter, business cards, brochures) this contributes to the effectiveness of the conferences and press releases.

Press relations: This is a special form of public relations aimed at journalists, but respecting their deontology and independence. It is a case of getting information into the media which is rigorous and favorable, designed to draw attention to one person, an idea, an activity, an organization, a service or a product. The main objective is to obtain press publicity, space in the media to be seen, read and heard by customers or prospective ones. To do this, the company can produce a press release giving clear, precise information, limited in time and devoid of any flamboyance. They can produce a press handout which offers the media a complete package of documents and information (factual aspects, anecdotes, quantified data, etc.) which they can use all or part of. Another means of action is to get several journalists together during a press conference, or offer to give certain of them an interview with the aim of supplying more personalized information. Finally, certain

390 Aerospace Marketing Management

companies organize one-off press trips, with a group of journalists invited for one or two days to a congress or to visit a factory.

7.2 Sponsoring

Sponsorship can support public relations activities. This form of event-related communication is certainly the most visible, outside the media.

Corporate philanthropy or patronage: This concerns art, culture or great social causes in specific areas. It consists of support for a cause, with no return or exploitation, and staying well in the background of the event, with no possible control over the way it is treated by the mass-media. This sort of publicity aims at improving the image of the company in the medium or long term.

Boeing, the aeronautics manufacturer is very active in this way with aid for schools and universities, contributions to programs and organizations linked to health, and in the financing of artistic and cultural events (Long Beach Symphony in California, the Seattle Symphony, etc.).

Figure 12-46. Air France participating in the ECPAT International campaign[25]

Similarly, Air France supports the Restaurants du Cœur (free meals for the homeless) by transporting free of charge musicians and artists who participate in fund raising. The company has also shown a video on board its aircraft, presenting the Mécénat Chirurgie Cardiaque Enfants du Monde which is an association devoted to treating children with cardiac deformities in France who cannot be operated on in their home countries. It has also offered and sold teddy bears on board its planes with the profits going to End Child Prostitution Pornography and Trafficking International (ECPAT), an international organization fighting against sexual and commercial exploitation of children (Figure 12.46).

Sponsoring: Sponsorship is financial backing given by a company to a cultural or sporting activity in return for increased exposure or some expected benefit in terms of image. Here there is a commercial spin off with any pictures being widely exploited. Any goodwill generated by the event sponsored, must be recuperated to a maximum extent. The action is short term, rationally developed and under partial or total control.

In order to develop its awareness or reinforce its image, the company can project key values through sponsoring, e.g. competitiveness, dynamism, etc. In this way the Pizza Hut company put a gigantic logo on the Russian Proton launcher for the Zvezda module of the International Space Station (ISS), and helped to finance the launch. Similarly, the watchmakers Breitling have, since their foundation in 1884, always been associated with important aeronautical events. For example it sponsored the successful Breitling Orbiter round the world balloon attempt, with Bertrand Piccard and Brian Jones. CNN broadcast pictures to the whole world.

Sponsorship is like a witness for the public and marks out the difference from competitors. There are positive repercussions and effects on customers but also on distribution and the sales team via the enhanced image that it creates. This form of communication has the disadvantage of not lending itself to objective measurement of the effects – because of the presence of other simultaneous actions – with the exception of TV sponsoring where numerous measures and tests exist.

8. *SALES PROMOTION*

Sales promotion covers all the stimulation techniques aimed either at the industrial customer, the distribution network or the end customer. Unlike advertising, promotion acts on behavior and its effects are measurable in the short term[26]. Sales promotion is direct, which is why it is known as hard selling as opposed to advertising actions which are soft selling. In order to maintain its image, the company must keep an eye on the coherence between promotional actions and other communication activities (ad campaigns, public relations operations, etc.). Promotion in this way is often used as a support for direct marketing actions or to back up merchandising[27] in tour operator sales outlets, for coordination, presence or for getting exposure (Figure 12.47). Promotional actions combined with advertising increases the effectiveness of the communication.

Promotion can be:

- one off: it could be a special invitation onto a stand during an air show, giving a sample or lending an example for the launch of the product (helicopter, business jet, etc.) ;

- repetitive, repeated frequently throughout the year: these are then bonuses, reimbursements, etc.

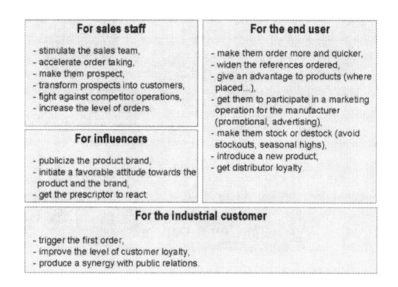

Figure 12-47. The main objectives of promotional activities

Promotion can include a whole program of activities over a long period and is not just limited to one off events. Setting up sales promotions means defining:
- the breadth of the promotion (level of stimulation),
- the targets chosen,
- the media to be used for the operation (postal, inserted in publicity material, etc.),
- the schedule and the length of the operation.

FREQUENT FLYER PROGRAMS

Frequent Flyer Programs (FFP) are competitive, promotional tools aimed at making customers (often Business class) loyal by offering them advantages as a function of how much they travel (generally calculated in miles traveled). The principal of these programs is to make a route attractive by concentrating on elements other than the price (or the price/quality ratio), in such a way as to make the traveler feel gratified by the gifts (miles accumulated in his name even if it is a company paid, professional trip). In fact, several studies in the United States have shown the important role played by FFP's in business customer buying decisions.

To encourage customers to persevere with the program, there are very often free flights given out using a decreasing sliding scale: the first free flight is offered for example after 30,000 miles, the second after only 20,000. With the development of yield management, the airlines have been forced to reconsider the way in which free flights are attributed, in order that they are not made on the highest frequency flights. The rules covering accumulation of miles have also been reviewed and the way free flights are taken has been tightened up. To compensate for these restrictions on the one hand, and to spread the costs over several partner companies on the other, the FFP's have been extended to include other services and products not concerned with air travel (banking, car rental, hotels, telephone services, department stores, etc.). For the alliances between airlines[28], each time that a person who is part of the FFP, travels on an approved flight run by an alliance member, the miles are counted. In this way miles amassed with American Airlines Aadvantage, British Airways Executive Club or Cathay Pacific Marco Polo Club, can for example be used on any other member company of the OneWorld alliance such as Iberia Plus with Iberia, Canadian Plus with Canadian Airlines, Finnair Plus with Finnair or Frequent Flyer with Qantas.

Launching promotional operations above all means verifying coordination in terms of delays in the exchange of information between the technical and sales departments in order to ensure availability of supplies, production and the sales materials. Once the operation has been carried out, its profitability needs to be evaluated by comparing the quantitative and qualitative sales. The surplus orders generated relative to the cost of the operation (publications, routing, publicity, bonuses, samples, etc.), must be calculated.

The difficulty lies in choosing which budget item the promotion can come under: reimbursement offers, special offers, games and competitions, distribution of samples, invitations to shows and payment of costs incurred, etc.. For example, the latter can sometimes be considered as promotion and sometimes as public relations.

Several sales promotion techniques stimulate the market in the short term. Basically, these are sales with bonuses, trial offers and samples, games and competitions and sales with price reductions.

Sales with bonuses: This is a very general technique used to gain new customers or obtain their loyalty. There are two sorts, direct bonuses which are offered at the time of purchase, and indirect ones which are given later against proof of sale. "Direct bonuses" are mainly used in the supermarket sector and very little in aeronautics and space, they work by offering something which may or may not be associated with the product:

- on-pack bonus is attached to the product: e.g. to promote sales of special air tickets, extra services could be added for the same price. For example the travel agent could offer reduced prices for hotel accommodation (whether they belong to the company or not) or special conditions for car rental.

- "giraffe" or "nested" offers give more of the particular product for the same price, and for industry, usually concern consumables. To promote a new service, a telecommunications company could offer 4 hours communication for the price of 3 hours. There is a financial advantage. The company effectively spends about 25% of the cost price more, but there is not a 25% reduction in the sale price. The customer however, remembers the figure 25%, rather than the amount to which it applies.

Deferred bonuses are mainly used to get customer loyalty and increase income from sales. This customer bonus is only awarded for several purchases. For example an airline offers miles as a function of tickets sold, giving a boost to company priorities. This could be:

- improving load factors in a slack period. In this case it would be stated that the normal gain of 1,000 miles is increased to 1,200 for this period,
- a new destination almost unknown to potential customers.

The drawback of this loyalty system for the customer company, is that their employees are encouraged to travel more, and into the bargain the miles are usually not in the company's name.

Trial offers and samples: They are used to promote new products and services. This means encouraging customer trials, getting them to try out the product or service, and appreciate the quality, performance, etc. For a helicopter or corporate aircraft manufacturer, this technique involves lending an example to the prospective customer company in order to obtain an order. In general, trial and sample techniques are the most effective for increasing the number of buyers ; but they are also the most costly.

Games and competitions: As for the other operations, it is always necessary to verify that a game or competition comes within the law, which can vary from one country to another. Competitions consist of asking customers (simple) questions concerning the company's activities. The winners who have got the correct answers get a prize. Luck does not come into it, and so it is not possible to budget ahead of the operation, because the number of winners is unknown. This is why the prizes are relatively cheap (T-shirts, watches, pens, etc.). Games (and competitive games), involve luck, like draws or lotteries, and they cannot in any way be linked to an obligation to buy, meaning that a customer can participate having copied out the rules. An operation of this sort was offered to Air France passengers on long-haul flights by the JetPhone company. For the launch of its on-board service, JetPhone produced a game where the prizes were three tickets for Paris-New York on Concorde, including accommodation in a four star hotel for two people, transfer by limousine and $800 to spend on Fifth Avenue (Figure 12.48).

For games, the company knows in advance the number of winners and so can budget for them, which is why this method is used much more than competitions. Games and competitive games are seen as attractive all over the world, but they are generally not used in business to business marketing except in shows to get people

12. Selecting Media

to come onto a stand or answer questions by e-mail. The visitors give their business card which serves as proof of participation.

In its loyalty program on Internet, KLM has set up a game called " 3-2-1 click, go " in tandem with humorous postcards showing the different destinations served by the company. The visitor to the site clicks on the card and sends it to anyone he chooses. By doing this, this visitor is automatically added to the list for a competition to win air tickets, whose destination will be determined by the 3 postcards which are selected the most (number of clicks). Participation in this game is accompanied by a questionnaire which allows the company to constitute a prospective customer profile, database. The whole point of the operation is its " viral/contagious " nature: the senders and receivers of postcards are both part of the operation and could for example recommend a visit to the site to friends ; a cheap way of publicizing the site.

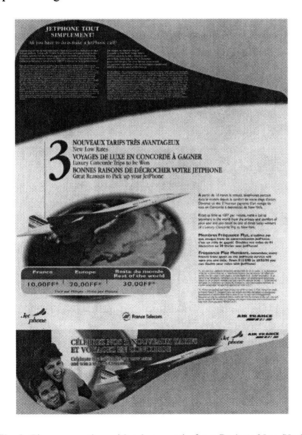

Figure 12-48. The JetPhone operation with a luxury trip from Paris to New York by Concorde as the prize

Reimbursement offers: These are used especially for the launch of a new product in order to encourage the customer to try it.

THE EXAMPLE OF SPOT IMAGE'S PROMOTION FOR THE LAUNCH OF SPOTVIEW

For the launch of SpotView in the Geospot product range, Spot Image wanted to present the functions and encourage orders of this image map which can be integrated into the customer's geographic information system (GIS) and give geometrically corrected satellite pictures in standard map form.

The SpotView operation was in two parts:

- a reimbursement offer to get the most orders for the demonstration CD-Rom, thus encouraging the initial purchase, and combined with this, a free poster with an exclusive SpotView picture ;

- a game with no obligation to buy, designed to boost CD-Rom orders and enrich the database of prospective customers (activity sector, domain, geographic information system equipment, type of image processing system used, etc.). The first prize was a trip to Kourou to see an Ariane rocket launch (Figure 12.49).

Figure 12-49. SpotView promotion with a reimbursement offer and a game

Price reductions: They group together the techniques aimed at reducing costs for the customer, and it is sometimes difficult to separate this technique from those sales conditions which last the whole year. Price reductions are used both to build loyalty as well as to penetrate a market. They are often aggressive and are mainly aimed at defending positions in the face of cheaper competitor activities. The effect on sales is often very brief and disappears at the end of the operation. Price reductions must be made carefully due to the often negative effect in terms of image and the risk of being outdone by competitors. In addition, it can give the impression that the supplier usually makes a large profit if he can drastically cut his prices. Price reduction techniques such as couponing (given directly to the customer, published in the press or sent by post or e-mail) can be used in travel agencies for example, where the coupon's reduction can be taken off the next purchase. This technique is less expensive than a simple price reduction because the cost is limited to the percentage of customers who take advantage of the offer. This type of operation is most often used by tour operators and leisure-tourism companies.

Creation of traffic: This is mounting operations designed to send end customers to distributors, e.g. travel agencies which distribute company products in their own towns. These therefore are operations which work in the supplier's interest (airline companies or leisure tourism suppliers) and the distributor's.

Gifts: This includes all small useful objects – calendars, key rings, diaries – and various other company presents. There are two categories of small advertising goods according to whether they are practical objects connected with work (office supplies: pens, post-its, rulers, calculators, diaries, calendars, etc.) or not (publicity clothes, smart cards, etc.). These gifts have an emotional value and are amusing; they will not be the factor which settles the order, but they can create good relations and their effectiveness comes mainly from the way in which they are offered. They must never be seen as an obligation to buy.

Settling advertising expenses ("co-op"): This consists of participating in customer's or distributor's advertising expenses. It is a technique used when setting up large operations, such as when an airline orders new aircraft, etc. When the manufacturer's name appears in the company's advertising message, this technique is similar to co-advertising.

The different promotional techniques presented are more or less relevant depending on the company's marketing strategy: penetration or loyalty building.

Other elements contribute fully to the communication policy and to strengthening the company image such as:
- the design of products, packaging, and the company premises
- the carrying out of surveys for customers: partnership, understanding of the activity sector, the market, marketing support
- the maintenance: deadlines, quality, taking into consideration production constraints, safety, warranty
- recruitment ads, which not only present a message but are also a vehicle for the company image (visual identity code, logo, etc.).

In any case, all these measures will not contribute to effectiveness of the communication policy unless they are coherent with desired objectives.

NOTES

1. For update listing of trade shows, see the following websites: airshow.com, expobase.com, reedexpo.com, tradeshowweek.com.
2. Photo taken by J. Sykes, published in *Air & Cosmos*, (2000), p 28, 20 October, n°1767.
3. Ad published in particular in *Aviation Week & Space Technology*, (1999), October 15, p. 93.
4. Photo taken by Ollivier, published in *Aviation International News*, (1999), vol. 31, 10, 14 June, p. 2.
5. Photo by Ollivier taken from Eurocopter AS350B3 Ecureuil Helicopter Flown by Hervé Jameyrac, published in *Aviation International News*, 1999, vol 31, n°10, June 14, p 1.
6. Sticker given on Dassault's stand at Paris Air Show.
7. See Chapter 14.
8. See Chapter 2.
9. See Chapter 9.
10. Photo taken from Pascal Chenu see site www.multimania.com/airbus/beluga.
11. See Chapter 15.
12. See Chapter 4.
13. Illustration taken from the www.jcdecaux.fr site, world leader for billboards, especially in airports (see www.skysites.com)
14. See *Orient Aviation*, (1999), Security Alert, vol. n° 6, n°7, May, p 16-17.
15. Picture from the Airbus Military Company website: airbusmilitary.com
16. Adapted from Airbus Military Company
17. Adapted from Airbus Military Company
18. Adapted from Airbus Military Company
19. Adapted from Airbus Military Company
20. Adapted from Airbus Military Company
21. Picture from the Airbus Military Company website: airbusmilitary.com
22. Picture from the Airbus Military Company website: airbusmilitary.com
23. Picture from the Airbus Military Company website: airbusmilitary.com
24. Adapted from Airbus Military Company
25. Advertisement published in the *Air France Magazine*, (2000), n°36, April, p 97.
26. See Chapter 11.
27. See Chapter 9.
28. See Chapter 15.

REFERENCES

Aubert, V., (1999), Airbus-Boeing: des minutes qui valent cher, *Le Figaro Economie*, p 46, 16 June.

Herrera, E., (2000), Applications de l'e-business au secteur aérospatial, Mémoire Mastère Spécialisé Marketing et Communication Commerciale, October.

Le Picard, O., Adler, J.-C. and Bouvier, N., (2000), *Lobbying, les règles du jeu*, Paris, Les Éditions d'Organisation.

Lehu, J.-M., (1996), *Praximarket*, Paris, Éditions Jean-Pierre de Monza.

Molière (1664), *Tartufe*.

Nicovitch, S. and Cornwell, T.B., (1999), An Internet Culture ? Implications for Marketing, *Journal of Interactive Marketing*, 12, 4, 22-33.

Patri, G., (2000),Meteor-Airbus A400M: Two British Decisions in Favor of EADS and BAE Systems, *Revue Aerospatiale*, June, n°169, p. 5-7.

Chapter 13

BRAND MANAGEMENT

Today, brands are of increasing concern to business professionals in the aeronautics and space sector, as well as being the subject of numerous surveys and research. Most of the time, brands are analyzed in terms of the consumer market. However brands are of major importance in the Business to Business field. Thus the different brand concepts and tools need to be examined from the two marketing perspectives. Vehicle of the company's strategy, brands can be managed in very different ways in terms of brand name creation, visual or sound identity (logo, jingle) and slogan.

1. BRAND FOUNDATION

A brand is generally defined as the combination of elements which makes it possible to identify a product and to differentiate it versus its competitors. Among these elements, the name comes first. Then the symbol, drawing, design, slogan or jingle identify the brand on visual or sound materials.

The brand can be defined as set of strong associations, which can be transmitted and can influence behavior (Leuthesser, 1988). Unlike a particular communication strategy which is generally short-term oriented[1], branding is a more lasting tool. Furthermore, it is very important to understand the complementarity relationship between the product/service and the brand which is an abstract mental translation of the latter: "the brand only exists in the customer's mind"(Kim, 1990) (Table 13.1).

Product	Tangible, physical characteristics
Brand	Abstract, imagination field

Table 13-1. Complementarity between brand and product

Moreover, the brand assures a double continuity in time and in space.

In space: Enabling customers to more easily recognize the supplier's identity whether:
- From one geographical area to another,
- From one distribution circuit to another, for example, from travel agencies to on-line travel operators,
- From one point of sale to another, within the same geographical area.

Even if the brand name is written in a foreign language, its visual identity code (logo, colors) allows customers to identify the company (Figure 13.2).

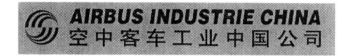

Figure 13-2. Brands assure a link in space: Airbus logo and its Chinese transliteration

In time: Products and above all services have a limited lifespan[3]. The brand is therefore helpful in maintaining the company's long-term position by facilitating purchase switching:
- For the same product, from one time period to another,
- For a product/service at the end of its life cycle to a new product/service of the same brand (brand outlives the product),
- From one ad campaign to another.

Again taking the analogy of the product's life cycle (comparable to that of individuals), it is possible to use another analogy between a family and a brand (Table 13.3).

Brand	Family
Product	Family member

Table 13-3. The brand, ensuring the lifetime of the product

1.1 Brand mechanisms

Awareness, a quantitative tool for evaluating the brand

Brand or company awareness can be defined as an individual's level of knowledge of the company or brand in question. This is a key concept, since it is through knowledge of the brand name that a customer asks for one brand or another. The different levels of awareness have been identified (Aaker, 1991):

13. Brand Management

- Brand recognition: when questioned, the consumer does not refer directly to the brand, but if he is reminded, he is capable of describing the product categories concerned;
- Brand recall when the brand is known: when questioned about a product category, the consumer spontaneously refers to the brand, but among other competing brands;
- Top of mind when the brand is present in the mind of the buyer: this is the first brand he refers to for a particular product or service category.

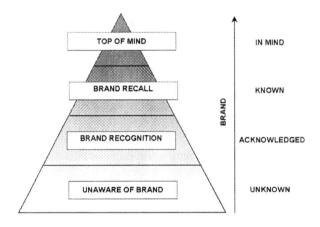

Figure 13-4. The awareness pyramid[4]

While awareness measures the level of knowledge of a brand, attention measures the existence and importance of the information relating to the brand stored in the memory. Attention can be broken down into two components, re-memorization and recognition (Alba *and al.*, 1991). Re-memorization is measured by brand recall, and in particular, by measuring top of mind, while recognition is measured using brand recognition. The ability to spontaneously remember brands (brand name recall) is much lower than the ability to recognize brands (brand recognition). These two measures are obtained by questioning buyers about the brands that they know in the product category.

Although these measurements can apply to the industrial field, they are not so useful: it is part of the responsibilities of a professional buyer to know the different suppliers (and their brands) capable of meeting the needs of the company.

A study carried out on 30 brands in four different industrial sectors thus showed that all the brands examined had a considerable awareness level among professionals in the sectors considered (from 95-100%) (Malaval, 2001). On the other hand, supplier brand awareness with the general public is much less discriminating: the same study reveals a very wide range in industrial brand awareness with the general public (from 0-100%). Certain brands communicate directly to end consumers. To strengthen its reference status with professional customers the company can exploit this brand awareness. Strong brand awareness among professionals or consumers is not however enough to optimize the brand. The evocation still needs to be positive and the image projected by the brand must be better than competing brands.

Image, a qualitative tool for evaluating the brand

To understand the "advertising efficiency" of the brand, its qualitative content, as well as the associations accompanying it, need to be analyzed. Image specifically measures the quality of the ties established between different kinds of information stored in the brand, in other words, the quality of associations linked to this brand. Brand image results mainly from the positioning[5] chosen and the communication policy created by the marketing department, as much in qualitative as quantitative terms. The overall image is an expression of the perceptions and feelings of individuals with regard to the company or to the product. A positive or negative image depends on incomplete and biased information (however, it is perceived as complete and objective by the buyer) and is likely to influence the buying decision. Image analysis is directly operational too in business to business marketing. While buyers have a rational approach, they are often sensitive to image such as the supposed permanence of the supplier, his professionalism, a classification in a trade magazine, expertise in a particular sector, his nationality, quality and reliability of the products or services offered.

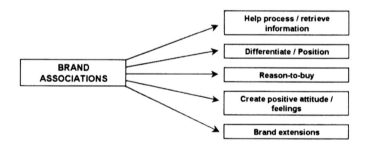

Figure 13-5. The different parts of the image[6]

Brand image can be analyzed by distinguishing the image of competing brands, users, the product, and the manufacturer. However, brand image depends mainly on its quality as perceived by industrial customers (Chernatony and McDonald, 1992), in other words, its capacity to supply the characteristics that the customer wants. To improve perceived quality, marketing departments ask the research and development people for product improvements so as to get closer to their ideal product. Innovation (Schmalensee, 1982) and the design of new products (functions, exterior and interior design, product maintenance, etc.), cannot be disassociated from the image: they provide its content.

Brand image is made of associations which can be defined as "anything linked in the memory to the brand" (Aaker, 1991). Associations are comprised of:

13. Brand Management

- Concrete elements such as products or services, which are "branded" along with their attributes and the product categories they are attached to,
- More abstract elements such as those expressed in the company's and competitors' various communication measures.

The associations can be defined in terms of uniqueness (association not shared with other brands), strength (association more or less linked to the brand), cohesion (convergence between associations), and favorability (positive impact of associations). A brand is strongest when it is successful in these four characteristics.

Loyalty: main objective of a brand strategy

To win over new customers is generally more expensive than to maintain the existing customer pool. In fact, for a prospective customer to become a real customer, the brand must catalyze a first sale, which will often involve important sales backup or promotions. This expense can be decreased after the first purchase, assuming that the customer has been satisfied by the experience. Marketing expenses to trigger the first sale are recouped in direct proportion to customer loyalty.

Brand loyalty also means greater confidence, favoring cooperation between the customer and the supplier brand. Increased loyalty allows the two companies to exchange information, to develop specific projects together, and to confidentially test new products coming from the supplier. Thus loyalty can be defined as a favorable attitude towards a brand resulting in repeated purchasing of that brand over time.

In business to business, the loyalty of a company to its different suppliers is especially strong: a supplier who has given satisfaction for 5, 10, or 20 years is difficult to shift[7]. The customer company gets to know the supplier brand, and is used to working with its particular methods and culture. Trust becomes mutual when the two companies come to an agreement and work together. Loyalty is in fact the result of constant investment in the customer (awareness, image, and quality of the products).

Awareness, image, and brand associations enable to better understand how the brand functions (Figure 13.6).

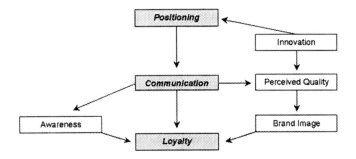

Figure 13-6. Simplified diagram showing how the brand functions (Malaval, 2001)

However the latter factors are useless if they are not part of a larger plan: brand management. Brand management consists of giving life to the brand, from its birth through to its development and evolution (birth of related brands, brand extensions, brand suppressions, etc.) [8].

1.2 Brand functions for the company

Generally speaking, the functions of the brand for a company in the industrial sector are identical to those of a company in the consumer market. In both cases, the two principal brand functions are positioning and capitalization. From a managerial point of view, these two global and essential functions of the brand correspond to the different objectives set for it; these can be divided into three main categories (Figure 13.7).

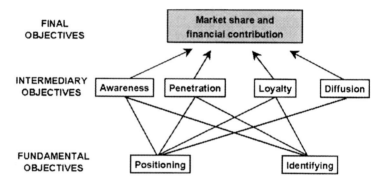

Figure 13-7. Brand's main objectives for the company

Fundamental objectives

Develop positioning: Through positioning, a specific place is allocated to the brand in the minds of the clients, differentiating it from its competitors – providing, of course, that the brand brings "added value" to the product it designates. Brand value lies in the ability to continually influence customer buying behavior, make customers loyal, and encourage them to ask for new products. Customer preference for a particular brand depends on the product's intrinsic characteristics and/or the overall characteristics of the brand. For example, Swissair airline is appreciated by passengers in particular for its seriousness and exactitude.

Make the product or service more visible to the client: This is the very first function of the brand. To make products or services offered more visible, the company will use different means: communication, media, easily identifiable visual identity (on packaging, and vehicles, etc.) [9]. These measures aim to increase brand efficiency, making it more visible, recognized, appreciated and purchased.

13. Brand Management

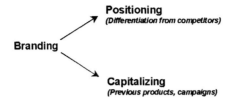

Figure 13-8. Fundamental objectives of the brand for the company

Intermediary objectives

Increase awareness: For the customer company to make the purchase, the supplier brand must be known. Creating awareness is thus an intermediary objective, an essential prerequisite to the purchasing act.

Increase penetration rate: This concerns winning new customers. These may be completely new clients for the brand in question or existing clients who also adopt a new product under the same brand name. The increase in penetration rate may be achieved by marketing action focused solely on the brand, solely on the product or a combination of both, with follow-up by the sales force in the field.

Assure greater brand loyalty: This involves consolidating the position already held by the brand and /or the product in the customer company. Investment made by the brand in terms of image and quality aims at producing greater customer satisfaction. The greater the customer satisfaction, the higher the likelihood that a new product under that brand will be chosen – whatever the category of product concerned.

Widen product diffusion: For branded products to be purchasable by the customer company, they must be available in the particular geographical area, at the distribution channels, and at the points of sale where the customer usually buys. The customer company must be able to procure the products or services proposed for each and every one of their localities (national or international). Delivery conditions should also be compatible with the company's activity. As a result, one of the intermediary objectives of the brand is to maximize the value availability in order to transform buying intention into purchase.

Mobilize internal human resources: The supplier brand also has the role of federating and mobilizing the internal resources of the company. It serves as a benchmark and a banner around which the personnel (staff and shareholders) rally, reinforcing cohesion. The brand is a concentrate of the fundamental and distinctive values made visible – on company vehicles, work clothes, or promotional items. Within and out of the company, a well-appreciated brand stimulates a certain warm attachment. Engineering students are thus more attracted to an aeronautics company that has prestigious brands, as is the case of Boeing or General Electric.

Final objectives

Increase market share: Very often, the supplier brand has no way to influence market evolution, all brands considered. Generally speaking, it is impossible to expand the global value of the market, although exceptionally some actions taken collectively by a particular industrial sector – or by a certain supplier brand in a more individual way – influence the market downstream and thus develop the global market. A brand can, however, increase the company's market share, providing the intermediary objectives mentioned above are achieved. Market share thus represents the combined result of the different intermediary objectives. Expressed in volume, then in value, it is finally translated by a financial contribution.

Increase financial contribution: The financial contribution results from the quantitative leverage brought by the brand to the company's communication and promotional efforts, and also from the difference between the acceptable market price and company cost price. By increasing sales, the cost price tends to become more competitive because of the increase in production quantities. Simultaneously, by its perceived value, the brand can support a higher price than an ordinary basic product. The weighted average price may be increased in the case of multiproduct industrial brands by improving the structure of inter-product sales. The objectives of the brand incorporate three financial aspects: the level of sales, the cost price, and the selling price.

1.3 Brand functions for the customer

Consumer goods context

- The first brand function is to identify the goods and services offered by the company. The brand's visual identity code ensures the coherence of different communication media, making recognition easier for the company's target customers. For an airline, this can range from tickets, to a distributor in the city, signs at airport counters, identification on aircraft, and different forms of advertising;
- Identifying the product in today sea of advertising is the second brand function. The end consumer can thus identify the company as well as its products and services. This function facilitates communication and in particular advertising through the traditional media, such as television, billboards, etc as well as the Internet. Brand identification and in turn merchandising are particularly important in multibrand points of sale, when the end customer has a choice between different suppliers.

In addition, as has already been seen in the chapter on the marketing of services[10], the service provided needs to be made tangible in some way for the customer; in the case of an airline, the passenger should be able to memorize the identity of the company (airline logo engraved on cups and dishes, toys for children, and the magazine provided to passengers, Figure 13.9).

13. Brand Management

Figure 13-9. Royal Air Maroc Magazine[11]

- The brand also acts as a guarantee, helping to build customer confidence. It is a kind of contract specifying reciprocal roles and obligations. If the customer is satisfied with the service, the strength of the brand, as well as how well it is memorized and the positive feelings it evokes all help to build customer loyalty.

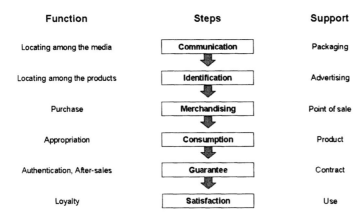

Figure 13-10. Brand functions for the customer

On the consumer market, the brand has other more subjective functions such as facilitating social integration within a group. In fact the brand expresses a sociological need to belong: by buying a Breitling watch, the end customer projects a certain image to those around him. He belongs to an informal group of customers who appreciate quality watches and are interested in aeronautics. At the same time, the consumer builds his self image. The brand is just one way for consumers to set themselves apart, affirming their personality and originality (Figure 13.11).

Figure 13-11. The brand: satisfying the consumer's psychological and sociological needs

As far as airlines are concerned, passengers often seek originality. To be different, they will use foreign suppliers. This is equally true of French consumers or Germans, who choose an American airline, as for Americans who choose to fly with a European company. Then there are certain brands such as Virgin, which project an anticonformist image compared to the main airline brands.

Finally, the brand also has a symbolic function (luxury brands, cult brands..) for the individual customer (Figure 13.12).

Figure 13-12. Ad for the Breitling brand[12]

In the same way, in the business to business sector, this symbolic function is linked to the corporate image that a company wishes to project. Drawing media attention to certain prestigious, sophisticated technological or ecological purchases, to a certain extent fills this symbolic function. By allowing customers to express themselves, brands thus contribute to identity.

Business to business context

There are four essential brand functions (Figure 13.13):

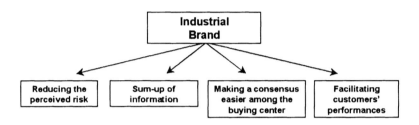

Figure 13-13. Main industrial brand functions for the customer company

Reducing risk: The brand guarantees the know-how of the supplier company. The members of the customer company's buying center can be reassured by the different guarantees provided by the supplier brand (quality control). Furthermore, the supplier brand ensures the traceability of components integrated into the final product, clarifying the issue of responsibility. As the brand provides continuity from one product to another, the customer company is more confident about follow-up, from technical assistance to the availability of spare parts. In business to business a particular form of co-branding has developed whereby the supplier brand appears right next to that of the finish product. The consumer brand/supplier brand combination improves the sales of the finished product. Such products benefit from the technical and technological guarantee of industrial brands and the high degree of quality associated with their production. The customer company benefits from the image and the renowned expertise of the supplier brand, thus reassuring its own clientele. This co-branding can be seen in the air transport sector, where the passenger is made fully aware of the aircraft brands used by the airlines.

Concentrating information: The brand allows the buyer to clearly identify the product or service, making it easier for him to choose from among the range offered. The brand facilitates memorization while concentrating the information associated with it (past performance, slogan, sales arguments, values…). Given the often highly technical nature of business to business purchases[13], the brand concentrates information, saving the buyer time and effort. The condensed information provided by the brand includes rational factors relating to the product and delivery as well as intangible elements such as the nationality of the supplier company, the personality of its managers, etc.

A consensus of decision: Given differences in responsibilities, training and motivation, the managers of the buying center do not respond to the same sales arguments. This can result in internal conflicts, such as strongly differing views about which suppliers to choose. Making the brand essential is one of the most commonly methods used to build consensus among the buying center members.

Performance facilitator: Among the reasons influencing the managers to choose a supplier of a strong brand, we can find:
- The brand's capacity to innovate and encourage innovation within the company,
- Their capacity to assist in the design of company products,
- Their capacity to make changes in processes or materials acceptable to those working with them,
- Their capacity to establish a consensus of decision,
- Their ability to accompany the company in the assurance of quality,
- Their assistance in production and maintenance,
- The commercial weight that these supplier brands represent,
- Their warrantee of durability in the long term.

In general, the industrial brand is especially appreciated by the customer company as a performance facilitator (Figure 13.14). This assistance can be technical, commercial or operational.

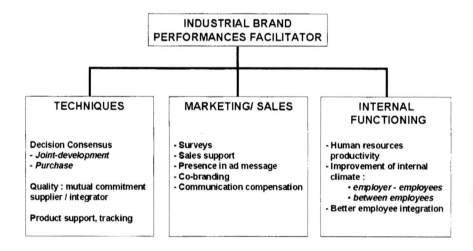

Figure 13-14. Industrial brand: triple role of performance facilitator

13. Brand Management

Facilitating technical performance

The word "technical" englobes three possible types of supplier brand/customer partnerships:

- Collaboration between the research and development departments in design and new product launch. Snecma engines manufacturer is a good example of this: thanks to considerable investment in research and development, this company has positioned itself with this particular type of partnership in the aeronautics and space sector. In the same way, BFGoodrich has contributed to the improvement of aircraft performance by its know-how in design and production of landing gear.

- Collaboration between technical departments aims at optimizing the production process by adopting new materials and/or new production methods. The supplier company improves technical performance by facilitating decision-making and consensus among the decision-makers in the customer company. For example, the prestige of the Rockwell Collins brand contributes to legitimizing the use of its data management systems on aircraft. In the same way, Latécoère, manufacturer of fuselage sections and doors has participated in this type of technical partnership with different aeronautics constructors such as Airbus, Bombardier, Lockheed, Dassault[14]

- Collaboration between the maintenance department of the supplier company and the production or maintenance department of the customer company reinforce loyalty. By incorporating the notion of regular investment in equipment maintenance into strategy, a company like Pratt&Whitney gains the confidence of its industrial customers – which are in turn more likely to stay loyal to their brand and even recommend it.

- Over and above the functional quality of the material, the industrial brand makes tracking possible. In the case of dysfunction or accident for example, the supplier brand can trace back a certain component or spare part to the producer.

It is easier to gain accross-the-board support for a brand that has demonstrated rigorous follow-up of quality control (maintenance) and its technical commitment (development).

Facilitating commercial performance

The aim of the supplier brand is to ensure that the customer company product is more easily sold than if it were commercialized under a competing supplier brand. This may be achieved by the juxtaposition of promises made to the final customer, by a close collaboration in communication campaigns and/or in the distribution channel. Thus, Boeing has a shared communication strategy with General Electric, based on the Boeing 777 and the GE90-115B engine (Figure 13.15). This co-advertising allows each of the companies to optimize their offer.

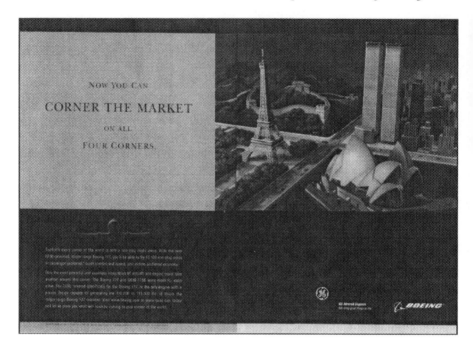

Figure 13-15. A communication partnership between Boeing and General Electric[15]

- Image partnership: this implies that the industrial customer benefits from the renown and positive reputation already gained by the supplier brand. The partnership of Intel® with PC manufacturers can be summed up in this way: "The demand for your products is good, but it will be even better if the Intel® brand appears on the PC as well." The Intel® brand communicates a certain promise to the consumer: a technical superiority of its microprocessors. By improving the performance of PC the supplier brand stimulates consumer demand. Co-branding thus contributes to both the customer's and the supplier's commercial performance.

- The partnership can cover both the product itself and the communication strategy. Thus, the supplier brand sometimes participates in the design or financing of the customer company's advertising campaign. This is the case of Airbus with the Biman Company, Bangladesh Airlines[16].

- Commercial performance is also facilitated through the sales force, particularly during events such as trade fairs. In the aerospace sector, it is common for an equipment supplier to allow its distribution partners to share its stand and invite their own customers.

Facilitating the operating performance of the customer company

From the supplier's viewpoint, this means improving the running of the customer company without directly acting on technical or commercial performance. By facilitating operating performance, higher productivity may be achieved along with a better working atmosphere and improved customer reception. This type of partnership concerns mainly professional service brands.

- Productivity improvement characterizes computer equipment and software suppliers for example. Through know-how and experience, a partner such as EDS (Electronic Data System) helps to improve its customer's management performance.
- A better working atmosphere can influence the efficiency of customer company teams. For example, thanks to catering companies like Compass, Sodexho-Marriott, the customer company can demonstrate to its personnel the efforts made to assure them comfort during lunch and rest breaks. In addition to improving the working atmosphere, this type of partnership contributes to the operating performance of the customer company.
- Improvement in the quality of receiving clients and visitors to the customer company is developed in the aim of increasing final customer satisfaction. This in turn gives a more positive image of the company, perceived as more efficiently run and results in higher loyalty.

AERONAUTICS AND SPACE BRANDS AND PERFORMANCE FACILITATION

In the aeronautics sector, aeronautics and space brands facilitate performance at the technical level. This is also true of parts manufacturer brands, which have become performance facilitators as far as in-house operations are concerned, increasing their customers' productivity and helping to improve working conditions.

However, aeronautics brands are not yet really involved in facilitating customer performance at a commercial level. Until now, there has been no brand strategy comparable to that of Intel® (Intel Inside), Lycra® (Du Pont de Nemours), Gore-Tex®...The struggle between suppliers and integrators for the largest added value in the final product is going to change the balance of power. Constructors will need to develop pull strategies, in other words, to influence final demand. The ultimate goal will be reaching the end customer and influencing his choice of a specific aircraft when he goes to purchase a ticket.

Thanks to branding, the supplier is no longer a simple provider of goods or services, but rather a partner, whose aim should be to improve customer performance. To achieve this, the supplier must rely on tangible product elements as well as intangible elements related to the brand. These two factors are difficult to separate.

2. *SPECIAL CHARACTERISTICS OF THE INDUSTRIAL BRAND*

Unlike consumer brands, industrial brands are generally not seen or are never purchased by end users. Concerning essentially professional customers, industrial brands can differ greatly depending on "purchaseability" and visibility by end consumers.

2.1 "Purchaseability" levels of the industrial brand

For an industrial brand, "purchaseability" can be defined as the ability of end consumers to buy products or services separately from the final product purchase. The glass bottle, for instance, can only be obtained by end consumers when wine or orange juice is purchased. The same is true for an Airbag system, which cannot be bought separately from the car purchase. Given the diversity of industrial brands, there are different levels of "purchaseability" (Figure 13.16).

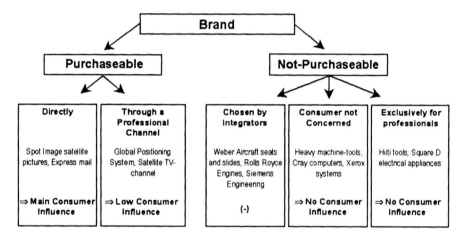

Figure 13-16. The criteria of "purchaseability" and "non-purchaseability" of industrial brands relative to final consumers

"Purchaseability" depends mainly on the product category's technical complexity and on the supplier strategy concerning non-professional customers (Figure 13.17).

Figure 13.17 shows that technical complexity restrains purchaseability, especially when:
- requirements in sophisticated equipment are important;
- there are some special conditions to use the equipment;
- a highly specific technical skills are required;
- there are severe regulations.

The decision to sell on the consumer market can be stimulated by opportunities for mass distribution, and thus potentially greater sales volumes. However, logistics difficulties, concern about maintaining a high level of quality when producing large quantities, the ability to maintain a satisfactory profit level for both suppliers and distributors and additional regulatory restrictions, can all hinder such a decision.

13. Brand Management

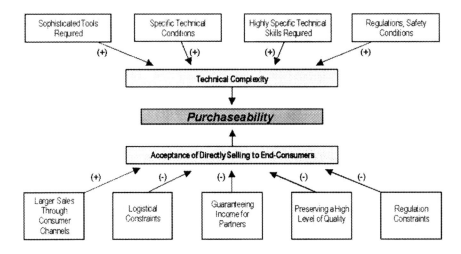

Figure 13-17. "Purchaseability" factors

2.2 The visibility strategy

Industrial brands can be visible by end consumers thanks to communication actions and the visibility strategy on the product itself. Brand visibility depends on whether the brand can be printed ("printability") or not. "Printability" can be defined as the ability for an individual to read the identity of the producer on the product when he purchases it or during the normal use of the product. The visibility depends also on the choice of the supplier and of his client to let the supplier brand appear on the final goods or not. As the diagram below shows, "printability" (Figure 13.18) depends on four main factors which are the product's physical support, the identification longevity, the visual access to the product and product transformations after delivery.

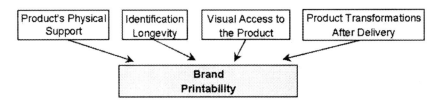

Figure 13-18. Main factors explaining the brand "printability"

The supplier brand visibility strategy involves to have its brand visible throughout the whole supply chain and especially on the final product. The aim of a this strategy is to increase awareness, to better differentiate its products and services and to allow customer to identify them during and even after their use. Brand visibility involves high marketing stakes for both suppliers and industrial customers. The latter can accept the presence of the supplier brand or not. The possibility of developing a long-term partnership with this supplier or benefiting from the supplier's investments in terms of communication incite customers to accept the visibility of their main suppliers' brand. On the opposite, he can refuse the visibility of his supplier's brand because of the fear of dependence on this supplier and the fear its products look more and more like those of competitors (Figure 13.19).

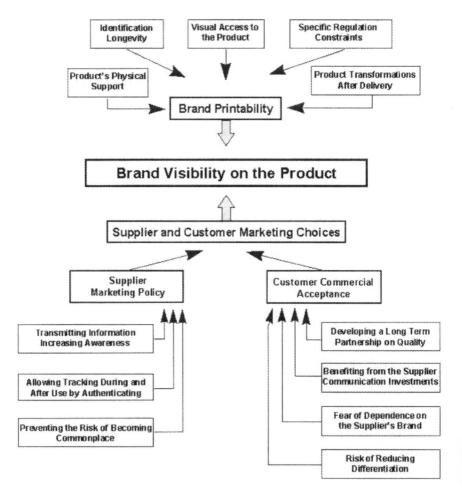

Figure 13-19. Factors explaining the brand visibility on the product

A visibility strategy for industrial brands involves being seen by all of the players in the economic chain, right through to the end customer. Backed up by a communication strategy (Figure 13.20), this can be a means of putting pressure on the industrial customer by developing downstream demand for his product.

13. Brand Management

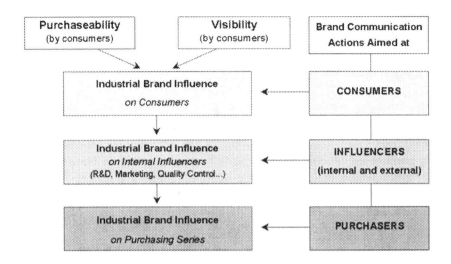

Figure 13-20. Industrial brands and a communication policy targeting the consumer market

With this strategy, the industrial supplier aims to transmit information (identity of the supplier, authenticity, traceability...), increase renown, and strengthen image (Figure 13.21). The goal is to influence final demand, creating a more balanced power relationship between the industrial supplier and the industrial customer.

Figure 13-21. Visibility of the engine brand: Rolls-Royce[17]

Airbus used this type of policy on the North American market when it ran advertisements aimed at the consumer public. Different scenarios presented the A340's competitive advantage: in first and business class, the seats are arranged two across thus minimizing discomfort for the passenger in the middle...

2.3 Airbus: "Setting the Standards"

The Airbus brand policy

The meaningful brand "Air-Bus" designates the product itself as well as the group that designs it. Airbus thus plays the role of both product-brand and corporate brand (Figure 13.22). Literally "a bus for the air", the brand has the advantage of being immediately understandable in English speaking countries. The initial aim of the company was to design, manufacture, and sell a 300 seat aircraft. This project provided the name for the first aircraft, the Airbus A300, upon which the brand policy was built.

⑨ AIRBUS INDUSTRIE

Figure 13-22. The Airbus logotype

The programs that followed were given names based on the Airbus corporate name and the first name of the first A300 aircraft. Thus the Airbus portfolio includes Airbus 310, Airbus A319, Airbus A320, Airbus A321, Airbus A330, and the Airbus A340.

The A380, the latest aircraft in the Airbus range, is coherent with the product family. The number chosen, to replace the temporary name, A3XX, situates the aircraft at the top of the portfolio (A380, A340, A330) In addition, it reminds the customer that the plane can be contained in a square 80 m by 80 m. Finally, in Asia, a key market for the A380, eight is a lucky number.

The name of each Airbus in the range varies depending on the technical characteristics and the chosen options (long or shortened models, etc.) such as the Airbus A310-300, Airbus A330-200 or the Airbus A340-600, etc.

The Airbus brand is often highly visible on the aircraft and can be seen for example on the rear fuselage or on the front door. Even inside the planes, passengers are aware of the brand, which is featured on the safety instructions and mentioned in the pilot's welcome speech.

The Airbus brand is synonymous with the aeronautics industry and high technology. It is perceived as an innovative and reliable brand that habitually shakes up the sector. Airbus also symbolizes the European capacity to design, mobilize, and win. Accordingly, for the 25th anniversary of the first flight of the A300 on 28th October 1972, Airbus launched a worldwide media campaign addressing the general public and highlighting its role in the construction of the European Community (Figure 13.23). Furthermore, the campaign recalled the brand's solid contribution, through its competitive innovations, to the modernization of the worldwide aeronautics industry.

13. *Brand Management* 419

The campaign was launched:
- In different European newspapers such as the Sunday Times in Great Britain, Le Monde and Le Figaro in France, Die Welt and Focus in Germany, Corriere Della Sera in Italy, El País in Spain, De Standaard and Le Soir in Belgium,
- In more international magazines such as The Economist WW, Aviation Week & Space Technology, Feer, Newsweek, Fortune and Asiaweek.

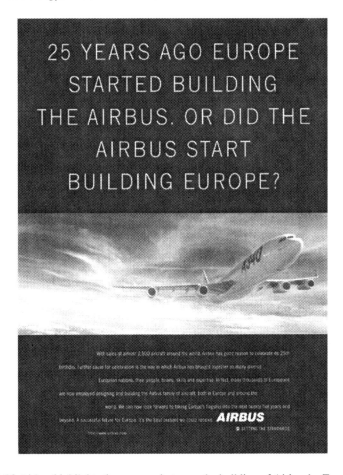

Figure 13-23. Airbus highlights the synergy between the building of Airbus by Europe and its own contribution to building Europe

On a more regular basis, Airbus's communication policy is based essentially on product communication oriented to professionals in the aeronautics industry. Airbus communicates mainly in technical journals or in specialized aeronautics magazines such as Aviation Week, Flight International, Airline Business, Air Transport World, or Air & Cosmos. The aim is to reach populations that are not directly targeted by the sales department, but who are influential in making decisions in airline companies and governmental bodies.

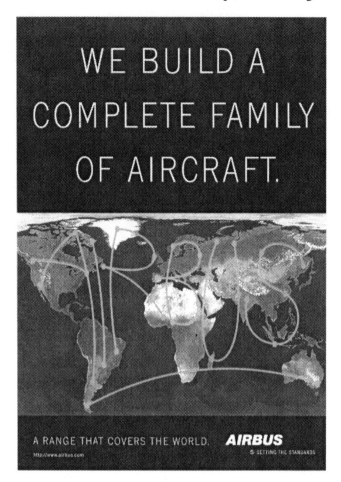

Figure 13-24. An example of Airbus brand corporate communication: the Airbus "signature" is obtained from different destinations in the world

Airbus brand policy also targets a wider public including private individuals and political decision makers. In addition to advertisements in national daily newspapers or in general economic magazines, Airbus uses press relations. For new orders or the signing of new contracts, the media generally features the expected positive effects on the European economy in terms of activities, employment, or the trade balance.

In general, television is little used by aeronautical builders. However, Airbus launched a major campaign on CNN with ads featuring the inside of the four-engine Airbus A340. A humorous tone was deliberately chosen, targeting the passenger in the middle. In this way, Airbus has tried to raise passenger comfort standards of long distance flights in North America by recommending a new arrangement of seats in the cabin.

13. Brand Management

Airbus: facilitating performance

Faced with solidly entrenched competition, Airbus, right from the beginning, developed original and innovative aeronautics programs. The result? In less than 30 years Airbus has gained control of half of the world civil aviation market. Today most of the world's major airlines are Airbus customers. The strength of the brand lies in its capacity to innovate and listen to customers. Thus the aircraft are designed in consultation with customer companies so as to define the expected cost, comfort, and performance requirements.

Taking into consideration customer expectations at the earliest stages, Airbus continually strives for new, better solutions. For Airbus, the goal is to establish the new market standards, as the brand slogan specifies "Setting the Standards".

Airbus's contributes to its customers' performance in many ways. First, there is the homogenous design of the cockpits, facilitating pilot training and making it easy to change from one aircraft in the range to another. Aircraft maintenance is facilitated thanks to the ergonomic design of equipment, and of course cost reduction, better knowledge of parts, and maintenance operations, etc. From a technical point of view, Airbus seeks to optimize aircraft performance and reliability, thereby reducing costs from stopovers, or holdups due to technical reasons.

For example, the Airbus A340 can take off filled to capacity in difficult weather conditions. This makes it possible to avoid reducing the freight transported by the aircraft, which can achieve its flight even in temperatures above 40°C on the ground, as is the case in Pakistan for example. This argument was featured in a campaign in which the Airbus A340 was presented as the highest performing aircraft "under the sun", in both a literal and figurative sense.

Figure 13-25. Airbus, technical performance serving airline companies and passengers

By improving technical performance, Airbus seeks to better satisfy airlines and their passengers. For example, the long distance range of the A340 eliminates long, costly, and annoying stopovers and leaves passengers feeling less tired at the end of their journey. In this way, Airbus helps the airline companies to improve their service and thus their sales performance.

Airbus has carried out a number of studies on the lines used by its customers or on passenger satisfaction. The airlines have specific and up-to-date information on their market and can improve their service. In addition to these marketing tools, the very image of the Airbus brand can reinforce that of customer companies by serving as a guarantee. The brand projects positive associations such as comfort, reliability, quality service, innovation, etc.

For example, Biman company in Bangladesh used the Airbus brand in its own communication campaign in Far Eastern Economic Review, insisting on the quality and interior comfort of the A310-300 making the flight particularly pleasant. This use of the Airbus brand is totally coherent with the Biman slogan, "Your home in the air".

Figure 13-26. Airbus: a guarantee brand facilitating Biman sales performance

The Airbus brand and in particular the A310-300 were also used by the airline Austria, in its communication policy. In its press campaign, the Asian company, Dragonair features the modernity of its fleet made up of A330's and A320's integrating the highest level of service and technology. Singapore Airlines, in its communication campaign, highlights its service presenting the A340 with all of its advantages (non-stop range, comfort, etc.). In a campaign targeting travel agencies, Air Lanka presented itself in 1994 as the first Asian airline to offer the comfort of the "most modern commercial aircraft in the world", the A340. The Spanish company Iberia and the British company Virgin Atlantic have also featured the Airbus A340 in their campaigns as a guarantee of quality and service: the technological advancement of Airbus ensures safer and more comfortable long distance flights.

Through its innovative policy and its commitment, Airbus has become a brand that significantly improves the performance of its customers both on a technical and commercial level.

3. INDUSTRIAL BRANDS CLASSIFICATION

3.1 *According to the use of goods*

Business to business brands can be classified according to how the industrial goods are used. From this perspective, there are three categories of business to business brands:
- Entering goods, or those integrated into the final product like for example Air Cruiser seats, Cordant components, Honeywell electronic systems;
- Production goods, or those used in the production process such as Siemens Engineering aircraft steps, aircraft such as the Boeing 737-400 or the Airbus A320-200 which enter into the "production process" of airline companies;
- Facilitating goods, which indirectly contribute to the production process such as IBM computers, Falcon or Fairchild business aircraft for the transportation of company directors, for example.

3.2 *According to the international brand policy*

Brands can also be classified according to their international policy. The globalization of markets have led aeronautics and space brands (Airbus, Boeing…) to follow a global brand policy (although this does not necessarily apply to all aspects of the marketing mix).

Other brands follow a local brand policy, using differentiated marketing strategy closer to the target market (positioning, cultural proximity…). Generally, parts suppliers use a local policy. This is the case of Zodiac[18] with its brands, Air Cruisers, Sicma Aeroseat, Weber, etc. This strategy focuses on one brand adapted to the local market (leaving room for greater creativity in the choice of an evocative or meaningful name for the brand, for example).

	"Identical"	"Adapted"	"Suggested"	"Indicated"	"Not Used"
Use of corporate brand name	Same in every country	Translated and written in different languages	Local subsidiary brand names are kept but the graphic style of the parent company is used	Local brand names are kept. The belonging to the global brand is indicated	Total local brand
Type of brand strategy used	Global brand strategy	Adapted global brand strategy	A combined global identity and local brand strategy	Local brand strategy	Local brand strategy

Table 13-27. Different strategies possible for the company brand internationally

In fact, there is no best strategy, given the diversity of possible options (Table 13.27) in terms of international presence, types of products or services offered, brands concerned (corporate brands, product brands, family name brands…) and targeted geographical areas. The advantages and disadvantages for the products as well as for the corporate name of either a global or a local brand policy need to be carefully balanced.

Today, business to business brand policies are increasingly sophisticated. In fact the trend is similar to what has happened over the last twenty years in the consumer market. More and more often, companies are turning to agencies specialized in the creation of brand names such as Name-it, Brand Institute, Lexicon Branding, Hundred Monkeys, Naming company, Nomen International, Demoniak, etc. This is particularly common in the wake of mergers or acquisitions. Generally there are two situations:

-the new name is created by combining the names of the two companies; for example, the merger of Hamilton Standard with Sunstrand, giving the corporate brand, Hamilton-Sundstrand

-a new name is created often using aspects of the preceding brand (logo, colors, names, initials, and syllables). This is the case of Detexis, from the merger of Dassault Electronique (DE), Thomson Electronique (TE) and XIS (system).

For aeronautics alliances[19], common companies are often created. The structure is often given a new corporate name. This was initially the case for Airbus and for Eurocopter. Joint ventures can also use a name created from the combining of partner names, in particular when there are few partners. Thus the defense electronics company, Thomson-CSF, renamed Thalès, joined an alliance with Raytheon. They called their common company, Thalès Raytheon Systems (TRS), world number one in air surveillance and detection with close to 40% of the market.

3.3   According to brand origins

Family named corporate brands

The history of the brand often explains the present company name as well as certain of their models. Brand names and logos are thus the result of different strategies, of discoveries and technical advances, which have marked the history of the companies, some of which are named after their founders. Family name brands often go back a long way; they are created from the company founder's name and maintained over the years. On the other hand, the logos of many original, founding companies have changed over time, with technology and the creation of new sectors.

In civil aviation, Boeing McDonnell Douglas, which is the result of several mergers and buyouts, owes it name to three respective founders: William Boeing, James McDonnell and Donald Douglas. First called Bloch (the genuine family name of its founder Marcel Bloch), the company later took the new name chosen by its founder, Marcel Dassault. The planes have their own names: Falcon for the civil aircraft, Rafale, Mirage for the military aircraft. The name of the company has changed from Avions Marcel Dassault-Breguet Aviation, to Dassault Breguet Group and finally Dassault Aviation. Dassault aircraft adopted the four-leaf clover logo favored by the founder. In the same way, parts suppliers such as Latécoère or Rolls-Royce have the name of their founders or co-founders, Pierre Georges Latécoère, Charles Rolls and Henry Royce.

While the founder's name can be used for the brand at the outset, it can also be used partially or totally to designate certain aircraft, even a whole range. This is the case of aircraft brands using their founders' names such as Andreï Tupolev, Ospiovitch Sukhoï, etc. The range uses certain parts of the name like, "LL", for Lirunov, "IL" for Ilyoushine, "YAK" for Yakovlev. This is the case for the "TU-144", Tupolev or the "AN-24", Antonov. McDonnell Douglas also use the name with DC for "Douglas and Co", while Lockheed use "L" and Vickers-Armstrong, "VC".

ANTONOV (*Oleg Constantinovitch*)	ILYUSHINE (*Sergeï Vladimirovitch*)
BELL (*Larry*)	LATECOERE (*Pierre-Georges*)
BOEING (*William*)	LOCKHEED (*Allan & Malcolm Loughead*)-MARTIN (*Glenn L.*)
BOMBARDIER (*Joseph-Armand*)	MCDONNELL (*James*) DOUGLAS (*Donald*)
BREGUET(*Louis*)	MESSIER (*George*)-BUGATTI (*Ettore*)
DASSAULT (*Marcel Bloch* pseudonym)	MIG (de *Mikoyan Gourvitch*)
DE HAVILLAND	RENAULT (*Louis*)-MORANE
DEWOITINE (*Émile*)	ROLLS (*Charles Stuart*) ROYCE (*Frederick Henry*)
DORNIER (*Claude*)	SIKORSKY (*Igor Ivanovitch*)
FOKKER (*Anthony H.G.*)	TUPOLEV (*Andreï Nikolaevich*)
GRUMMAN	YAKOVLEV (*Alexander Sergeevich*)

Table 13-28. Some aeronautical family brand names (and founder given names)

Evocative product brands

The old family name brands are usually immersed in the pioneer image surrounding the beginnings of aeronautics. With a long history, these brands are very well known and have developed a strong image over time. More recent brands such as Airbus, founded in 1967, owe their renown to the quality and success of their products.

The companies in this sector, quite apart from their corporate names, have aimed to make their product names easy to remember. Their policy has been to derive their brand names from the corporate name or from something sounding highly technical. Some companies have pushed this reasoning even further and developed names which are easily recognizable not just by the aeronautics professionals concerned. In this way names have been thought up which are easily remembered by the media and the general public. Rather than a "dry" reference like RAF-84XTIN, companies have chosen more meaningful names.

For example, in the military aeronautics sector, the names of predators have been used, like Aerospatiale Cougar (AS 532), Panther (AS 565 MB/SB) or Super Puma (AS 332 L1). The most recent helicopter is known as "Tiger". Aerospatiale then adds an abbreviation after the brand name to designate the type such as HAP for support/protection, UHU for multi-role, or HAC for anti-tank. The predator approach has also been used for the British Seacat, Tigercat, Python or Sidewinder (rattlesnake) missiles and in the Dassault Jaguar fighter aircraft. It should however be noted that predators and animal names are also used in civil aviation with Dassaults' Falcon or Rolls-Royce's Viper 630 engine.

Warrior words are especially common names for missiles, witness the American Trident, Harpoon, Tomahawk, the French Arpon (harpoon), the Franco-British Scalp along with the European, Eurofighter. Brands are also taken from the weather and words for movement such as in the Franco-British Storm Shadow missiles (Matra BAe Dynamics), the British Skyflash and Thunderbird missiles, the American Tornado and the French Dassault, Mirage and Rafale fighter aircraft.

During the Gulf war, the media coverage given to the Boeing-developed American Patriot missiles was facilitated by the almost universal meaning in the name and the fact that it was easy to remember. Whether it be for civil or military brands, there is always the question of the language of the brand name. As a general rule, helped by the ubiquitousness of English, American aeronautic equipment keeps its name or sales name regardless of where it is sold; but this is equally the case for the French Aerospatiale helicopters with their Ecureuil (squirrel), Colibri (hummingbird), Dauphin (dolphin), Super Puma, Fennec (military version of the Ecureuil), Super Five, Panther and Cougar brands.

These strongly evocative brand names give the products an extra dimension, setting them apart and making them easy to remember. Indeed this branding policy is often accompanied in the military sector as much as the civil one, by a brand support and promotional communication policy. While engine and parts manufacturers deal essentially with professionals within the sector, plane builders also communicate with the general public. This can be directly through TV campaigns, PR, sponsorship or by the visibility of their brand on the aircraft, or indirectly through press relations or by customer companies citing their brand etc.

The following table (Table 13.29) is a classification of aeronautical brands as a function of the product category (planes, helicopters, missiles/rockets and equipment) and brand nature (family name brands, importance etc.).

	Planes	Helicopters	Missiles/rockets	Equipment
Family name brands	Breguet, Boeing, McDonnell, Douglas, Lockheed, Dassault, Tupolev, Antonov, Ilyoushine, Yakovlev, Mikoyan Gourevitch (*Mig*), de Havilland, Grumman Messerschmidt, Dornier, Fokker	Bell, Sikorski, Grumman, McDonnell, Douglas		Messier-Dowty, Messier-Bugatti, Latécoère, Pratt & Whitney, Rolls-Royce, Lockheed-Martin, Renault-Morane
Meaningful brands	Airbus, Étendard, Caravelle, Concorde, Awacs (*Airborne Warning and Control System*), Aerospatiale	Eurocopter	Sidewinder, Iris-T (*Infrared Imagine Sidewinder Tailed Controled*), Slam, Pac-3 Patriot, Exocet, Matra BAe Dynamics (*Matra: mécanique aviation traction*)	Microturbo (Labinal), Turboméca (Labinal)
Brands which evoke natural phenomena	Mystère, Mirage, Tornado, Rafale	Typhoon, Thunderbold, Tempest, Hurricane	Meteor, Skyflash, Storm Shadow, Thunderbird, Aster, Hellfire	Airesearch (Garret)
Warrior brands	Eurofighter, Spitfire, Corsair, Crusader, Intruder, F5-Starfighter, Flying Fortress, Walkyrie	Apache, Commanche, Sea Knight, Kiowa Warrior	Trident, Arpoon, Harpon, Scalp, Tomahawk	
Animal name brands	Jaguar, Faucon, Falcon, F22Raptor, F15Eagle, Vampire, F5-Tiger, F20-Tiger Shark, F18 Hornet, F14-Tomcat, Fockewolf	Ecureuil, Fennec, Tigre, Panther, Puma, Cougar, Cobra, Frelon (hornet), Gazelle, Lynx, SeaHawk, BlackHawk, SeaDragon, Alouette, Super Stallion, Mosquito OceanHawk, RescueHawk, SkyHawk	Sidewinder (rattlesnake), Python, Seacat, Tigercat	Viper (Rolls-Royce)
Divine or mythological brands	Mercure, Hercule		Ariane, Titan, Hermès, Apollo	

Table 13-29. A classification of some of the aeronautical brands

Product-brands with a "technical" dimension

Quite apart from brand names coming directly or indirectly from a founder's name, there are numerous aeronautical brand names with an important technical dimension. This often involves the use of letters and numbers and their various combinations. For Boeing and Douglas, the letter F added to the aircraft brand name means that it is a " freighter " or cargo plane, e.g. B747 F or DC7 F. The Boeing 747 SR (short range) is a model destined mainly for Japanese internal flights allowing a large load (500 passengers) to be carried over a distance of about 650 km (frequent medium range journeys). Conversely, the Boeing 747 SP (special performance) with a shorter body, can carry 331 passengers non-stop for distances of up to 9500 km. The American armed forces classify their aircraft using one or two-letter prefixes to denote the category to which it belongs, e.g. A for attack (A-10 tactical support aircraft), B for bomber (B-52), C for carrier (C-5 Galaxy transport aircraft), F for fighter (F15). For its range of civil aircraft, Avro, British Aerospace (now BAE SYSTEMS plc) added the letters RJ, or regional jets, to this brand. One of their ad campaigns used the letters visually to show the positioning of its regional short haul aircraft (Figure 13.30).

Figure 13-30. A British Aerospace ad for the Avro RJ shorthaul regional aircraft range[20]

Letters and numbers are frequently used with aeronautical brands to show the chronological order of the generations of aircraft such as the McDonnell Douglas DC1, DC2, DC3, etc. or Dassault's Rafale A,B,C,D or the Ariane[21] launcher models 1 to 5. For its missiles, Aerospatiale uses code letters to indicate the range and so there is Trigat MP (Moyenne portée = medium range; 2000m) or Trigat LP (Longue portée = long range; more than 7000m). Both Boeing and Airbus have a brand policy using a combination of the first letter of the company name with a number (to which other numbers are sometimes added to denote different variants) to designate their aircraft range. Thus there are for example Airbus A300, A310, A319, A320, A321, A330, A340 and A380 or Boeing B707, B717, B727, B737, B747, B757, B767, B777.

Brands can come from technical abbreviations such as in the case of the Patriot Advanced Capabilities missile or Pac-3 and the Sea Slam derived from Stand-off Land Attack Missile. The names of other American missiles use similar abbreviations: Slam Er (Stand-off land Attack Missile Expanded Response), GBU-15 (Guided Weapons System), Jassm (Joint Air-to-Surface Stand-off Missile), Jdam (Joint Direct Attack Munition) etc. In the same way Boeing's E-3A Awacs come from "Airborne Warning and Control Systems".

The technical sounds are sometimes very evocative as is the case with Snecma's commercial brands: Microturbo and Turbomeca, given to engines, turbines and turbojets for helicopters and planes. Microturbo, the manufacturer of aeronautical starters and propulsion systems sometimes uses abbreviated names as brands e.g. TGA15 (turbo-générateur d'air) for their air turbo generator. In a similar fashion Pratt & Whitney have developed evocative names like Turbojet or JT3D Turbofan, for engines fitted in particular to Boeing 707's.

Explicit brands

These are brands whose names are explicit enough for the "customer-promise" to be directly understandable. These brands are effective in terms of memorization, recognition, and suggestion. Some examples are the waterproofing product, Skyflex®, from Gore-Tex®, the aeronautics constructor, Airbus, the airline, EasyJet, the company specialized in on-board films for aircraft, AirTV, or the alliance of airlines, Skyteam[22], among many others. An other example of explicit brand is that of NoiseGard® from Sennheiser which clearly express the product advantage that is, the better quality of sound and communication within the cockpit thanks to sophisticated systems used to reduce noise (Figure 13.31).

13. Brand Management

Figure 13-31. NoiseGard, a Seenheiser brand[23]

Brands formed from acronyms or made-up of initials

These are the brands whose letters correspond, for example, to the initials of long names. This is the case of Awacs, for Airborne Warning And Control System, EADS for European Aeronautic Defence and Space Company. This shortened form makes it easier to read and memorize the brand name. Many brands are better known by their shortened name than the complete original name such as BAe for British Aerospace or GE for General Electric.

Brands using specific endings

These are generally high technology and scientific brands. Among the endings used are: " on " (Textron), " el ", " ex ", " ax ", " ix ", " ium " (Astrium), " is " or " xis " such as in Detexis (Figure 13.32).

432   *Aerospace Marketing Management*

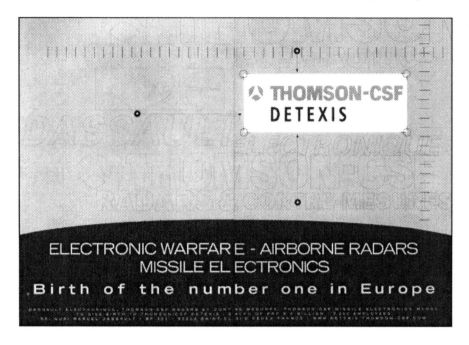

Figure 13-32. Ad presenting the new name of the merger between Dassault Electronique and Thomson-CSF: Detexis[24]

Airline brands

Most airline names are of national or geographic origin. However, some companies, especially the more recent ones, have taken different names such as Sun Express, Star Airlines, EasyJet, Air Liberté, Air Alfa, Lauda Air (family name), Virgin, Go, Futura and Reno.

Brands directly named after the nationality of origin	Brands evoking the geographical zone
Emirates Iran Air Kuwait Airways Oman Air Qatar Airways Royal Jordanian Saudia Syrian Arab Airlines Yemenia	Gulf Air Middle East Airlines

Table 13-33. Main airline brands in Middle Eastern

13. Brand Management

Brands directly named after the nationality of origin	Brands evoking the geographical zone	Meaningful or suggestive brands
Air India	Aero Asia	Airlink
Air Lanka	Air Pacific	Bouraq Indonesia (bird)
Air New Zealand	Asiana Airlines	EVA Airways
All Nippon Airways	Far Eastern Air Transport	Jet Airways
Biman Bangladesh Airlines	Garuda Indonesia Airlines	Kantas Airways
China Airlines	Trans Asia Airways	Mandarin Airlines
Indian Airlines		Merpati Nusantara (bird)
Japan Air System		Silk Air
Japan Airlines		Uni Airways
Japan Asian Airways		
Japan TransOcean Air		
Korean Air		
Malaysia Airlines		
Pakistan International (PIA)		
Philippines Airlines		
Royal Brunei Airlines		
Singapore Airlines		
Thai Airways International		
Vietnam Airlines		

Table 13-34. Main airline brands in Asia and the Pacific area

Brands directly named after the nationality of origin	Brands evoking the geographical zone	Meaningful or suggestive brands
Aerolineas Argentinas	Aero California	ACES
AeroMexico	Rio Sul	Aero Ejecutivo
Aeroperu	Transbrasil	AeroContinente
Air Jamaica		Aerolineas Internacionales
Cubana		Aeropostal
Ecuatoriana		Aerorepublica
LAN Chile		Air Aruba
Lloyd Aereo Boliviano		Austral
San Colombia		Avensa
		AVIACSA
		Avianca
		Aviateca
		Copa
		Dinar Lineas Aereas
		Lacsa
		Ladeco
		LAPA Lineas Aereas Rivadas
		Saeta
		Servivensa
		Taca
		Taesa
		TAM
		Varig
		Vasp

Table 13-35. Main airline brands in Latin America

Brands directly named after the nationality of origin	Brands evoking the geographical zone	Meaningful or suggestive brands
Air Canada Air Wisconsin Alaska Airlines America West Airlines American Airlines Atlantic South East Hawaiian Air Midway Airlines Midwest Express North West Airlines South West Airlines US Airways	American TransAir Canada 3000 Reno Air United Airlines (USA & Airlines)	Air Nova Air Transat Aloha Airlines Continental Airlines Delta Airlines Frontier Horizon Air Kiwi International Skyservice Sun Country Airlines Tower Air TWA (TransWorld Airlines) West Jet Airlines

Table 13-36. Main airline brands in North America

Brands directly named after the nationality of origin	Brands evoking the geographical zone	Meaningful or suggestive brands
Air Berlin Air France Air Holland Air Malta Alitalia Austrian Airlines British Airways British Midland Caledonian Airways Cyprus Airways Finnair Icelandair Istanbul Airlines KLM (Royal Dutch Airlines) Spanair (Spain) Swissair TAP Air Portugal Turkish Airlines	Adrian Airways Aer Lingus (clover) Balkan Brittania Airlines Crossair (Swiss flag) Deutsche BA El Al (Israel Airlines) Iberia Lot (Polish Airlines) Malev (Hungary) Olympic Airways (Olympic from Greece) Sabena (Belgium) Tarom Airlines (Romania)	Aeroflot Air 2000 Air Alpha Air Liberté Air Littoral Air One Air Tour Aviaco CityBird CityFlyer Express Condor Corsair (Corsica & Corsair) Easyjet Edelweiss Air Eurofly SPA Eurowings Flying Colours Futura International Go Airlines KLM City Hopper L'Aeropostale Lufthansa Luxair (Luxembourg & luxury) Monarch Airlines Pegasus Airlines Star Airlines Sun Express Virgin Atlantic Virgin Express

Table 13-37. Main airline brands in Europe

Brands directly named after the nationality of origin	Brands evoking the geographical zone	Meaningful or suggestive brands
Air Algérie Air Gabon Air Madagascar Air Mauritius Air Tanzania Air Zimbabwe Cameroon Airlines Egypt Air Ethiopian Airlines Ghana Airways Kenya Airways LAM Mocambique Nigeria Airways Royal Air Maroc South African Airways Sudan Airways Tunis Air	Air Afrique Comair (Comores)	Belview Airlines Lotus Air Sun Air

Table 13-38. Main airline brands in Africa

Brands directly named after the nationality of origin	Brands evoking the geographical zone	Meaningful or suggestive brands
Air China Air Macau China Eastern Airlines China North West Airlines China Northern Airlines China South West China Southern Airlines China Xinhua Airlines China Yunnan Airlines Hainan Airlines Shanghai Airlines Shenzhen Airlines Sichuan Airlines Xiamen Airlines Xinjiang Airways Zhejiang Airlines	Cathay Pacific Airways	Dragonair Swan Airlines

Table 13-39. Main airline brands in China

First and foremost, before companies or public organizations, airlines target individual consumers, and until fairly recently, consumers from the same country as the airline. This explains why most airlines (~60%) include geographic origin in the name whether the name of the country or the region. This is especially true for US and China.

Airline brands named after the geographic origin are especially important in South East Asia, China, Africa and the Middle East, while brands with meaningful names make up around 36% of all brands. The latter are especially common in

Europe and North America. South America on the other hand has a high proportion of airlines with initials for names. Overall, unlike for aeronautics constructors or parts manufacturers, family names are rarely used for airlines.

In conclusion, industrial brand policy is increasingly sophisticated both for the entering goods sector (Cordant, Gore-Tex, Sextant, BF Goodrich, Rockwell Collins) and equipment goods (Siemens Engineering, Embraer, Boeing, Airbus, Allied Signal, Messier-Dowty, Pratt & Whitney, Snecma...) as well as for professional services (UPS, Fedex, DHL, EDS, Smiths Industries, AirTV, Middle River Aircraft Systems...). This trend involves greater use of the media (including mass media such as television and billboards) and stems in part from the determination of supplier companies to acquire greater visibility. Aeronautics companies are paying more and more attention to creating a coherent identity, in particular through the logo, the slogan, and in general, respect of a visual identity code.

4. *VISUAL IDENTITY CODE, LOGOS AND SLOGANS*

A brand can only rarely be summed up by its name: various different elements contribute to its identity, which progressively develops throughout its life. In addition to the numerous evocations and associations, and tangible product or service attributes brought to mind by the brand, identity is also built on distinctive elements such as the logo, slogan or jingle. Representing the brand and communicating its basic values, these elements must be carefully managed starting with establishing an effective visual identity code.

4.1 Logos

The brand and the different elements displayed with it become so inextricably associated that the brand might be unidentifiable without them. The graphic unity of these different components form the brand's visual identity.

The latter is strictly defined in order to preserve coherency whatever the support used (poster, brochure, Internet site, company car, work-clothes, TV ad, etc.). A company's visual identity should reflect its values – in other words, its culture, personality, vocation and projects for the future – in a recognizable tone and style. The defined visual identity may contain up to four elements:
- a typographic transcription of the brand (name, slogan),
- a visual symbol (logo),
- a color code,
- an aural code (jingle or associated musical theme).

While obviously music is not visual, it is directly associated with advertising images and therefore is an integral part of the brand's identity. Visual identity should be long term to have a significant effect on company awareness and strength.

13. Brand Management

Evolving with time, it should follow the different phases of company development, reflecting any changes in company image (Brun and Rasquinet, 1996).

The logo constitutes the main element in brand identity. In the past, military colors showed who was friend or foe; heraldic arms communicated the identity of the family or country. Today, the logo has the same role, both in-house and in "commercial and marketing battles", each company adopts its own "crest" with color code, type face, motto, and slogan.

The logotype or symbol represents a complementary – often indispensable – element for building brand awareness and consolidating associations. The logotype makes the brand easier to recognize and quicker to identify; image processing rather than word processing is in fact easier for the brain.

Thus the logo should:
- Translate, over the long term, the image of the company (and its possible evolution),
- Be easy to spot,
- Be easy to read and to memorize, whatever the media support,
- Be accepted from the outset by the different populations connected with its use (federating role of the brand).

Certain logos also communicate the brand's activity: the Boeing logotype symbolizes flight around the world, thus evoking the company's aerospace activities (further Figure 13.40). Other logotypes are not as referential, but nevertheless clearly identify the company. This is the case of the General Electric logotype which includes its initials (Figure 13.40). Used alone, the logo identifies the company on packaging, vehicles or work clothes.

Figure 13-40. General Electric logotype

To build a solid image, the logotype should be the graphic transcription of the company, coherently combining the brand name, type face and graphics chosen. Different impressions will be created by juggling with the complexity, colors, modernism or classicism of the type face. Narrow or italic letters are dynamic; Helvetica, Arial or Universe type convey brand force and aggressiveness Times suggests balance, elegance and classicism. In the same way, different colors evoke different sectors of activity, concepts, and references depending on the culture. Colors may even have opposing associations: warm colors may evoke a friendly atmosphere – or danger.

Because of the relatively low cost of creating a logotype and its efficacy in terms of memorization, many industrial firms, even small ones, have developed strong visual identities. In fact, the logo is one of the first items of information communicated by the company to its clients or suppliers. A well-designed logo can strengthen a company's position, increase awareness and memorization of the company name, even reinforce its meaning and consolidate brand positioning.

Although it would seem relatively easy to create a logo, it does in fact require serious thought and often creates problems. When creating a logo one must take into consideration the fundamental values of the company, its competence and know-how, markets, partners, etc. The logo should not detract from the existing image of the company but rather reinforce it. When it is necessary to change symbols, there is not only the difficulty of harmonizing the new with the old, but also the problem of re-educating the various target populations (beginning with in-house). However, such changes can inspire new energy, communicating a new company dynamic; the brand identity should progress while preserving the heritage of the former symbol (Aaker, 1994) (Figure 13.41).

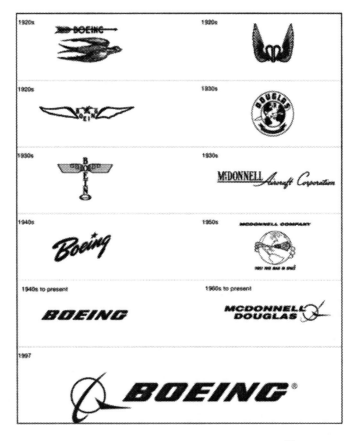

Figure 13-41. The evolution of the Boeing logos[25]

4.2 Slogans

Slogans are generally very concisely formulated in order to make memorization easier and facilitate automatic association with a product or service, the company and/or the brand advertised. When associated with a logo, a brand, or a company name, the slogan is the translation of the company's key message, its signature and even its "philosophy". Just as there is a variety of logos, there are many different types of slogans: from highly descriptive ones to the more general, from those that are almost aphorisms, to those that are quasi esoteric (Tables 13.42, 13.43 and 13.44). Slogans are often registered in the same way as the brands themselves.

Companies	Slogans
AirLiance	For airlines by airlines
Applied Signal Technology	Technology is not just our last name, it's our advantage
BFGoodrich	Creating value through excellence in innovation, quality and people
CFM	The power of flight
Cordant Technologies	When it better work, we make it work better
Crane Aerospace	Delivering more
Eros	Trust in the air
GE Aircraft Engines	We bring good things to life
Hamilton-Sunstrand	The aerospace power company
Honeywell (GE)	We make perfect sense
ILS (Lockheed-Martin)	The power to lift the world
International Aero Engines	Building on a tradition of excellence
Jet Aviation	The only global business aviation service company
Loral (Space systems)	Performance is the best strategy
Lucas Aerospace	A big company with big ideas
Martin-Baker	Another reason why flying's safer than driving
Messier-Dowty	The landing gear company
MTU Maintenance	Anytime. Anywhere. We care.
OAG	The world leader in business travel information
Pratt&Whitney	Smart engines for a tough world
Rolls-Royce	Advanced Power Systems for Tomorrow's World
Sagem	Technology in action
Shell Aviation	Shell is Flying
Singapore Technologies Aerospace	Taking you higher
Sita	The sky is not the limit
Tag Aviation	A higher altitude
Thiokol	Aerospace and industrial technologies
Thomson-CSF / Thales	Securing your future
TRW Aeronautical Systems (Lucas Aerospace)	Passion for excellence
Turbomeca	Powered by passion
Via Inmarsat	Broadband your horizons

Table 13-42. Some examples of slogans of aeronautical and space equipment manufacturers

Companies	Slogans
Airbus	Setting the standards
Alenia	Technology powered by imagination
ATR	The front-runner turboprop
Casa	The things we do, we do the best
Avro (BAe)	The regional jets of choice
Bell Helicopter	Today's world demands
Boeing	Forever new frontiers
Bombardier Aerospace	Ask us for the world
Embraer	Working together to reach a higher standard
Israel Aircraft Industries	Innovation is our business
Eurocopter	Redefining satisfaction
DaimlerChrysler Aerospace	Engines for the world
Northrop Grumman	Technology's edge
Raytheon	Expect great things
Dassault Aviation	Your air force
Eurofighter Typhoon	Superiority for pilots

Table 13-43. Some slogans of constructors

Companies	Slogans
Abu Dhabi International Airport	Many happy returns !
Aéroports de Paris	The shortest way to your satisfaction
Air France	Making the sky the best place on earth
Air France Industries	Winning the hearts of the world
Atlas Air (cargo)	Dedicated to your success
British Airways	The world's favourite airline
British Airways Engineering	The difference is a world of experience
Cologne / Bonn	...Here we go
Copenhagen Airports A/S	The rising star of northern Europe
Fort Lauderdale-Hollywood International Airport	The easy come, easy go airport
Hartsfield Atlanta International Airport	Hub to the world
Lufthansa Flight Training	The perfect way up
Lufthansa Technik	More mobility for the world
Macau International Airport	Gateway to China
Miami International Airport	All the right connections
Munich International Airport	Service nonstop
Swiss Air	World's most refreshing airline
Vienna International Airport	Europe's best address
Williams Gateway Airport	The perfect climate for airlines

Table 13-44. Some slogans of airlines and airports

When a brand name does not adequately describe the activity, it is sometimes necessary to add an "explanatory" slogan in baseline. For example, CFM signed with "The Power of Flight", referring to the engine propulsion systems.

Slogans enrich emotions and associations induced by the brand name, facilitating memorization of information concerning activity and values. The slogan is also a complementary support capable of rectifying the image of a brand. Compared to the brand name, which is a real summary of information, the slogan is generally placed in a discreet position – in support – to recall the essential message. Its life is usually shorter than that of the brand name. Because of its discreet role, the slogan can communicate new messages for the brand (new activities, mergers, and new strategies). To avoid changing slogans too often, a general signature is mainly used in order to translate the basic spirit of the company.

Slogans are usually placed under the brand name or to the side. Certain are deliberately placed at a slight distance to give the effect of a brand echo. When logos are communicated in an audiovisual form rather than a written one, the slogans reappear on the screen or on voice-off to reinforce the contents of the brand and its message. Some industrial brands also use jingles or music as a signature.

4.3 Jingles

Jingles are a musical or audible signature added to the brand, logo or slogan. The aim is to facilitate memorization of the brand, its logo, slogan, and message by the memorization of the musical accompaniment.

Although this technique is used less in the industrial sector, it has become quite popular in business to business with the development of industrial brand mass communication to the general public (via television, radio, etc.). It is mainly used, however, by those companies with a mixed clientele (general public/professionals). Service companies such as temporary work agencies, banks and insurance companies also use jingles. Such is the case for IBM whose audiovisual campaign has used the same music for the last few years.

The classic example is Intel whose five notes are systematically played in each advertisement every time the brand is mentioned – that means not only when it is an Intel campaign but also those of its clients like Compaq, Hewlett-Packard, Dell, etc. As the memorization rate of a jingle is high, Intel's reputation and brand awareness grows accordingly. The use of the Intel jingle results from an agreement between the company and its clients to mention the presence of Intel components in their products.

4.4 Visual identity code

Because of the diversity of elements which are associated with a brand and its evolution (changes, abandon, new creations, etc.) it is vital to have general guidelines containing details of just how the various entities linked to the brand should be used. These guidelines, which are a sort of internal reference document, help the company to coordinate the elements contributing to brand image. They define in a concrete manner how the brand should be written, the reproduction of the logo, any size and placement constraints regarding the logo, slogan, brand-name, and the jingle. In general, these guidelines serve as a visual identity code (even though they may include reference to a jingle). It is of particular importance in the aeronautics and space sector where companies are set up around the world.

Visual identity codes or normalization guidelines define a standard usage and application of the logo, be it on the product, packaging, letter-heading, vehicles or any other visual support used by the company such as Internet. By respecting the established code – whatever the means of communication – a greater global brand coherence is perceived by the different target populations. Thus use of the Boeing logo or slogan must be authorized by the company. Specific usage rules are given in the Boeing Corporate Identity Code[26] as to color, form, and layout of the different graphic elements.

The visual code specifically covers:
- the make-up of the logo and the different elements it contains,
- the position of these elements, their size and relative proportions,
- the overall size of the logo and the smallest permissible size when printed,
- the black and white version specifying the nuances of black,
- the exact composition of the colors used (single or four-color),
- the possible different backgrounds on which the logo can be used, and how it should be adapted (colored background, four-color, very dark background, etc.),
- the exact position of the logo on headed paper. The type face and any modifications which may be necessary (spacing, style), size, line spacing, incorporation of the company address and corporate name,
- the exact position of the logo and the details to appear on business cards; the typography and use of color for envelopes, letter paper and bills are also defined,
- the characteristics of the logotype (position, color, typography, size..) defined for all possible applications: on vehicles (stickers...), press ads, television commercials, signs, banners, flags, neon signs, street signs, totems, exhibition hall and stand decor, etc.

The visual identity code enables a company to reinforce its visual identity, making its communication policy more effective. There is greater consistency between brands, as well as a rallying of the company's different activities. Internationalization is facilitated by the adoption of a universal "language" and by the consolidation of a sense of belonging within the company. By controlling the image more closely, the company is better recognized by the market and improves its reputation.

5. *LATÉCOÈRE: TECHNICAL PARTNERSHIP AND ITS OWN PRODUCTS*

5.1 *The rise of Latécoère*

Founded in 1917 by Pierre-Georges Latécoère in Toulouse, the Société Industrielle d'Aviation Latécoère started by making warplanes such as the Salmson

2A2 or the Breguet XIV. At the end of the First World War, the company became involved in civil aviation with a special interest in extending its aircraft range.

These efforts allowed Pierre-Georges Latécoère for example to make the first air crossing of the Pyrénées when he flew from Toulouse to Barcelona one and a half months after the armistice. In 1926, another project close to Pierre-Georges Latécoère's heart was achieved with the crossing between France and South America. In this way, the company designed and developed more than 83 aircraft study projects, made 33 prototypes (most of which were the largest of their era) and set up 11 series production aircraft for commercial airlines or military seaplanes for the Navy. Latécoère aircraft earned France 31 world records: in 1930 for example, Mermoz made the first seaplane crossing of the South Atlantic in a Latécoère Late 28.

LATECOERE

Figure 13-45. Latécoère logotype

In 1939, after having taken part in the French war effort and in the face of mounting clouds on the political horizon, the company gave up most of its factories. In this way it reduced its production potential and focused on activities which the enemy could not use. At the end of the Second World War, the company, which had in fact secretly continued its research and hidden aircraft parts, was in a position to offer the type Lionel de Marmier 75 ton Latécoère seaplane reinforcing its world lead in heavy lift aircraft.

Figure 13-46. Following World War I, the Latécoère France – Spain – Morocco line was born

5.2 Latécoère, technical performance facilitator brand

Latécoère, intermediate equipment supplier for manufacturers

With the restructuring of the French aircraft industry in the fifties, the Latécoère company began to cooperate with various firms, by constructing tools and aircraft parts. The accumulated know-how and experience of the company had, for this period between 1950 and 1970, allowed it to become the specialist in environmental tests and mechanical trials and tests, as much for materials as for various structures. In the nineties, the company began to turn more towards the international scene, becoming the respected partner of the main international contractors such as Airbus, Boeing, McDonnell Douglas, and Dassault as well as participating in the main world aeronautical programs in the electrical-electronics field and on-board and ground equipment.

In 1998 the Latécoère company had developed a large amount of know-how and acquired great experience in electronics (cabling and wiring looms, automatic testers, equipment systems and avionics, etc.) and composites (carbon fibers, Kevlar®, metal-metal bonding, etc.). It was one of the world leaders for human centrifuges for pilot training, astronaut and medical research.

With a turnover of 160 million euros, the Latécoère company is now one of the leaders in:

- Studies and production of aeronautical structures and electronic assemblies: EADS, Dassault Aviation, Hispano-Suiza, Rohr, Boeing, Lockheed, Northrop Grumman, etc.
- Studies and production of satellite structures, on-board equipment and ground equipment for space activities: CNES, EADS, Alcatel Space, Matra Space, SEP, BAe, etc.

While it is often true that family names do not indicate any precise information but just recall the founder's name, the case of Latécoère is very different. In fact there are few family names as evocative as Latécoère. This word, which recalls one of the most famous names in aviation, evokes memories of the beginnings of aviation and symbolizes the world of aircraft. The brand is thus known to the general public mainly because of its previous activities of designing and constructing aircraft. However, the skills of the company have evolved and the brand, whose image and renown are still well recognized by the general public, is today in fact known by professionals and the aeronautical world:

- For its expertise and design activities in the production of structures (fuselages),
- As well as for its complete, high tech equipment (centrifuges, etc.).

In fact, Latécoère is an intermediate and finished equipment supplier brand. The brand's objective is to contribute to its customers' success by enabling them to

improve the performance of their products or processes. Thanks to a sustained effort in research and development and active collaboration with different partners, Latécoère has developed very high level equipment for:
- Aircraft structures: fuselage, wing and nacelle elements,
- Equipment and systems: looms and electrical assemblies, on-board equipment, human centrifuges, air combat restitution systems, ground equipment.

High aeronautical standards and requirements necessitate the design and production of high quality aircraft structures. Any contribution to the technical performance of constructor customers, necessarily involves the development of new solutions which allow a gain in assembly time or facilitate maintenance, but equally improve aircraft performance (weight, resistance, etc.). And this is how Latécoère has developed numerous technical partnerships, allowing it to design and offer high tech and high performance equipment.

The fuselage, wing or nacelle parts produced by Latécoère, have been used in the following programs:
- Aircraft from the Airbus range A319, A320, A321, A330, A340-500, A340-600,
- Military aircraft ATL2, Lockheed C130, Rafale,
- Business aircraft Falcon 50, Falcon 900, Falcon 2000,
- McDonnell Douglas aircraft MD 90 and MD 11,
- The Eurocopter helicopters.

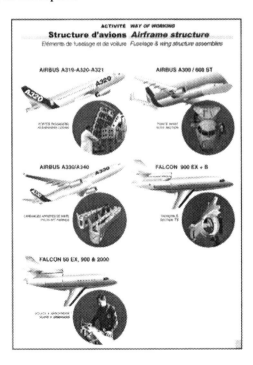

Figure 13-47. Some fuselage elements produced by Latécoère

446   *Aerospace Marketing Management*

As an equipment manufacturer of aircraft structures, the Latécoère brand is also associated with on-board equipment such as the software, limiting battery charging on the Airbus, pilot arming straps, the looms for the CFM 56 engines, the radio cables on the Falcon 2000 or the adapters for test programs. It is also present in ground equipment such as the cryogenic booms for Ariane 4 or the on-board/ground links for Ariane 5.

Latécoère, supplier of stand-alone equipment: centrifuges

Quite apart from the specialization in aircraft sub-assemblies for the main contractors, the brand also develops stand-alone equipment for sale to constructor customers as well as airline companies, like for example the training systems for air combat restitution (SEMAC) and in particular the human centrifuges.

Figure 13-48. Latécoère equipment

In parallel with the development of aeronautics and aircraft performance, the pilots have been submitted to higher and higher G forces. Under high G loads, the pilot may lose consciousness as for example in air combat maneuvers. To deal with this risk and to preserve the plane and its pilot, it is necessary to train them in centrifuges capable of recreating G forces using specific systems simulating flight changes and accelerations. In the same way, medical research laboratories need human centrifuges to study the effects of high G forces, to determine their long-term consequences and develop means of protection against these effects.

The centrifuges developed by Latécoère are safe and reliable systems, which allow extreme tests to be carried out while at the same time permanently monitoring the pilot's well being. In this way pilots can train themselves to recognize and understand their resistance and physical reactions. In addition, the airlines have a device for safety purposes and training which is capable of simulating the most difficult situations encountered by pilots.

5.3 Sales performance facilitation

With a view to improving the performance of airline companies, Latécoère has developed a passenger video camera system, adaptable on Airbus as well as Boeing. This consists of on-board cameras called Landscape Video, which are designed to film, on board the aircraft and in real time, the landscape passing below. The passengers in this way "see" from their seat the journey unrolling and the various geographical features. This innovation from Latécoère helps the airlines to distinguish their offer and contributes to their own innovation. By offering views of the exterior to civil airline passengers using the " landscape video camera system " the company has a plus product. With this invention, Latécoère has shown that it is concerned with improving its customers' sales performance[27].

6. ZODIAC: MANAGING A BRAND PORTFOLIO BY SECTOR

6.1 History

In 1896, Maurice Mallet, an aeronautical engineer of the late 19th century, founded the aerodrome management company Mallet, Melandri and Pitray. In 1908, he created a dirigible balloon company specializing in the study, construction and commercialization of aircraft. In February 1909, after the launch of the first balloon, the brand name Zodiac was registered along with its logo showing a dirigible superimposed over the signs of the zodiac (Figure 13.49).

Figure 13-49. Logo of the Zodiac brand, registered in 1909 by the French Dirigible Balloon Company (Société Française des Ballons Dirigeables)

In 1911, the company – renamed Zodiac – diversified into monoplane and biplane manufacture. In 1934, a Zodiac engineer named Pierre Debroutelle had the idea of using material normally employed on dirigibles to build the first inflatable floater canoe: the inflatable boat was born. Because of its great practicality and high performance it had numerous applications in the military field, and also in the leisure sector. Doctor Bombard used this inflatable boat to cross the Atlantic Ocean in 1952 (just fishing and using the water from the rain and the ocean). The brand quickly became well known, liked, and trusted. The product was used in situations where the forces of nature had to be confronted and safety and performance were key factors. Allied with this was a pioneer spirit and innovative image.

Although for most people today, Zodiac is associated with inflatable boats (over a million boats sold worldwide) the company continues its aeronautical developments. In going back to its roots, the company is re-confirming its vocation of mastering air and water (Figure 13.50).

Figure 13-50. The new logo adopted in 1978 and the emblem of the 100th Anniversary of the company: symbol of continuity of the pioneering spirit combining the old logo (in the form of a dirigible) with the latest one (Z)

13. Brand Management

6.2 Activities of the Zodiac Group today

The group has a turnover of about 1 billion euros, 80% of which is generated in the foreign market. It employs 6,500 people and has three main branches of specialized activity:

- Aeronautical equipment: associating aeronautical activity targeted at constructors (civil and military aviation, helicopters, engines), and civil and armed services such as emergency evacuation systems, airbraking systems, de-icers, non-rigid fuel tanks, composite materials, super-insulators for satellites, scientific and weather balloons...
- Airline equipment: including activities concerning cabin equipment such as passenger seats, technical seats, modular systems, on-board sanitary systems for civil airline companies...
- Marine-leisure: oriented mainly to the non-specialist market (consumers), three divisions:

    - marine: ranges of inflatable and semi-rigid boats,
    - swimming pools: ranges of freestanding and sunken pools, spas, cleaning robots...,
    - leisure: ranges of beach articles, inflatable, promotional and leisure items such as canoes and kayaks.

Figure 13-51. World leadership areas of the Zodiac group

Zodiac and its different subsidiaries are world leaders in:
- Aeronautical equipment, world leader in emergency escape slides for civil aviation and floatation systems for safety-service helicopters (Air Cruisers, Aerazur), parachute systems (Aerazur, EFA, Parachutes de France, Pioneer), non-rigid fuel tanks (Amfuel, Superflexit, Aerazur, Plastiremo),
- Airline equipment, world leader in passenger seats for civil aircraft (Sicma Aero Seat, Weber Aircraft) and world leader in on-board sanitary systems (Mag Aerospace),
- The marine-leisure sector, world leader in inflatable boats (Zodiac, Bombard Jumbo), in inflatable water-sport items (Sevylor, Baracuda), free-standing pools (Debes and Wunder, B. Kern, Muskin, Europool) and pleasure-boat safety rafts (Zodiac, Bombard).

Almost two thirds of the group's activity concerns professional customers for whom Zodiac designs and commercializes different systems and aeronautical equipment. The internal and external growth strategy has enabled the group to widen their offer, making the best of technology synergies.

6.3 Brand policy

The group's brand policy is dictated by the renowned corporate brand Zodiac for the marine-leisure mass market branch. In this domain, the brand name Zodiac is visible, appearing directly on the products. In the other activity sectors, oriented towards professionals and industry, the identity of the different subsidiaries is preserved, in conformity with the company policy that brands are created independently and sign different productions. Nevertheless, the group's name is also mentioned. This is the case of emergency slides used in the recent Airbus A319, A320 and A321 and Boeing B777 programs. They are manufactured both in the United States by Air Cruisers and in France by Aerazur (Figure 13.52).

Figure 13-52. Emergency slides

This is equally true for airbraking systems such as those used in military and civil parachutes, engine braking, airdrops, ejector seats and also runway stop barriers. These systems are produced in Europe by Aerazur-Parachutes de France, and by Pioneer in the United States (Figure 13.53).

Figure 13-53. Example of a Zodiac parachute

In the airline equipment domain, Zodiac has been able to consolidate its position as world leader thanks to its association with the well-known brands of the subsidiaries Sicma Aero Seat in Europe and Weber Aircraft in the United States. Airline companies in this sector are constantly looking for ways to innovate and improve passenger comfort and service to differentiate and widen their offer. Seat dimensions, ergonomy, functionality (electric controls, head and foot rests, video, electronic games, telephones) become significant marketing sales points[28]. The brands in the Zodiac group offer personalized passenger seats for tourist, business or first class as well as technical seats for the pilot and crew (in conformity with the new 16G norm which defines an impact resistance of 16 times the force of gravity).

Over and above the supply of the equipment itself, Zodiac offers solutions which help their clients to improve their technical performance. The gains in performance, safety, and comfort make their offer more attractive commercially. As world specialist in non-rigid composite materials, Zodiac is pursuing its program of developing new products and new technical solutions. It benefits from the synergy generated by its merger with specialized companies. Studies on technology and the utilization of high performance materials have enabled Zodiac to design products meeting the highest standards, both for the aeronautical and the marine-leisure branch. The company thus has a technological advantage in terms of design and creation of new complex materials and expert know-how in the use of aluminum in aeronautic products.

Zodiac invests in CAD, robotics, new production techniques and processes (laser cutting, heat welding, molding, numeric machining, optimization of structure calculations, etc.). In this way, clients have access to the many technological advantages gained by the group in its different domains: nautical, space, civil and military aeronautics: Thermobandage (an industrial heat-assembly process), which

uses a new-generation material Strongan Duotex made of a high-tech plastic-coated polyester support. Brand management within the Zodiac group operates on the following principles:
- Use of Zodiac as the corporate, federating brand:
  - product brands: Sicma Aero Seat, Air Cruisers...
  - process brands: Thermobandage, Strongan Duotex.
- Use of French and American brands mainly on the different respective markets.

NOTES

1. See Chapter 11, Communication policy.
2. Source: Airbus Documentation
3. See Chapter 6, Innovation and Product Management.
4. Source: Aaker, D.A., (1991), Op. Cit.
5. See Chapter 4, Market Segmentation and Positioning.
6. Source: adapted from Aaker, D.A., (1991), Op. Cit.
7. See Chapter 2, The Individual and Organizational Purchase.
8. See further, Zodiac: managing a brand portfolio by sector.
9. See Visual identity.
10. See Chapter 7, Marketing of Services.
11. Source: *Royal Air Maroc Magazine*, (1998), March-April, n°88.
12. Ad published in *Aviation Week & Space Technology*, (2000), 16 October, p 9.
13. See Chapter 2, The Individual and Organizational Purchase.
14. See further Latécoère: technical partnership and its own products.
15. Ad published in *Aviation Week & Space Technology*, (2000), 9 October, p 46-47.
16. See further Airbus: "Setting the Standards".
17. Photo extracted from a Cathay Pacific commercial documentation.
18. See further Zodiac: managing a brand portfolio by sector.
19. See Chapter 15, Alliance Strategies.
20. Ad published in *Air Transport World*, (1999), June, p 43.
21. See in Chapter 6, Innovation and Product Management, the development "Different product generations: military aircraft".
22. See Chapter 15, Alliance Strategies.
23. Ad published in *Air Transport World*, (1999), June, p 64.
24. Ad published in *Journal of Electronic Defense*, (1999), vol. n°22, n°2, February, p 14-15.
25. Source: Boeing website, www.boeing.com/news/releases/1997/histlogo.htm.
26. Company's visual identity is presented on-line on its website: boeing.com.
27. See in Chapter 6, Innovation and Product Management, the development "Latécoère's innovation: on-board video systems"
28. See Sicma Aero Seat in Chapter 2, The Individual and Organizational Purchase, in the development "Senior marketing" and Weber Aircraft in Chapter 6, Innovation and Product Management, in the development "Weber Aircraft: real innovation".

REFERENCES

Aaker, D.A. and Lendrevie J., (1994), *Le management du capital de marque*, Paris, Dalloz.

Aaker, D.A., (1991), *Managing Brand Equity*, New York, Free Press.

Alba, J.W., Hutchinson, J.W. and Lynch, J.G., (1991), Memory and Decision Making, in *Handbook of Consumer Behavior*, Englewood Cliffs, N.J., Prentice-Hall, p 1-49.

Brun, M. and Rasquinet, Ph., (1996), *L'identité visuelle de l'entreprise au-delà du logo*, Paris, Éditions d'Organisation; Behaeghel, J., (1990), A Corporate Identity... to be Identified, *Gestion 2000*, p 91-106.

Chernatony,L. (de) and McDonald,M., (1992), *Creating Powerful Brands*, London, Butterworth-Heinemann.

Kim, P., (1990), A Perspective on Brands, *Journal of Consumer Marketing*, vol. n°6, n°4, p 63-67, Fall.

Leuthesser, L., (1988), Defining, Measuring and Managing Brand Equity, Cambridge, MA, *Marketing Science Institute*, Report n°88-104.

Malaval, P., (2001), *Strategy and Management of Industrial Brands*, Norwell, Kluwer Academic Publishers.

Schmalensee, R., (1982), Product Differentiation Advantages of Pioneering Brands, *The American Economic Review*, vol n°72.

Chapter 14

BUILDING LOYALTY : MAINTENANCE, CUSTOMER TRAINING AND OFFSETS

In addition to marketing actions, companies in the aeronautics and space sector have developed many tools to increase the loyalty of customer companies including maintenance, training and financial compensation (offsets). These tools play an important role from the definition of the product/service mix to the development of customer loyalty.

1. MAINTENANCE

Etymologically, the term 'maintenance' comes from the Latin 'manutenere' (manu, hand and tenere, to hold) which referred to the action of maintaining military troops in condition for combat. Since then, its meaning has evolved and 'maintenance' today has a more technical and general meaning. It involves the necessity of keeping equipment such as an aircraft or helicopter operational. Generally speaking it means providing assistance and technical support to a product that has already been sold.

In fact, after-sales service comes into play right from the purchase: in fact it is one of the sales arguments used during the negotiation. Beginning with the Letter of Intent (L.o.I.), which shows the customer's interest in an aircraft (model, quantity...), the sales process continues with the Memorandum of Understanding (M.o.U.). The latter specifies customer expectations as far as the aircraft is concerned and is accompanied by payment of a reimbursable deposit. It is at this phase that after sale-service comes into play and then is integrated into the Purchase Agreement. Thus before delivery of the aircraft, maintenance training for aircraft personnel, the organization of maintenance services and the availability of spare parts (Figure 14.1) have already been established.

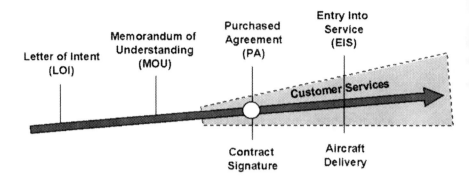

Figure 14-1. The role of maintenance in the sales cycle of an aircraft

Maintenance is based on human skills and is a customer-oriented quality procedure (Figure 14.2). In fact, quality can be defined as the properties and characteristics of products which allow them to satisfy the customer's implicit or expressed needs. By adopting a quality procedure, the company seeks:
- first, on the production level, performance quality in terms of technical characteristics, reliability and durability,
- then, on the marketing level, quality to meet customer standards.

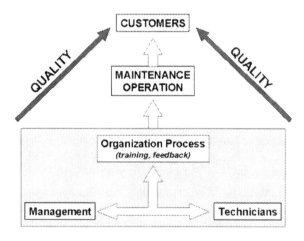

Figure 14-2. Maintenance at the heart of customer-oriented quality procedure[1]

The quality procedure is based on analyzing customer satisfaction, training, team involvement and a quality audit of the organization and its procedures (Figure 14.3).

As a function of the results of customer expectation surveys, the marketing department draws up specifications in partnership with technical departments. It thus

14. Building Loyalty : Maintenance, Customer Training and Offsets

determines the desired quality for the production process (relationship 1). The company then measures the results obtained and compares them to the requirements of the marketing department. This is the evaluation of the company performance, in other words, the quality obtained (relationship 2).

Depending on the company's communication policy as well as the visibility of the brand and products, customers perceive a certain level of quality: the perceived quality (relationship 3). In this way, the company's marketing department can measure customer satisfaction by verifying any gaps between perceived and expected quality. This information is very important so as to be able to take the necessary corrective measures and improve global performance (relationship 4). Customer perceived quality depends on three factors:
- the customer's own evaluation of the service provided,
- the supplier company's communication policy on its own quality,
- communication on the results of the customer satisfaction survey which can put into perspective or support an individual judgment.

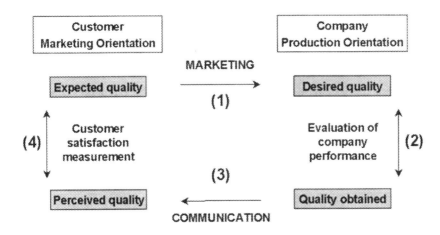

Figure 14-3. The quality cycle[2]

The increasing technological complexity of equipment as well as requirements in terms of quality, productivity, costs and reactivity have considerably favored the development of maintenance. The aeronautics support services market which is already large is expected to grow with the expansion of the worldwide fleet[3]. Based on data, in particular from the ICAO (International Civil Aviation Organization), Boeing estimates that the total maintenance market for civil aircraft over the next 20 years will be close to 50 billion dollars annually[4]. There are different categories of technical support in aeronautics (Table 14.4).

Activities	$ Billion
Heavy airplane maintenance *(airframe repairs, airframe painting, interior refurbishment, incorporation of airworthiness directives and service bulletins, aging aircraft repairs...)*	18,5
Airplane servicing *(line maintenance, ground support services, aircraft fueling, preflight inspection, preparation for departure, on-wing engine servicing, engine removal or replacement...)*	13,8
Airframe component repair *(landing gear, avionics components...)*	12,3
Engine repair / off wing *(engine breakdown, buildup and engine component repair)*	8,4
Airframe & engine repair parts *(OEM proprietary parts, vendor parts and standards, spare OEM components)*	8,1
Major airplane modification *(freighter conversions, re-engining, cockpit conversions, in-flight entertainment systems upgrades...)*	1,8

Table 14-4. Breakdown of the global civil aviation maintenance market[5]

The maintenance market can be broken down into:
- Heavy maintenance (major visits): These visits require immobilizing the aircraft from one to seven days (type C verification) or from 3 to 7 weeks for a complete general overhaul (type D verification) depending on the type of aircraft. Mainly carried out by the company's in-house departments (75%) this type of maintenance is increasingly handled by outside service providers[6].
- Light maintenance: This involves checking all of the essential parts of the plane (type A verification). Among the traditional upkeep operations are ground support services, fuel provision for the aircraft, pre-flight inspection, maintenance of the aircraft's engine (on-line maintenance). Aircraft are in service and operations are often handled by company personnel or by a specialized service provider. Certain airlines which handle this maintenance internally are trying to extend their activity to heavy maintenance.

Planes have a special characteristic, they are "new" throughout their lifetime[8]. For obvious security reasons, airlines cannot afford to wait for a breakdown[9]. Thus planes benefit from constant and scheduled maintenance ensured by the operating airline as well as diverse independent verification organizations.

Methods and procedures are defined by the constructor, accepted and approved by the authorities and carried out by operators and control organizations[10]. A test plane makes it possible to establish the aging of the aircraft (in terms of structure, mechanical, electrical and thermal wear) and the necessary number of visits to maintain the plane in total safety.

14. Building Loyalty : Maintenance, Customer Training and Offsets

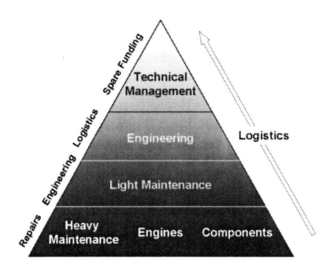

Figure 14-5. Aeronautics maintenance: from repair to management of components including engineering and logistics[7]

Depending upon the use of the aircraft (frequency of takeoff and landing cycles, flight hours, age...), verification visits for components differ. However, every five years aircraft undergo a "major visit": the plane is entirely taken dismantled, component by component, and put together again. Each component is changed or verified until it is like new (Table 14.6).

Aircraft type	Maintenance		Heavy Maintenance
	(A) All important parts of the airplane	(C) Simple Operations "Out of Service" Status during 1 to 7 days	Check D Full Overhaul "Out of Service" Status during 4 to 6 weeks
B747-100/200/300	650 FH	18 months	72 months
B747-400	650 FH	18 months	80 months
B737	300 FH	18 months	126 months
B767	600 FH	6000 FH / 18 months	72 months
B777	500 FH	5000 FH / 12 months	

Aircraft type	Maintenance		Heavy Maintenance	
	(A) All important parts of the airplane	(C) Simple Operations "Out of Service" Status during 1 to 7 days	Intermediary Check Technical Inspection & Refurbishment	Check D Full Overhaul
A320	500 FH	3500 FH	6 years	8 years
A340	500 FH	18 months	5 years	10 years

Table 14-6. Aircraft maintenance: a visit every 500 hours[11]

"The equipment reports" are also analyzed, in other words the record of incidents observed by the crew during the flight or technicians during pre-flight visits of the planes on the ground. Between two "major" visits, all parts are examined by specialists, from cabling to the landing gear, including the fuel pump and the engines:

- Avionics systems: the central information and control equipment of the plane (navigation, automatic pilot, flight control, radio communication) are constantly checked (in particular the alarm panels and engine control screens). A general verification is carried out every 15 to 18 months.

- Landing gear: tires are frequently replaced (owing to their intense wear, a thousand times greater than that of car tires), the brakes and the landing gear legs are checked daily. Every 10,000 cycles the landing gear benefits from a complete overhaul.

- Structure: every 5,000 hours (15 to 18 months), the fuselage and the wings are analyzed. Every 10 years the cellule is serviced. The windows are removed, checked, re-polished and re-set.

- Engines: every 10,000 hours, the engine is removed. The rotating parts are carefully examined (Figure 14.7), the engine is then bench-tested before being reset. The combustion chamber is inspected at the half of its lifetime.

Figure 14-7. Engine maintenance[12]

Once the plane has been completely overhauled, remounted, and then refurbished with new seats, it is flight-tested. For optimal security, maintenance is based on tracking: all operations are recorded, signed by the technicians and countersigned by their superiors. The parts used are recorded and must come from a well-known and regularly inspected supplier: if there is any problem, it is possible to find the record of any operation, as well as any part installed and the technician responsible for the operation.

All of these operations and precautions are so costly that, at the end of its lifetime, the plane will have cost as much in maintenance and repairs as the purchase.

A KEY TO AIRCRAFT SAFETY (Piernaz, 1999)

Incidents which occur on board aircraft, as well as anomalies or defective parts found during ground maintenance, are recorded and analyzed. The safety and control organizations, as well as the constructors are informed of these problems. Thus, other airlines are regularly informed of the necessity to carry out changes on their fleets (at the risk of being deprived of their right to fly if they do not conform, certificate of airworthiness). In this case, for greater security, they advise new procedures or the replacement of parts by safer more resistant components. International measures can be applied to entire fleets, reinforcing the security process essential to the operation of aircraft. These new measures apply to maintenance procedures. Certain of them are applied in an emergency context following an accident. Even though the aircraft remains the safest means of transport with 0.05 deaths per 100 million passengers-kilometer transported, safety is the number one priority of the authorities, the airlines, and the constructors. Procedures are increasingly strict, technologies are more and more safe (greater engine reliability, better control of pressurization technology) and crews are better trained. Safety is based on worldwide collaboration between the different players in the sector.

Following an accident with the Swissair three-engine MD11 in Halifax, Canada, in September 1998, the safety and control authorities carried out an analysis of the causes. Verification of the wreck showed signs of thermal damage, in particular on certain electrical wiring of which the insulator was destroyed. Investigations carried out by the Office of Transport Safety of Canada were followed by those of Civil authorities and aeronautics constructors. The FAA (Federal Aviation Authorities of the United States) took preventive action and accepted to question certain of its procedures.

Thus, it ordered the inspection and replacement of two switches regulating the light intensity of the MD11's control panel owing to possible overheating. It also requested that this procedure be applied to the electrical cabling situated above the front exits of the aircraft. In addition, the FAA recommended to Boeing and Airbus and all constructors of regional aircraft the replacement of the insulation of the walls of the aircraft in all of the 12,000 aircraft in the world. Two materials were recommended: fiber glass and Curlon polyacrylonitrile fiber (manufactured by Orcon), both of which are polyimide coated. The FAA took the following measures:

- development 6 months later of a new certification of insulating material for better fire resistance,

- requested constructors to replace the thermal and sound insulator inside the cabin and outside the aircraft with more resistant materials,

- obligatory inspection and possible replacement of the electrical cabling installed above the front exits/doors of the MD11s,
- inspection of cabling and fuse panels situated in the cockpit of all the MD11's in service.

Long considered as just an expense for the company (maintenance represents around 12% of an airline's expenses), maintenance is now perceived differently by companies who see it as:
- favoring the profitability of the industrial investment;
- an integral part of the quality policy;
- the logical and even essential next phase in the fabrication or installation of equipment;
- a commercial asset completing the project range and generating greater profits...

In light of ever greater competition and the generalization of products, suppliers are increasingly highlighting service and proximity to their customers, thus strengthening their long-term relationship.

1.1 The different forms of maintenance

Maintenance meets the needs of industrial buyers concerned about operating safety, reliability, availability and maintainability (the capacity of a product to be maintained or repaired). Among the different forms of maintenance, the trend is towards the development of preventive and predictive maintenance.

Curative maintenance: Also called corrective maintenance. This is the primary form: intervention takes place once the failure has occurred. Today, this form of maintenance is less and less used because of the problems it creates (production shutdowns, quality degradation, delays, negative image...).

Systematic maintenance: Rather than relying on occasional preventive maintenance, this involves systematic preventive maintenance based on a schedule established according to wear criteria: operating hours, quantity produced, energy consumed, miles traveled, number of takeoffs and landings... Costs are lowered thanks to better knowledge of "the behavior of machines" and the failure rates of components, using for example wear measurements and auto-diagnostic measures based on data from sensors...

Preventive maintenance: This consists in intervening before the failure occurs using operative maintenance (cleaning, lubrication, greasing). Since the exact lifetime of parts and components is not known, it is necessary to periodically examine equipment according to pre-determined criteria so as to reduce the probability of failure. Nevertheless, visits are sometimes useless and costly: certain parts are changed long before their failure. Yet this form of maintenance is preferred as it is easier to bear this extra costs than costs due to failures.

14. Building Loyalty : Maintenance, Customer Training and Offsets

Predictive maintenance: Especially adapted to sophisticated equipment, this consists in determining failure conditions and predicting the breakdown thanks to continual follow-up of operating parameters. This requires the purchase, installation, and continual use of complex surveillance tools: thermal imaging, infra-red temperature control, interferometers, oil analyzer by spectrometry, ultrasound, sound level control, lasers... Spectrometry has improved maintenance productivity. Its use makes it possible to determine particles present in the lubricant. Any abnormal rubbing is thus quickly observed provoking a more in-depth verification.

When a company acquires equipment such as a machine tool or an aircraft, it can entrust maintenance to its in-house departments and outside service providers and, more generally a supplier of the same equipment (Table 14.8).

Disadvantages	Advantages
Losing know-how Dependence on a service provider Less stimulated to improve equipment productivity Paying an annual maintenance fee for little used services and increasingly reliable materials	Freedom to focus exclusively on production Right from the purchase of equipment, the global lifetime cost of maintenance can be calculated Clearly establish the responsibilities of the maintenance and productions teams Free the company of in-house maintenance teams which are often too numerous when everything is working normally and insufficient when there is a problem

Table 14-8. Main advantages and disadvantages of external maintenance for a customer company

The maintenance market is evolving with the development of outsourcing (Table 14.9). Apart from aircraft builders, most of manufacturers are very important on the maintenance market owing to their specialized know-how and technical expertise.

	Airlines in-house	Airlines / focused subsidiary	Independent	Manufacturer	Value (billion $)
Heavy maintenance	75%	15%	10%	-	7,2
Light maintenance	80%	15%	5%	-	4,7
Regional jets	10%	10%	40%	40%	0,6
Military transport	-	-	40%	60%	1,6
MRO / Maintenance Repairs & Overhaul (avionics, systems)	-	55%	5%	40%	5,4
MRO / Maintenance Repairs & Overhaul (engines)	-	25%	15%	60%	10,5
Global market share	31%	25%	12%	32%	30,0

Table 14-9. Market share in aircraft maintenance (in value) [13]

464   *Aerospace Marketing Management*

For aircraft engine maintenance, an airline can directly deal with the specialized structure of the engine manufacturer such as GE Engine Services, Rolls Royce, Pratt & Whitney, SNECMA Services or MTU Maintenance. For an airline, one of the advantages of handling maintenance internally is the development of expertise which is likely to become an independently profitable branch in the long term. This is the case of Swiss Air (SR Technics), Air France (Air France Industries), Lufthansa (Lufthansa Technik) and Singapore Airlines (SIA Engineering). Thanks to this expertise, they can offer their services to competing airlines (Figure 14.10).

Figure 14-10. Example of SR Technics, the profit making maintenance center of Swissair[14]

Certain companies are specialized in maintenance. An example is Aviation Sales Company, the world leader for the re-distribution of spare parts and the North American leader on the maintenance market (Figure 14.11).

14. Building Loyalty : Maintenance, Customer Training and Offsets

Figure 14-11. Aviation Sales Company, a specialist in maintenance and aircraft parts[15]

On the maintenance market, there are the subsidiaries of the main airlines, original equipment manufacturers, as well as specialized service providers (Table 14.12).

In addition to maintenance, associated activities such as spare parts stock management are also increasingly externalized. So as to offer world service in terms of availability of spare parts, airlines that have established complementary networks, have joined together to form a common company: AirLiance (United Airlines, Air Canada and Lufthansa Technik). Airline alliances tend to favor the centralizing of the maintenance activities of partners[16].

In recent years, aeronautics maintenance has evolved along with e-business. Many players have taken an interest in the growing maintenance market such as aerospan.com (associating the SITA and the American company AAR specialized in spare parts stock management) and Aeromanager (Rolls-Royce, ANXeBusiness, Data Systems and Enigma). In addition, certain airlines have joined together such as in AeroXchange or they have joined with manufacturers and maintenance providers or spare parts companies to create specific sites (PartsBase.com) [17].

Rank	Company	Country	Maintenance	Type	Turnover (million $)
1	GE Engine Services	USA	Engine	Manufacturer	5000
2	UAL Services	USA	Fuselage Engine	Airline	2100
3	Lufthansa Technik	Germany	Fuselage Engine	Airline / Indep.	1938
4	Air France Industries	France	Fuselage Engine	Airline / Indep.	1400
5	American Airlines	USA	Fuselage Engine	Airline	1300
6	Delta Air Lines	USA	Fuselage Engine	Airline	1200
7	Japan Airlines	Japan	Fuselage Engine	Airline	1100
8	Rolls-Royce	UK	Engine	Manufacturer	1100
9	Pratt & Whitney	USA	Engine	Manufacturer	1050
10	SR Technics	Switzerland	Fuselage Engine	Airline / Indep.	845
11	Sequa / Chromalloy	USA	Parts repairs	Independent	745
12	Alitalia	Italy	Fuselage	Airline / Indep.	715
13	All Nippon Airways	Japan	Fuselage Engine	Airline	700
14	US Airways	USA	Fuselage	Airline	650
15	Northwest Airlines	USA	Fuselage Engine	Airline	640
16	KLM Eng. & Maint.	Netherlands	Fuselage Engine	Airline / Indep.	624
17	AA&ANZES	Australia/NZ	Fuselage Engine	Airline	600
18	Snecma Services	France	Engine	Manufacturer	560
19	British Airways Eng.	UK	Fuselage	Airline	500
20	MTU Maintenance	Germany	Engine	Independent	451
21	Sogerma Maintenance	France	Fuselage Engine	Independent	410
22	Honeywell	USA	Engine	Manufacturer	400
23	Qantas Airways Eng & Maint	Australia	Fuselage Engine	Airline	400
24	Iberia Airlines Eng & Maint.	Spain	Fuselage Engine	Airline	390
25	Sabena Technics	Belgium	Fuselage Engine	Airline	384
26	FLS Aerospace	Denmark/UK	Fuselage Engine	Independent	383
27	Korean Air	South Korea	Fuselage Engine	Airline	360
28	BFGoodrich Aerospace Airframe Maint.	USA	Fuselage	Independent	360
29	Standard Aero	Canada	Engine	Independent	360
30	Celsius Commercial Aerospace	Sweden	Fuselage Engine	Indep/ Manufacturer	360
31	SIA Engineering	Singapore	Fuselage Engine	Airline / Indep.	335
32	South African Airways Technical	South Africa	Fuselage Engine	Airline	330
33	Aviation Sales	USA	Fuselage Engine	Independent	325
34	Air Canada – Technical Operations	Canada	Fuselage Engine	Airline	315
35	Gulfstream Aerospace	USA	Fuselage	Manufacturer	280

Table 14-12. The 35 leading civil aircraft and engine maintenance companies [18]

Maintenance has thus become increasingly sophisticated. It has become an integral part of the marketing mix as well as the marketing information system.

1.2 Maintenance: a tool for the marketing-mix

For suppliers, maintenance is an essential part of the mix in that it can serve both as:

A guarantee of credibility: Maintenance serves as a guarantee reassuring the customer about the reliability of his relationship with his supplier, who commits to following up his product.

A differentiation factor: Considered to be one of the five essential activities of the company in the value chain, after-sales service is often a crucial, competitive advantage, in particular in the framework of differentiation strategies. Maintenance becomes a commercial argument when the supplier offers a free basic service: "a product identical to that sold by the best competitors but with the added advantage of after-sales service". Maintenance sets the product apart and personalizes it. Technical assistance can be adapted to customer priorities (number of visits, more or less systematic replacement, advice depending on the intensity of usage of equipment...).

Increasingly suppliers offer their customers a system rather than a product by itself, including different associated services and in particular maintenance. To strengthen this service orientation, right from the design stage, manufacturers integrate maintenance such as by facilitating the identification of parts and access to elements that need to be maintained (easy access system, easier assembly and dismounting). Given the total cost of maintaining a fleet (salaries, purchase of parts and consumables, stock management...), it has become essential to rationalize the maintenance of different aircraft.

This is why Airbus has tried to maximize the commonality of different aircraft, in other words the number of identical or very similar equipment used whether in the cockpit, the engine area or the passenger zone. This commonality creates savings:
- a smaller number of parts thus easier stock management;
- savings in the training of maintenance personnel[19].

Overall, maintenance is a way of ensuring customer loyalty: "the systems provider", or the product maintenance supplier who is involved in the heart of his customer's activity develops a common experience which is intangible and cannot be calculated in terms of numbers. If the customer wanted to change suppliers, he would lose not only the value of the equipment but also the acquired knowledge and mutual adjustments.

1.3 *Maintenance, a tool for the marketing information system*

For suppliers, the fact of regularly intervening on its own brand of equipment or that of other suppliers is also an important source of information (Figure 14.13). Thus maintenance makes it possible to gather 3 main types of information: about the customer, the product and the competition.

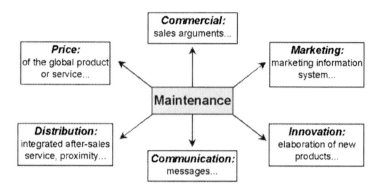

Figure 14-13. Maintenance: the main uses of information gathered

About the customer: By providing the customer with maintenance service, the supplier is frequently in contact with his customer's personnel, developing the technical rather than the commercial aspect of the relationship. This proximity favors the exchange of information, allowing the supplier to better understand the reasons for latent dissatisfaction, better know decision makers as well as the culture and internal organization of the customer's company, and even learn about new projects. Thanks to this information, the supplier can better personalize his product or service for the next commercial negotiation with his customer. Thus customer surveys allowed Airbus to better understand customer expectations in terms of productivity. This favored the development of the idea of commonality between the components of planes allowing easier stock management and lower costs.

About the product: Any failures provide maintenance operators with information: origin of the failure, poor or excessive use, frequency... This information can help to improve the product intrinsically or can be used as the basis for better usage instructions. The analysis of latent needs can be used to recommend possible innovations to the research and development department.

About competition: Maintenance is an opportunity to observe in situ competing products and when maintenance is multi-brand, the company gathers even more useful information about its competitors: evolution of products, reliability, degree of customer satisfaction... By handling a repair, the maintenance department knows the exact condition of competing equipment, its main characteristics, its most common failures and the general problems that it causes. Thus the sales department can be alerted and briefed about the main advantages and disadvantages of competing equipment.

The challenge lies in coordinating, organizing and transmitting information gathered about products, customers and the competition and then using it efficiently. This involves first cross-checking the different information collected so as to highlight the weaknesses and strengths of the product/service and competitors. The weaknesses observed can be used as a basis for the product plan which will try to correct the imperfections. The observed strengths can be used as a basis for better product communication and sales arguments.

AIRBUS' AFTER-SALES MARKETING FUNCTION: FOUR MAIN OBJECTIVES

At Airbus after-sales service, the marketing department has established four main objectives.

• Studying the needs of airlines: this involves determining the types of needs of airlines to make a useful segmentation. Results optimize the organization of after-sales service by major geographical zone and by type of airline.

• The study of customer airline satisfaction: this is based on several tools including a questionnaire which is deliberately simple (Figure 14.14). A gauge represents the level of customer satisfaction like the gauges that are found in the cockpit of aircraft.

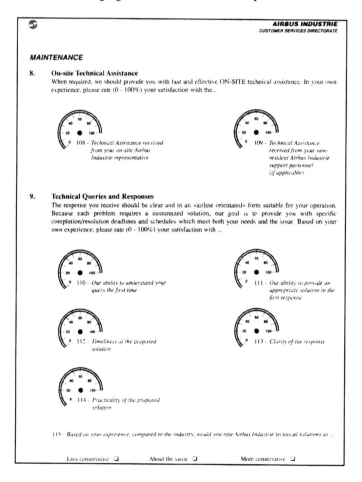

Figure 14-14. Extract from a customer satisfaction questionnaire

The Customer Satisfaction Improvement Program (CSIP) was created to improve after-sales and in particular maintenance. Thanks to the questionnaire, customer satisfaction was measured using 12 criteria (Figure 14.15).

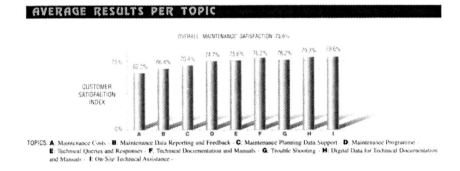

Figure 14-15. Summary of the results of the customer satisfaction study[21]

Like for the other areas evaluated, the surveys results highlight the main reasons for satisfaction as well as the main areas that need to be improved (Figure 14.16), thus facilitating the action plan.

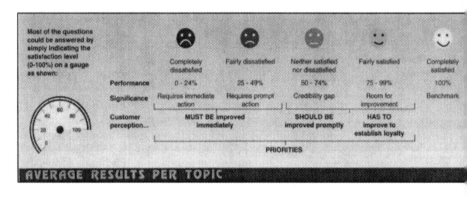

Figure 14-16. The priority level of actions[22]

These surveys are regularly carried out for maintenance, training, customer relations, flight operations support, the warranty and equipment. Based on these elements, a global evaluation of Airbus customer satisfaction is carried out.

• Product communication presenting the results of the customer satisfaction study: It is important to give customers the results of the survey in which they have participated. The "customer satisfaction improvement program newsletter" summarizes the results of studies carried out and is sent to the airlines to present the quality of service at Airbus and the decisions made in terms of action priorities.

14. Building Loyalty : Maintenance, Customer Training and Offsets

In addition to this newsletter, there are two other supports: Fast (Flight Awareness Security Technical) and The Link. Also accessible on the Internet via Airbus online services (AOLS), Fast is a thirty-two-page bi-annual review which requires a lot of work but which is very appreciated by customers. In fact, in each issue, several subjects are dealt with in depth and the latest improvements in products or processes are presented.

Figure 14-17. Communication media: Fast and The Link[23]

The Link is much lighter (6 pages) creating an additional bridge with customer airlines, keeping them informed about evolutions in customer service. Videos, presentations, brochures and CD-ROMs provide information on the customer service from maintenance to training and including technical engineering.

- Economic intelligence: this involves studying the actions of competitors so as to compare and improve services offered. In addition, this makes it possible to crosscheck the information gathered.

Thus starting as an essentially technical function, maintenance finds itself in the rear guard of marketing by integrating the customer satisfaction factor into the objectives of different departments.

2. CUSTOMER TRAINING

The complexity and the high technology of numerous industrial products make customer training increasingly necessary. Explaining the function, structure, environment and specific characteristics of the product to the user, is the primary objective of training. For example, aircraft are fitted with more and more intelligent Instrument Landing Systems (ILS), or collision avoidance systems capable of communicating with ground-based infrastructures. However the truth remains that 70% of all aeronautical incidents are caused by man (not understanding the machine, lack of crew communication, misunderstanding ground control, etc.). Airlines therefore invest heavily in crew training, and in addition to flight simulators, this is also centered around culture and psychology in order to improve crew efficiency (Niedercorn, 2000).

PILOT TRAINING: A MAJOR AND ESSENTIAL EXPENSE

In general, the cockpits of different aircraft models have their own particular characteristics. This means that pilots are qualified to fly only one particular model, and are obliged to obtain another 'type' of qualification to fly something different.

This separation of pilots within airlines according to type of aircraft flown, leads to the need for standbys who can replace a sick crew before the actual flight. Another consequence is that pilots with different type ratings coexist inside the company, and internal promotion is by changing category of aircraft. This results in a chain reaction with several pilots changing group. For a company which uses four different types of aircraft, the arrival of a new recruit can lead, by the process of internal promotion, to seven pilots moving (Figure 14.18). Pilots, in general change aircraft every three or four years: between 25 and 33% of the crews change group and do not remain operational in their previous group.

Training takes between 25 and 30 days, which means five to six weeks when the pilots are inactive. Such group changes have important financial consequences for the company which, over this period must:
- finance the training,
- continue to pay the inactive pilot,
- pay the pilot who replaces him during the training period.

Therefore for the company, crew training is a very expensive affair.

14. Building Loyalty : Maintenance, Customer Training and Offsets

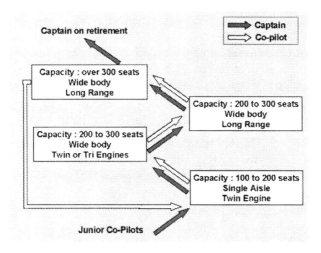

Figure 14-18. The traditional pilot promotion scheme[24]

The area covered by training is gradually increased to include other domains and other players in the buying act (decision makers, influencers, etc.). Training is no longer just becoming familiar with the products, but with techniques, the company, the market and even management.

2.1 Different training objectives

Training does not just have a pedagogical role, it is also an important communication tool. When a company decides to set up a training program it generally has five objectives.

Training, a tool for assuming customer loyalty: Training the influencers and users, familiarizing them with the new product concept, new functions and new technology increases customer confidence when it is finally launched. With better knowledge of the company's products and a positive attitude towards them, the people having undergone training will be more likely to work with this supplier. The company, by improving the level of competence of its customers, is looking to the long term and loyalty.

Training, a tool for quality control: If training can assure customer loyalty, it can also be a quality control tool. The instruction given by the company ensures a better level of quality control vis-à-vis customer production (maintenance, user skills, etc.). By so doing, the company is controlling not only the quality of the work but also its own image in the eyes of customers and end users.

Training, a tool for influence: The company, communicates on its products via its training and also transmits its culture. By training prospective, actual and future customers, the company becomes the reference point which will tend to be recommended. Training is thus a useful tool to influence customers.

Training, a communication tool: Training gives valuable information about the products, and the image of the company is transmitted along with it. The practical usefulness, rigor and quality of the training makes a large contribution to reinforcing the image of the company and its reputation with precise targets in the professional arena. Similarly, sales training, which companies sometimes give their customer's sales team, is a form of communication which has a prestigious and corporate aura about it. The companies which undertake training usually announce it in their corporate documents and this indirectly gives them credibility as a professional expert.

Training, as an element in the sales action plan: While training can be used for communication, it can also be a determining sales argument which reassures the customer at the moment of purchase. Tailor-made training – sales techniques, quality, production management – creates a sort of partnership with the customer and differentiates the product by adding value. Training is often part of sales policy by acting as a lever behind the scenes.

2.2 The contents of training

A technical content

In order for training to attain fixed objectives, it must have an actual technical content which allows it to perfect trainee's knowledge by passing on know-how. Successful training must therefore be useful, even essential for the customer, and if it is of a high quality it should be objective and answer the participant's expectations. All this means that the programs must be up to date.

Social recognition

Social recognition is another essential factor for the success of training. Giving a diploma, certificate or equivalence rating to participants will allow them to use this qualification later on in their professional careers. The social value comes from the prestige attached to a training course organized by a recognized and respected market supplier. The latter's acknowledged know-how is transferred to participants.

This dimension is also important for those in charge of the customer company and who authorize their employees to take the training course. They save money on their training budget, and could even get improved staff motivation thanks to this supplier organized session.

14. Building Loyalty : Maintenance, Customer Training and Offsets

Favorable conditions

While training must be given in a serious and lively fashion, it must also satisfy the participants. This means modern and numerous teaching aids, a suitable site (training rooms, training production lines, pilot sites, etc.), and an impeccably organized and carefully planned environment (meals, evening events, local visits, etc.).

In order to be credible and taken seriously, training must make use of a wide range of teaching skills and form a coherent part of the company's general strategy, and this normally must be paid for. Product training however, with a sales vocation is given free of charge by companies.

Certain companies have adopted an original approach and give " training credit points " to customer companies: this functions like a refund which instead of being financial becomes a " training " one. For a given volume purchased, customers are given training credit points which they can then use as they wish: this could be by rewarding an employee by letting him take the training course. There is therefore a double loyalty device here:

- firstly in the short term, assuring customer loyalty with the training credit points ;
- secondly in the long term by making the employees who have taken the course loyal, in that they will remember it regardless of where they end up in the future.

TRAINING, AN ESSENTIAL PART IN THE AIRBUS STRATEGY

The Airbus training department undertakes to ensure the successful entry into service of all aircraft delivered. This means ensuring that the airline companies have all the requisite information and skills for safe and efficient use of the aircraft. In addition, the training department ensures help with customer technical support and supply of information updates (Figure 14.19).

Figure 14-19. Training in the cockpit[25]

In the interests of saving the airlines money, Airbus has developed the concept of commonality which, thanks to fly-by-wire technology allows them to offer cockpits which closely resemble each other on all aircraft of the new generation (A320 family: 1318/A319/A320/A321 and the A330/A340 family). The cockpits have the same characteristics except for on the engine side, where variations obviously exist according to the number present. In this way transitions within the same family are very rapid (less than 2 days), and also between families (maximum 12 days). There are numerous advantages as much for the companies as for the pilots.

- Firstly, transition qualification from one aircraft to another is easier and faster thanks to Cross Crew Qualification (CCQ). This was approved by the FAA in 1994 for all new generation Airbus aircraft. It consists of qualifying experienced crews on a new type of aircraft (e.g. the A318) using training which covers the differences between this and the aircraft for which they are currently qualified (e.g. A320) with no need to go through a full training course which would be a New Type Rating. In addition, pilots moving up from the A320 to the A330 or A340 need only be trained on the specific characteristics of these aircraft, and this " Difference Training " considerably reduces the cost and equally, improves the productivity of the crew.

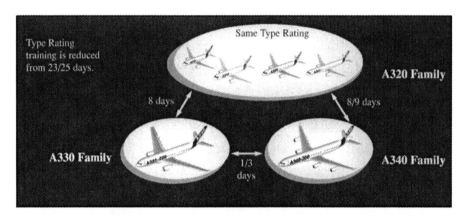

Figure 14-20. Efficient, customer oriented training[26]

- In addition, the principal of commonality allows simultaneous qualification on several aircraft and Mixed Fleet Flying (MFF). The same pilot can combine short, medium and long haul flights, and can therefore be on standby for more than one type of aircraft. Non-productive time is thus reduced. Together, CCQ and MFF give greater pilot mobility and productivity for the company: they spend more time flying (and less training), and fly other types of aircraft easier which gives flexibility. Crew fragmentation is reduced and the process of promotion is eased thanks to open training, facilitating integration. The company can optimize its planning schedule and the pilots can take their rest periods spread over several types of aircraft.

Air transport regulations state that a pilot who has flown for 12 hours must rest for 24 hours before flying again. If this pilot is on the Cayenne – Paris route (8 hours flying time) he

14. Building Loyalty : Maintenance, Customer Training and Offsets 477

will be obliged to stop for 24 hours even though he has not reached his 12 hour quota. MFF allows him to do 4 extra Paris – Bordeaux (1 hour flying time) flights before resting. These extra 4 hours bring with them 8 hours rest which becomes part of the original 24 hour rest period. This increase in the number of takeoffs and landings gives a decrease in the number of breaks in the pilot's sleep rhythms and eliminates the need for Recency Training (designed to ensure a certain minimum number of takeoffs and landings over a 90 day period).

- Overall, these possibilities give the companies the potential to make large savings (several million dollars each year) on training costs, equivalent to a reduction of 25 to 40% in fuel consumption.

- General pilot safety is also improved thanks to the commonality between their previous experiences and the new procedures that they have acquired. Each time they change from one aircraft to another the pilots increase their skill and experience (Figure 14.21) by getting to know more airports (national flights alternating with intercontinental ones).

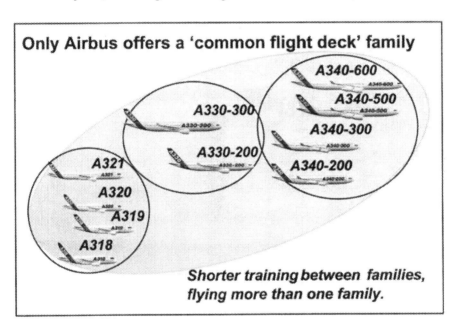

Figure 14-21. One family for nine types of aircraft from 100 to 440 seats[27]

In the same way as maintenance cited earlier, training takes place even before the sales contract has been signed. During the negotiation phase, visits to the customer allow an evaluation to be made of their exact needs in terms of training. When the actual contract is signed, the training clause 16, sets out the future training needs of the company (organization of company personnel training, contents of lessons, adaptations to the specific needs of the company and its personnel, teaching aids to be used, etc.). Some time later there will be a Training Conference to present the training program for crew, mechanics and hostesses (general administration, practical details like accommodation, transport, trainee spending, etc.).

This initiates different phases of the training schedule:
- maintenance training for the technical staff (which begins 6 months before the delivery date),
- flight crew training (which starts 3 months before),
- cabin crew training (1 month before).

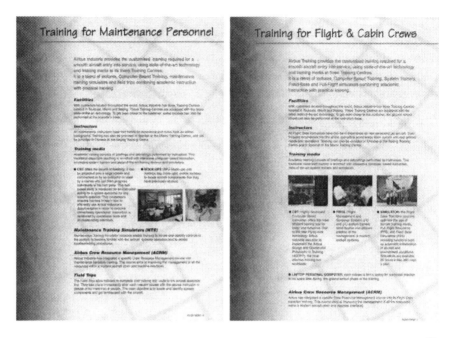

Figure 14-22. Presentation of training for maintenance personnel, flight and cabin crew[28]

When the aircraft is delivered, training continues with the pilots being accompanied, and the presence of ground support staff.

Training tools

The basic training for pilots comprises several key stages. For example, the training period for a New Type Rating on an Airbus A320 is 25 days, made up of:

• phase 1: theoretical training (11 days): general functioning of aircraft ;

• phase 2: normal operations training (6 days): how to fly the plane under optimum safety and efficiency conditions without any emergency operations ;

• phase 3: abnormal, emergency operations training (6 days): how to react to emergency conditions ;

• phase 4: evaluation (2 days): is the pilot capable of safely and efficiently flying the aircraft ?

• phase 5: actual flight (1 flight) of one and a half hours, accompanied by an instructor.

14. Building Loyalty : Maintenance, Customer Training and Offsets 479

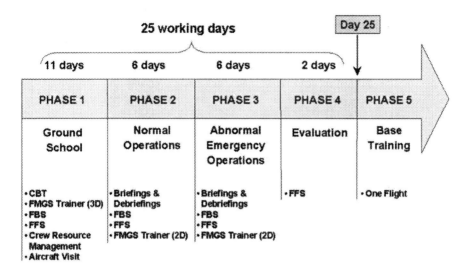

Figure 14-23. Description of A320 training[29]

In order to be able to offer training courses, Airbus has developed different tools such as:

- Computer Based Training (CBT) used by the trainee to learn via a computer the functions of the various instruments in the cockpit ;

- Training routines in 2D and 3D on the Flight Management Guidance System (FMGS), which gives real time information about the state of the aircraft. It permanently and automatically processes pilot commands, navigation (trajectory control) and optimizes the performance of the aircraft;

- The Fixed Base Simulator (FBS) and the Full Flight Simulator (FFS), which are flight simulators exactly reproducing the cockpit of the aircraft. All the instruments are linked to a computer which can simulate a flight as well as a certain number of emergency situations. They are for teaching the pilot how to control the plane under all circumstances. The FBS does not move, whereas the FFS sits on hydraulic rams which simulate the movement of an aircraft in flight. The simulators used by Airbus are produced by Thomson Training & Simulation (Thales). They can simulate a flight with CFM or IAE engines, the different radar tools used, and almost 350 malfunctions. Numerical communication and command control from the instructors seat, improves the training situation. A hydrostatic system with 6 different levels allows all types of normal aircraft movement (Figure 14.24).

Figure 14-24. The Thales flight simulator used by Airbus Integrated Company in its training centers[30]

The training sessions last different lengths of time according to type of aircraft and initial pilot qualification (Table 14.25).

	Basic training on A320, A330, A340	Difference Training					
		A330 ⇒ A340	A340 ⇒ A330	A320 ⇒ A340	A320 ⇒ A330	A340 ⇒ A320	A330 ⇒ A320
Fixed base simulator *(hours)*	42	0	0	9	9	6	6
Full flight simulator *(hours)*	21	6	0	12	9	12	12
Total simulator hours	63	6	0	21	18	18	18
Length of training *(working days)*	25	3	1	9	8	8	8

Table 14-25. The training programs according to type of aircraft[31]

Airbus have developed three training centers in Toulouse, Miami and Beijing where there are different combinations of simulators: from the FFS to FBS, the Cockpit System Simulator (CSS) and the Maintenance Training Simulator (MTS) (Table 14.26).

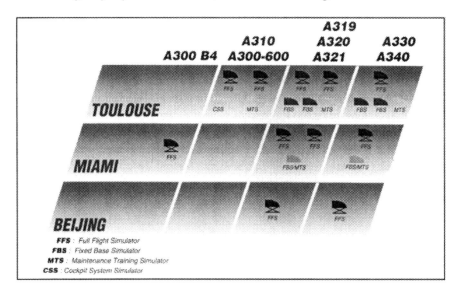

Figure 14-26. The types of simulator used in the different Airbus training centers[32]

2.3 The main types of training

Open to individuals from the buying center, the courses give prospective customers, influencers, advisors, decision makers and buyers, programs which are adapted to their needs and profile:
• technical training: usually linked to the products or equipment ;
• sales training: linked to the direct sales use of the product or to customer based development of business opportunities ;
• management training: linked to improving customer management skills through acquired expertise;
• quality training: linked to products and their use.

Objectives	Types of training				
	Technical	Management	Commercial	Marketing	Quality
Loyalty	+	/	/	/	+++
Control	+++	/	++	+	+++
Influence	/	/	/	+++	+
Communication	/	++	++	+++	+++
Commercial	/	+	++	++	++

Table 14-27. Different training objectives

There are basically two types of training course:

In-service training for customer operators: This is teaching personnel who are already operational, and doing jobs which are identical or very close to that covered by the course. For example engine manufacturers like Pratt & Whitney teach maintenance staff of customer companies. Aeronautical manufacturers like Boeing or Airbus train pilots in their training centers, to use new aircraft or new functions.

Initial training: This is teaching young persons as part of their professional apprenticeship programs. Examples are aeronautical manufacturers who participate in the training of engineers via conferences, lessons or specific teaching aids (software, technical booklets, documentation, catalogues, etc.). Spot Image goes to the extent of offering post graduate diploma courses (DESS) in space remote sensing.

GDTA: CUTTING-EDGE TRAINING[33]

Founded in 1973, GTDA (Groupement pour le Développement de la Télédétection Aérospatiale) is at the heart of the Toulouse (France) aerospace complex, one of the major sites for space technology in the world. The GDTA center offers training in the use of space data from earth observation satellites. More than 5,000 experts from 144 countries have been trained in this way since 1980.

The breadth of courses and who comes to the GDTA

More than 200 specialists cover all the areas and specificities of remote sensing:
- users from all domains: geology, cartography, town planning, environment, oceanography, land development, agriculture and forestry, etc.
- satellite and remote sensing instrument specialists ;
- experts in the numerical and analogue processing of images and in the Geographical Information System (GIS).

The courses are aimed at:
- decision makers and managers who want to know what space imagery can offer their business and who want to evaluate whether it is worth investing in this source of information,
- engineers and technicians from different areas wanting to acquire sufficient expertise to use the techniques by themselves or be able to follow and use work carried out by specialist groups,
- researchers who need to be able to follow the rapid evolution of technology to be able to design and develop new applications,
- teachers and instructors who want to introduce remote sensing into their courses.

Several training courses are offered:
- short courses, from one to five weeks, such as " Monitoring plant resources " or "Space cartography " (Figure 14.28),
- general courses,
- courses covering applications (specializing in a particular area),

14. Building Loyalty : Maintenance, Customer Training and Offsets

- courses covering " tools and methods " (Geographical and Remote Sensing Information System, processing and use of satellite images),
- a course in remote sensing, CETEL, leading to a post graduate diploma (DESS) (Figure 14.29),
- made to measure training courses and seminars, organized for different international organizations.

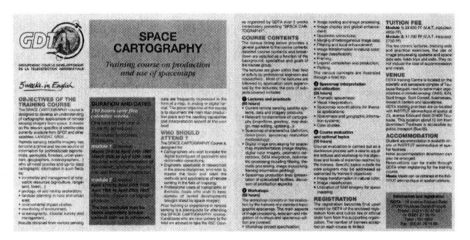

Figure 14-28. Documentation concerning "Space Cartography" course organized by GDTA[34]

The GDTA training program

The GDTA courses cover different levels and are complete and adapted to the needs of professionals. All satellites and associated probes are described and compared for potential users of these techniques. Spot Image regularly uses the training services of the GDTA with several objectives in mind:

- to assure loyalty, by familiarizing users with satellite image processing and working methods, and by letting them know what services are available;
- influence, by working with the decision makers and trying to ensure that they also are familiar with space technology ;
- for communication with researchers within the scientific community and customers who know about the products ;
- that of information for distributors and their trainees: the training center allows them to collect together information concerning user needs and concerns ;
- finally commercial: training, by transmitting technology and product information, alerting potential customers. By showing why satellite imagery is useful for the activities of different organizations and jobs, the GDTA maintains, defends and develops its partners openings and markets.

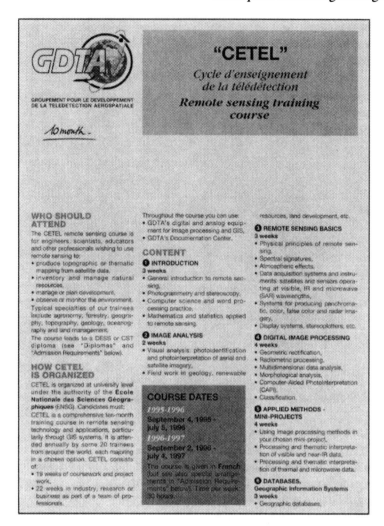

Figure 14-29. The long CETEL (10 months) leading to a post-graduate diploma (DESS) in remote sensing from the GDTA[35]

---

3. OFFSET, A BUSINESS TOOL

3.1 Offset: a means of payment

At the end of the Cold War, most countries and especially those in NATO made big cut backs in their military spending, and faced with this financial shutdown,

14. Building Loyalty : Maintenance, Customer Training and Offsets

there has been a slump in the military market. Customer relations have been modified and companies have had to adapt. Offset (Hartley, 1998) has begun to play an increasingly important role in the launch of defense projects by aerospace companies. Used within the framework of international contracts, offset has recently taken on an increasingly important role in the defense equipment markets, especially the aerospace ones (Dusclaud and Soubeyrol, 1998).

Different terms are used to define offset, such as:

- Barter: This is a one-off contract covering the sale and payment which doesn't need any money to be exchanged between buyer and seller. It is characterized by an exchange, one product for another and by the equivalence of what is exchanged (100% rate of compensation).
- Countertrade: This is an operation which generates two contracts. On the first, the seller receives the traditional payment, and he agrees under a second contract to buy goods produced in his customer's country. The details of the two contracts are set out in a general contract (length of time, price, etc.) including penalties for non respect of clauses.
- Buy back: This is where the seller buys back part of the purchaser's production which has been made on the equipment originally sold. An exporter supplies a production machine to the purchaser who in return gives him part of its production as payment.

All these cases have negotiated sales conditions under which the supplier agrees to a certain amount of commercial reciprocity which is to the purchaser's benefit. Offset is a business deal under which the seller agrees to make purchases, transfers, services or any other operation in the customer's country, in exchange for a sale which is only concluded once these conditions have been agreed. Offset is therefore a commercial exchange in a specific context: it is one of the elements of the aerospace program worth several million dollars.

Offset consists of offering the customer nation an attractive solution. More than 130 counties have developed offset agreements at various levels: about 50 - 60% for Greece, Saudi Arabia, Belgium, Turkey or Spain, 100% for the United Kingdom, Australia, Netherlands, Switzerland and Finland and even 110% for Canada. Offset is therefore a method of payment based on of three elements

Reciprocity: In a traditional sales contract, the seller undertakes to supply goods in exchange for money from the purchaser. An offset contract obliges the seller to buy. In actual fact there are two sales contracts where each player is consecutively a buyer and a seller. For example, a national army orders two Northrop-Grumman Hawkeye surveillance aircraft with 100% offset. The first sales contract is signed between the army and the manufacturer. A second, binds the latter to the purchasing county's companies (Figure 14.30).

Exchange balance: The two sales contracts making up the offset must be balanced. There are no rules regarding offset rates resulting from negotiations and agreement between the parties, and in certain cases this could exceed 100%. For example, the French sales contract for 4 Boeing E-3F Awacs, mentioned 130% offset.

Monetary evaluation: Apart from bartering, this type of contract does not do away with money. All that is necessary is for the two parties to agree on a sum and on the real value of the goods.

Figure 14-30. Links in the "Hawkeye" contract

Offsets can be part of any business deal in the guise of " Offset Agreement ". Usually the purchase programs in which they are involved consist of two parts: direct offsets and indirect offsets.

Direct Offsets

Direct offsets consist of direct participation by the purchaser in the production of the goods to be sold. This is usually to do with goods and services exported to the selling country by the main manufacturer and/or sub-contractors, written into the purchasing program. This translates into operations for:

- local co-production: McDonnell Douglas for example, won the contract to renew the Swiss Army's fighter wing thanks to an offset agreement. The Swiss have assembled their own F18 aircraft with McDonnell Douglas playing the sub-contractor role, transferring know-how to the customer and ensuring a similar quality to the American made originals.

- sub-contracting: the airline company LOT purchased 8 ATR 72-210 aircraft and the contract was completed by a sub-contracting agreement between ATR and the Polish PZL company for the manufacture of wingboxes worth 100% of the

14. Building Loyalty : Maintenance, Customer Training and Offsets 487

initial contract. PZL therefore supplied wingboxes (which were not necessarily used on the ATR's for LOT) until the contracted amount was reached.

- or transfer of technology: the contract for 10 A320-200 for Vietnam Airlines was accompanied by transfer of know-how in the form of maintenance technician training by Airbus.

Indirect Offsets

This is investments made by the supplier in domains outside those of the acquisition program[36]. Normally these concern transfer of technology, granting of licenses, joint venture agreements, training offers, both at the local level and abroad, or the export of aerospace or defense products/services. It is up to the supplier to get the offset, and when he cannot achieve this completely, it may need to be referred to a third party.

Transfer of technology

The main point here is always that technology is transferred to allow the receiver country to be able to develop its own industrial production. The offsets are therefore medium to long term and are spread over several years (>5 years). The supplier can transfer technology in different ways:
- patents, licenses, software and data for local production,
- high level training for research and development, design and production,
- advanced technological resources and equipment,
- research and development programs linked to the project sold which will allow the purchaser to carry out his own maintenance.

Financial offsets

Financial offsets by debt conversion is a marginal and complex affair. It uses the financial markets: a country with credit in a debt ridden country will transfer this to a buyer who will use it to acquire goods, this is the Debt Commodity Swap, or to make an investment in the form of equity participation, and this is the Debt Equity Swap. The debtor country can instigate this conversion.

3.2 A business argument

Taken as a method of payment, offset is becoming more and more a sales tool used to clinch the buying decision, because in fact offsets give numerous advantages to the purchaser.

They allow the weight of the debt to be lightened and they limit the negative effects of a supply program on the country's balance of payments. In addition, offset allows new manufacturing opportunities to be created within the host country thanks to transfer of technology. In this way the country in question can develop its defense product's export market.

Offsets are often seen at the national level as being a method of payment which permits in addition (Rodriguez, 1991; Stahel, 1991):
- to create a defense or aerospace industry,
- to benefit from transfer of technology,
- to save and even create employment,
- to defend the country's currency.

For the purchaser, it is viable to combine these contracts with offset clauses for financial reasons (sales balance, reduced financing because of non monetary exchanges, etc.). In addition, the purchasing countries can come onto the international market at a reduced cost and can set up new contracts with other companies. Finally, the injection of technology allows the industrial fabric of the country to be reinforced (know-how, work load, etc.).

For the seller, the transfer of technology has certain drawbacks, the main one being that there is a risk of losing technological control and thus military superiority. The receiver country can become a competitor. In this way, the multitude of technology transfers between American and Israeli companies have allowed the latter to produce better anti-missile, missiles than those of their suppliers, or on-board detection systems which are just as good as the Awacs. In the face of these real risks, there is solid business sense: it is better to do offset than to sell nothing at all. At the very minimum, offset allows the seller to protect himself against customer insolvency and is a way of retaining and even gaining market share (Table 14.31).

For the purchaser	For the exporter
- Creates an equilibrium in the trade balance - Provides access to the international market - Technology transfer - Employment	- Reduces risk of non payment - Maintains or develops market share

Table 14-31. The advantages of offset for the buyer and the seller

For developing countries, offset is one of the possible ways in which they can improve their technological know-how. By having an industrial base and increased skills, they are better able to export, restrain imports, and reduce tension due to people out of work. For the offsetting industrial countries, the same arguments apply with in addition the need to stay technologically abreast of their competitors. A large part of offset agreements concern transfer of advanced technology using co-production, licensing and joint-ventures.

Offset agreements over the last few years have resulted in an increasing number of international partnerships (transnational joint ventures). Countries have tended to modify their offset policy in favor of indirect offsets on specific projects. As the export perspectives for military goods has fallen with the reduction in military spending, countries are opting for projects which will be more beneficial to their economy, in domains such as agriculture, health, the environment, construction and housing, telecommunications or electronics. Suppliers have now got the measure of offset and realized that it is a specific tool to be managed as such. They use specialized companies to identify possible offset projects for them, to define the main outlines and even sometimes to put them into effect. All this represents new opportunities for a large number of small companies, to extend their activities onto the world stage by taking on the management of specific projects.

NOTES

1. Source: Adapted from Reville, P.Y., (2000), Air France Industries.
2. Source: PIMS
3. See Boeing's *Current Market Outlook* and Airbus' *Global Market Forecast.*
4. See Boeing's *Current Market Outlook.*
5. Adapted from Boeing's *Current Market Outlook.*
6. Source: Boeing's *Current Market Outlook.*
7. Source: Reville, P.Y., (2000), Air France Industries.
8. Text based on Brosselin, S., (2000), in *Passager Aérien*, 2nd Quarter.
9. However a plane can takeoff with some equipment failures. This extremely carefully regulated situation is specified in international document: Minimum Equipment List.
10. See below: "A key to aircraft safety".
11. Source: Air France Industries, in Niedercorn, F., (2000), Dossier Industrie: Face au défi de la sécurité, *Les Echos*, 17 May, p 69-72.
12. Airbus photo
13. Source: *Airline Business*, (2000), October, p 62-73, Roland Berger estimates.
14. Advertising published in *Airline Business*, (2000), August, p 31.
15. Advertising published in *Aviation Week and Space Technology*, (1999), 25 October, p 20.
16. See Chapter 15.
17. See Chapter 7.
18. Source: *Airline Business*, (2000), October, p 62-73.
19. See below, Training.
20. Source: Airbus, Customer Services Marketing.
21. Source: CSIP Airbus
22. Source: CSIP Airbus
23. Source: *Fast*, (2000), September, n°26 and *The Link*, (1999), September.
24. Source: Airbus
25. Source: Airbus, *Customer Services Directorate*.
26. Source: Airbus.
27. Source: Documentation Airbus, *The Airbus Family Concept.*
28. Source: Documentation, *Airbus Customer Services.*
29. Source: Airbus.
30. Source: Airbus.
31. Source: Airbus.
32. Source: Airbus, *Customer Services Directorate.*

33. Cf. site gdta.fr.
34. Source: Documentation GDTA.
35. Source: GDTA.
36. See Chapter 10.

REFERENCES

Dusclaud, M. and Soubeyrol, J., (1998), *Enjeux technologiques et relations internationales*, Economica.
Hartley, K., (1998), *Economic Evaluation of Offsets*, The University of York, April.
Niedercorn, F., (2000), in Dossier Industrie: Face au défi de la sécurité, *Les Echos*, 17 May, p 69-72.
Piernaz, P, (1999), Les constructeurs d'avions face au feu électrique, in *L'Usine Nouvelle*, n°2677 bis, p 22-24.
Rodriguez, R.A., (1991), Offset, Buy-back, Technology Transfer, Joint Ventures in Cooperative Security, ISSC Seminar, Brussels, 26-27 June
Stahel, M., (1991), Offset, Countertrade, Technology Transfer in Cooperative Security, ISSC Seminar, Brussels, 27 June.

Chapter 15

ALLIANCE STRATEGIES

Alliances are one of the visible characteristics of the aeronautics and space sector. Originating in the seventies, the trend towards alliances has grown over the last thirty years, whether at the level of parts manufacturers, in particular the manufacturers of engines, constructors-integrators or the airlines. Given the importance of alliance strategies on a highly competitive aeronautics and space market, we have decided to present them analyzing objectives, the different types of alliances and the main applications observed in each activity sector.

1. TRADITIONAL FORMS OF COMPANY DEVELOPMENT

There are different development options for companies aiming to expand their activity. The three main development strategies (Garrette and Dussauge, 1995) are vertical integration, diversification, and internationalization (Figure 15.1).

Vertical integration: This is based on expanding activity upstream (supplier) or downstream (customer) of the supply chain. An example of this is a satellite integrator developing the diffusion of earth observation pictures or teletransmissions (telecoms, data) such as Astrium or Spot Image. This new activity which completes the original area of expertise downstream or provides a commercial outlet can be launched independently, with partners, or through the acquisition of a company already present in the sector (i.e. Boeing's acquisition of Continental Graphics Corporation, of Jeppesen Sanderson and Preston, to develop airline services: traffic management software, cartography, navigational software, meteorology). In the same way, an aircraft constructor like DASA could choose to develop engine activity downstream (MTU).

Diversification: This involves developing new sectors relative to the company's core area of expertise. An example is DASA, which is developing a new space activity with Dornier Satellitensysteme. This diversification can also depend on in-house expertise, the purchase of companies already established on the market or a partnership with other companies such as the alliance between Astrium and Thomson-CSF[1] Sextant for on-board multimedia services in aircraft.

Internationalization: This is a way of increasing activity by building a stronger presence in new geographic zones. This is often one of the aims of companies when they acquire firmly established local players, such as the purchase of Ratier-Figeac by United Technologies. Internationalization can also involve the forming of alliances with partners, as is the case of the alliance between General Electric and Snecma in CFM International, which has allowed each of the partners to expand their geographic range.

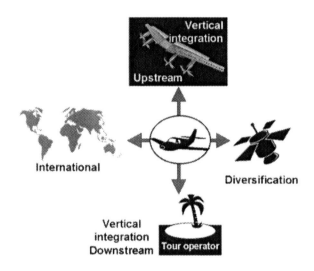

Figure 15-1. Main forms of strategic development

Thus companies can choose to expand their activity horizontally (diversification, internationalization) or vertically (upstream, downstream). They can achieve this goal through the purchase of new companies or the creation of a new division. Finally, as is increasingly the case in the aeronautics and space sector, they can choose to join an alliance.

2. SPECIFIC OBJECTIVES OF ALLIANCES

An alliance is a group of independent companies whose goal is to complete a project or work together on a specific activity through sharing of required know-

15. Alliance Strategies

how, resources and means (Garrette and Dussauge, 1995). It is an "unstable" form of relationship between companies as each one preserves his independence. In fact one of the strengths of alliances in comparison to mergers-acquisitions is that each of the partners is preserved in organizational and cultural terms. Furthermore, alliances do not have any specific legal definition or status. Nevertheless, they fall under the legal framework of contract law (an alliance is legally considered to be like a contract) and corporate law (either a joint venture, or a "GIE", Economic Interest Group). However, when a specific structure is created between the partners, this does not necessarily apply to all the areas of cooperation. Legally, it can uniquely cover sales or R&D for instance (even though if in reality the cooperation goes way beyond this). Thus a legal interpretation of the alliance does not provide a representative picture of its actual content.

Companies form alliances for very diverse reasons. They can be financial, in other words the desire to share the burden of investment costs. They can stem from difficulties in satisfying demand, whether in terms of products or geographical area, or for marketing reasons. International political issues can also come into play.

2.1 Financial objectives

The capacity to mobilize capital

From the point of view of financial partners, backers, or shareholders, a project that is driven by an alliance of companies is often seen as a safer investment. On the one hand, the financial players are reassured by the financial guarantees provided by the different partners. On the other, the credibility of the project, in particular if it is very innovative, is strengthened by the partners' commitment. Furthermore, the latter will not be in competition in this category of services, thus reducing the level of investment risk.

Risk sharing

The sophistication and diversity of technologies used in aeronautics construction, the space sector and air transport, have considerably raised the level of investment needed to launch a new project. By creating an alliance, the new partners share the financial risks associated with the investment, in particular concerning research and development costs, marketing and communication, as well as the costs of production equipment. By joining together, companies aim to combine and optimize their know-how, giving them more room to innovate than if they were acting independently. The costs and technical complexity of projects are such that, given the technological uncertainty, cooperation is often absolutely necessary between aeronautic companies, both upstream and downstream in the sector.

Lowering of costs

The synergies created by the alliance allow for assigning production phases as a function of previous upstream investments. Thus each partner can produce more for a given investment. Furthermore, the global quantities to be produced will be greater if the new sales objectives are reached. In this way an alliance tends to better amortize fixed costs and to more easily exceed the profitability threshold of a project.

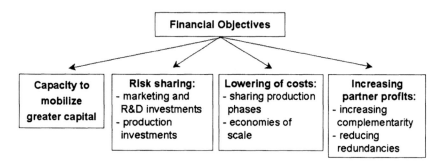

Figure 15-2. Different financial objectives

Improving results

If the same type of investment must be borne by the different partners, each on his own, there is no savings. That is why the success of an alliance fundamentally hinges on the degree of confidence between partners. Each must accept the idea that:
- on the one hand, he can no longer control all of the design and production operations,
- on the other, the exchange of know-how is required in the two alternative situations: the partner sometimes gains, thanks to what he receives, and sometimes gives, transmitting something to others.

That is why increasing profits depends on the degree of complementarity between partners. Thus improving profitability means:
-maximizing the level of complementarity of the partners,
-minimizing redundant costs.

2.2 Marketing and sales objectives

Marketing objectives can involve:
-the products/services that will be offered to customers,
-the geographical spread, that is the area in which the alliance will operate,
-the image obtained thanks to the involvement of the different partners.

As for the financial objectives, the degree of complementarity is the first factor on which the synergies in an alliance depend.

15. Alliance Strategies

Figure 15-3. Different marketing and sales objectives

Enlarging the product/service offer

In the equipment field, the sharing of technical, design and production resources, will make it possible to offer more products/services covering different market segments. This is the case both for aeronautics equipment with in particular the manufacturers of engines such as General Electric or Snecma as for constructors of the smallest planes to wide-bodied aircraft.

Improving the geographic spread

By joining forces, the partners in an alliance each benefit from the others' geographic locations. This is true of Airbus which uses the network of sites of its ex-partners, who are now merged in the EADS (Aerospatiale-Matra, which is French, DASA, which is German, CASA, which is Spanish) as well as the network of its British partner, British Aerospace. It is also often the case between part manufacturers and constructors from the United States and Europe. For example alliances have been created between Raytheon and Thalès in the defense sector or between Lockheed Martin and EADS (for the transformation of Airbus planes to radar or supply aircraft). The argument is even clearer for the airlines (Lawrence and Braddon, 1998). By joining an alliance, each can hope to enlarge his customer base geographically and offer his customers the same range and standard of services on all of the partner networks. In the case of Star Alliance, Lufthansa's regular customers will take advantage of the North-American network of its partner, United Airlines. Reciprocally, United Airlines will benefit from Lufthansa's European network.

Belonging to an alliance also facilitates the setting up of maintenance services. Thus the partner's delivery time is reduced both for new supplies and for spare parts.

Improving image thanks to easier communication

By joining forces under one flag, the alliance members greatly improve their advertising efficiency. In fact, each partner can cut down on or even eliminate investments under his own name as soon as he appears visibly as a partner in the new alliance. The new communication budget of Star Alliance for example is much lower than the sum of the individual budgets of the 13 airline partners up until now, meaning greater efficiency.

From a quality point of view, companies and individuals targeted by Star Alliance's ad campaign mainly remember the global aspect of the new player and the fact that his "regular company" belongs to this group. In turn, this legitimizes the new alliance for customers in each of the geographic zones.

The "competitive" alliance

Often the goal of the alliance is to create and optimize long-lasting competitive advantages that can withstand competitors both in terms of cost (increased production volume) or products (differentiation and/or expansion of the range thanks to combined skills and know-how). However, paradoxically, certain alliances are created between competitors, constantly shifting between opposition and cooperation. Alliances are sometimes established with the aim of limiting or even eliminating the real or potential advantages of competitors. For example, it is possible to keep in check the innovation potential of a competitor by providing him with ready-made technological solutions. Thus the competitor will not commit considerable resources to develop a competing technology. Furthermore, certain companies don't hesitate to ally themselves with partners, in particular to prevent the latter from joining forces with a dangerous competitor. This was certainly one of Boeing's objectives when it formed an alliance with Fuji, Mitsubishi and Kawasaki for the construction of the B777, thereby trying to limit the opportunities for these Japanese companies to ally themselves with Airbus.

In Israel, thanks to political and economic links with government authorities as well as with the heads of the airline, El Al, Boeing has maintained its market domination of the technology used. Thus the Israeli aeronautics industry is dependent on a vertical multinational, which provides technology and know-how. Israeli companies have limited autonomy, isolated as they are from other worldwide developments in aeronautics; consequently it has been very difficult for them to integrate European know-how. To arm itself against Airbus offensives, Boeing, which is linked to over 50 Israeli companies, has decided to invest in Israeli high tech and to publicize this fact. This enables Boeing to keep abreast of local start-ups and the latest technological developments, while limiting Airbus's access to the market (Rouach, 2000).

Sometimes it is very difficult for an outside observer to understand the nature of alliances and partnerships between competing companies given their sheer number and complexity. For example, in the helicopter sector, there are many partnerships, both at the level of aircraft and engines.

15. Alliance Strategies

Helicopters	Industries	Corresponding engines	Industries
S92	Sikorsky (United Technologies)	T700	General Electric
EH 101 - EHI	Westland – Agusta	RTM322 vs T700	Rolls-Royce / Turbomeca vs General Electric
NH90 - NHI	Eurocopter - Agusta – Fokker	RTM322 vs T700	Rolls-Royce / Turbomeca / BMW-Rolls-Royce vs General Electric / MTU
AS332 Super Puma	Eurocopter (EADS)		Turbomeca - Makila

Table 15-4. Industrial partnerships: from cooperation to competition[2]

These partnerships often involve a common structure, the capital of which is evenly controlled by the partners:
- Snecma (Turbomeca) is joined with Rolls-Royce in RRTM,
- Snecma (Turbomeca) controls 40% of MTR, where Rolls-Royce controls 20% and MTU (Daimler Chrysler) controls 40%. MTU also has a partnership with General Electric),
- Rolls-Royce controls 100% of Rolls-Royce America which is associated at 50% with Honeywell (former AlliedSignal).

Visually, it is possible to illustrate the competitive complexity of these agreements at the level of engines (Figure 15.5).

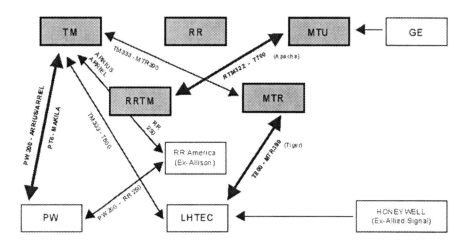

Figure 15-5. Partners often in competition[3]

In the aeronautics sector, such clusters of alliances are very common. For example, for each of its activities, the Italian company Finmeccanica is associated with different companies, certain of which are in competition (Revue Aerospatiale, 2000) (Table 15.6).

Co-companies / Joint ventures	Activity	% of capital controlled by Finmeccanica	Other partners
Eurofighter	Fighter	19.5%	BAe Systems (UK), EADS (Europe)
Augusta Westland	Helicopters	50%	GRN (UK)
AMS	Electronic defense	50%	BAe Systems (UK)
Gie ATR	Regional aircraft	50%	EADS (Europe)
LMATTS	Military transport aircraft	50%	Lockheed Martin (USA)
New MBD	Missiles	25%	EADS (Europe), BAe Systems (UK)
Astrium	Satellites	Under negotiation	EADS (Europe), BAe Systems (UK)

Table 15-6. Finmeccanica alliances

2.3 The international political stakes

Certain alliances are formed by public organizations motivated by the same objectives of lower costs and more efficient marketing. However, to the latter must be added political considerations such as preserving the independence of research and industry in space. The ESA, European Space Agency, is a case in point. Generally, one or more countries acts as leader by investing more money or trying to convince other countries of the value of the project:

• Financial objectives: like most private firms, one of the objectives is to cut costs. The aim is to make sure that countries such as Italy, Spain, Germany, Great Britain and France do not each spend money to develop quasi-similar projects.

• Marketing objectives: the first objective is to expand the direct potential market represented by the different partners and then to benefit from the networks of each.

Consequently these projects contribute to the building of international economic union such as the European Union, in particular through the exchange of research and the bringing together of companies.

3. DIFFERENT FORMS OF ALLIANCES

Presentation of alliances

Alliances in the aeronautics and space sector (Dussauge, 1992) have been very common for many years, involving parts manufacturers, builders of launchers,

15. Alliance Strategies

aircraft, missiles or helicopters and the airlines. The word "alliance" in fact covers a wide range of distinct realities including mergers, cooperative agreements, international programs, commercial agreements, the creation of joint ventures, a constellation of companies (Lorenzoni and Ornati, 1988), the transfer of licenses, and shared research. More specifically, an alliance is based on an agreement linking the partner firms together while preserving their respective independence (Garrette and Dussauge, 1995). Companies join together to pursue common goals while maintaining their strategic autonomy and own interests. Thus mergers and acquisitions are not alliances in that the new entities pursue coherent and homogenous strategic objectives, managed by a single power structure. When the same company forms numerous alliances for very different projects such as helicopters, missiles, aircraft and satellites, the boundaries of the company are sometimes blurred, weakening its very reason to be.

By definition, alliances involve continual negotiation and conflicts of interest between the different decision-making centers. For example, the French-British supersonic Concorde was developed thanks to a compromise between two different coherent projects: a medium range carrier on the French side and a very long range aircraft with weaker capacity on the British side. Ultimately, a hundred seats aircraft is not easily profitable over short distances, for while it can cross the North Atlantic it is not capable of making longer flights without stopovers (where the supersonic speed would have been a clear advantage) (Hochmuth, 1974). Conflicts of interest are natural in an alliance, given the independent nature of partners; in fact, the alliance's common objectives are really only partial objectives as far as each of the partners are concerned. Consequently, one of the partners can block a common project, even if it is commercially viable, if it creates direct competition for a product or service from his own portfolio of activities.

The (variable) vigilance of public authorities

The public authorities pay close attention to alliances, looking for signs of price fixing, concentration, or the establishment of a monopoly restricting free competition. The Fair Trade Commission in Japan and the Justice Department which oversees anti-trust laws in the USA survey alliances between competitors even more closely than mergers or acquisitions. The authorities are wary of alliances established with the primary goal of raising prices. In Europe (Lawrence and Braddon, 1998), the Treaty of Rome, which regulates mergers, also regulates alliances, except for those based on R&D cooperation, which are less controlled and often supported by the European Community under the aegis of such programs as Jessi, Eureka, Start or Esprit. This is also the case in Japan, where the Ministry of Industry and International Trade (MITI) supports research alliances. In the United States, the National Cooperative Research Act also authorizes R&D agreements if they are restricted to research and do not involve its common exploitation.

The diversity of alliances[4]

Alliances can be formed by non-competing companies:
- from distinct sectors: these are mainly inter-sector agreements, based on crossover know-how from different sectors. For example, an alliance between an aeronautic or space constructor and a computer company with the aim of developing flight simulators or general public video games. One of the objectives is to diversify the field of activity of each of the partners.

Figure 15-7. Agreement between non-competing companies

- alliances based on a customer-supplier relationship: these are vertical partnerships. Rather than producing its own equipment or relying on different suppliers, the company works with one specific supplier, participating with him in the design and production of required equipment. Purchase marketing generally leads to the development of this type of relationship[5].

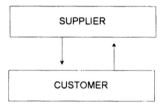

Figure 15-8. Agreement between customer-supplier companies

- alliances based on improved local market penetration: these are agreements between multinationals and local companies to facilitate the development of sales of products adapted to the local context. This often involves the creation of joint ventures in the form of indirect exports. The foreign partner, which often has more sophisticated technology and know-how, can distribute its product on a new market, leaving it up to the local partner to commercialize and develop the activity.

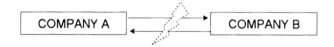

Figure 15-9. Agreement between potentially competing companies

15. Alliance Strategies

Alliances can be between:
- competing companies with equivalent structures (equipment, know-how...) or on the contrary, different but complementary structures;
- competing companies that cooperate so that each can benefit from the same product, which will be differently integrated into their respective portfolios. The aim being to make economies of scale (R&D, production) on the same component, which is then integrated into goods that are produced by both of the partners and which are often in competition;
- competing companies that cooperate more extensively to design and commercialize a common product (R&D, production, commercialization, after-sales). Commercialization of the common product can be carried out by a specialized sales structure, common to all the partners. This was the case for the Airbus consortium, each of whose partners were entitled to a share of turnover proportional to their participation. This commercialization can also be based on division of the geographic sales sectors of each of the partners.

Certain alliances are only based on a one-off project, others concern the entire company and last beyond the duration of the project, and still others give rise to mergers. It is thus necessary to classify them.

3.1 Tactical alliances

One-off alliances

These alliances are formed between partners on a one-time basis within a group with the aim of capturing a market. This type of alliance is especially common in project marketing[6]. The building of the new Peking airport was the object of a one-off alliance between BCIA (Beijing Capital International Airport), Aéroports de Paris and ABN-Amro. In this case the alliance comes to an end when the project is finished unless it is extended to the equipment usage phase (i.e. the operation of the airport in the preceding example).

Partial alliances

These only concern one part of an activity, for example the design or creation of a project or equipment.

For Airbus Military Company's launching of the turbo-propeller of the A400M military aircraft, the program brings together the usual partners such as Aerospatiale-Matra, Daimler Chrysler Aerospace, the Spanish CASA, BAe Systems, but also the Italian, Alenia, the Turkish, TAI, and the Belgian, FLABEL. The two latter partners are partially linked to Airbus via the A400M project. In addition, it is planned that the Italian, Finmeccanica, will participate in the program for the A380 wide-bodied aircraft, which will be a new partial alliance between this company and Airbus.

Figure 15-10. The partial alliance

3.2 Strategic alliances while maintaining the company's initial identity

For alliances to be successful, a common, motivating objective has to be set such as customer satisfaction. Then it needs to be verified that each of the partners understands and integrates the complementary strengths of the others. If the alliance works, each partner will see his value increase. It is thus important for the partners to share the same philosophy: "when one partners wins, everyone wins." In alliances bringing together companies from different countries, it is essential to create a climate of trust, in which each of the partners understands and respects the national and corporate culture of the others. An alliance is not only a signed agreement, rather it involves the full commitment of the staff and management of the companies involved. The creation of a specific organization lays the basis for a coherent framework in which objectives and respective commitments are clearly defined. The aim is for the organization to be able to make common decisions, evaluate the work of each partner, and favor the integration of partners, particularly the smallest ones. Successful alliances depend on on-going communication between all the players involved. Before moving towards an alliance strategy, management must question:
 - the pertinence of the alliance compared to autonomous development,
 - the choice of partners,
 - the organization of the cooperative effort: agreement, creation of a subsidiary...
 - the potential profit to be gained by the different partners (possible imbalances),
 - the estimated duration of the alliance (anticipating the rupture)
 - the operating mode for loyal cooperation while preserving the company's own strategic interests,
 - protection of its technology and know-how,
 - the effect of the alliance in terms of competitive intensity.

As soon as the objectives of the various partners differ, the alliance is threatened. Problems can include misunderstandings about commercial objectives, or lack of knowledge about the corporate culture of each of the partners, leading to greater incomprehension. These communication problems create serious functional weaknesses within the alliance. Another factor that can damage an alliance is too great asymmetry between partners and too different services in terms of quality, content, and regularity. In fact customers expect the same quality of service from all the partners, a standard which can be difficult to ensure. Sometimes lack of experience in an alliance or working cooperatively can also lead to failure.

Maintaining the company's initial identity

This case can be seen with airlines, which can use their own partner identity or the new group identity. Within Star Alliance, SkyTeam, One World, or Qualiflyer for example, each airline continues to communicate, produce and sell under its own brand. The airline alliance phenomenon developed in the middle of the nineties. In the wake of the Deregulation Act in the United States, the airline transport market became highly competitive, or the "open sky, open market". Alliances were created to deal with market growth, lower yields, fleet investments. Customers quickly learned how to look for the best deals. Faced with the same problems, the airlines tried to develop customer loyalty in a market of close to 1.5 billion passengers (which is supposed to triple over the next twenty years). The aim was for allied companies to increase their efficiency by creating economies of scale on long trips (cooperation in maintenance, catering, purchases..) while increasing the volume of activity. To this end, they had to increase their destination networks and sales by offering more attractively priced tickets. This is why many airlines developed worldwide alliances offering passengers more connections, access to private rooms in airports, and better loyalty programs. Thanks to code sharing, the airlines can offer:
- Reduced transfer time,
- Simplified boarding thanks to a single boarding card,
- Complete baggage follow-up,
- Accumulated loyalty bonuses.

In fact, code sharing is the most common form of alliance in the airline sector. It is a commercial tool that allows an airline to offer and sell tickets from other operators as if they were its own. For the sale of seats, the transporters add on the same flight, the two letters of their operator code and/or the flight number. Code sharing facilitates reservations on the Computer Reservation System. It makes it possible to share costs, increase revenue, and to protect against future competitors. Thus, one year after the opening of a Paris-Cincinnatti connection, Air France, thanks to code sharing with Delta Airlines, obtained a 96.5% capacity rate (a score that is generally only achieved after a connection has been opened several years). In fact, Delta Airlines offers 37 destinations leaving from Cincinnatti. In the first months, the crew must be prepared to answer the questions of passengers, who are sometimes disappointed when they discover that they are not on an aircraft of the airline they expected. Presented as a commercial advantage for the passenger, it is a powerful strategic asset for participating airlines, to some degree making the passenger a bit of a prisoner to the alliance. The airline market includes a few major alliances representing over 70% of world traffic:

- "Star Alliance": Air Canada, Air New Zealand, All Nippon Airways, Ansett Australia, Austrian Airlines, British Midland, Lufthansa, Mexicana, SAS, Singapore Airways, Thai Airways International, Tyrolian Airways, United, Varig;
- "SkyTeam": Air France, Delta Airlines, Korean Air, AeroMexico;
- "One World Alliance": Aer Lingus, Aerolinas Argentinas, American Airlines, British Airways, Canadian Airlines, Cathay Pacific, Finnair, Iberia, LanChile, Qantas;
- "Wings Alliance": Braathen, Continental Airlines, Kenya Airways, Northwest, KLM, Air China, Alaska Airlines, America West Airlines, Big Sky Airlines, Eurowings, Express Airlines, Garuda Indonesia, Gulfstream International, Hawaiian Airlines, Horizon Air, Japan Air System, Jet Airways, Malaysia Airlines, Mesaba Airlines, Pacific Island Aviation;
- "The Qualiflyer Group": Swissair, Crossair, Lauda Air, Air Europe, Sabena, Air Portugal, Turkish Airlines, AOM, Air Littoral, Lot (Polish Airlines), PGA (Portugalia Airlines), Volare Airlines.

Alliances and hubs

In fact these alliances serve as leverage for hub-to-hub connections. Delta alone represents 1885 connections in Atlanta. Air France alone represents 2208 connections in Charles-de-Gaulle. Through the alliance, the two partners represent 14978 hub-to-hub connections.

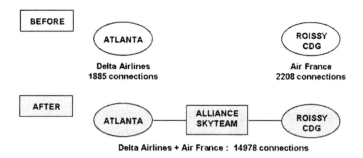

Figure 15-11. The multiplying effect of alliances

Hubs increase the international connecting flights available by organizing arrivals and departures in waves. In fact, the airlines have progressively restructured their traditional networks around hubs which enable the setting up of connection systems. Direct links have been replaced by connecting links within the hub (Figure 15.12). The additional passengers from these connections optimize aircraft capacity and eliminate the need for direct links between cities. Thus British Airways has its hub at London Heathrow, while Air France has its hub at Roissy Charles de Gaulle. For the airline, the "star system" rationalizes the network and is economically more profitable:

- operating costs are lower thanks to higher density traffic on the branches of the "star". The aircraft are filled to greater capacity. Furthermore, airlines can use bigger planes (the cost per seat decreases with increased transport capacity). They can also increase the frequency of flights, thus increasing the connections available within the hub and making the system more attractive to passengers.
- the concentration of certain activities (maintenance...) in hub airports allows economies of scale.

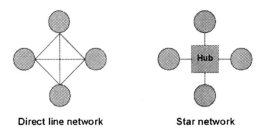

Direct line network   Star network

Figure 15-12. The evolution in networks towards a star structure via hubs

From the point of view of parts & engine manufacturers, these alliances:
- mean working with unified buying organizations,
- have a considerable effect if a market is lost or gained.

While alliances offer a less expensive and a wider range of services to passengers, they also can create a certain number of problems:
- they can transform the market into an oligopoly, reducing passenger freedom of choice between airlines,
- inconsistent quality of service,
- they can disappoint passengers because of code sharing policy.

STAR ALLIANCE: A WORLDWIDE AIR NETWORK

Star Alliance was created in 1997 to better satisfy frequent international fliers. Studies were carried out on the latter to determine their specific needs and requirements. The alliance includes 13 partners:
- Air Canada
- Air New Zealand and Ansett Australia
- All Nippon Airways (ANA)
- Austrian Airlines
- British Midland
- Lauda Air

- Lufthansa
- Mexicana Airlines (Compañia Mexicana de Aviacion)
- Scandinavian Airlines System (SAS)
- Singapore Airlines
- Thai Airways International
- Tyrolean Airways
- United Airlines
- Varig Brazilian Airlines.

Thus Star Alliance represents 296 million travelers a year, over 9,300 flights daily, 815 destinations and 130 countries in the world. The airlines joined together to offer what none of them individually could have provided:
- visible advantages, specific to each trip,
- a worldwide status and privileges,
- a non-stop trip,
- facilitated world access,
- a very comfortable environment,
- on-going research and anticipation of new customer needs.

Figure 15-13. Star Alliance

The network offers up to 7,000 world destinations and many world trips in first, business, and economic class. Itineraries are simple and attractively priced, calculated as a function of distance covered. This allows the customer to determine the duration of the trip (10 days minimum, 1 year maximum) and to choose the number of stopovers (3 minimum, 15 maximum). From a commercial point of view, Star Alliance ensures customers that whatever the airline, they will benefit from the same treatment, thus recalling the quality of service offered to privileged customers.

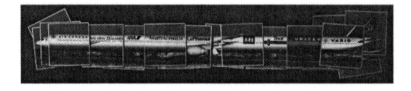

Figure 15-14. An ad campaign symbolizing the alliance between Star Alliance partners

Among those services offered to Gold or Silver Star Alliance customers are:
- access to private lounges, for working calmly or just relaxing,
- priority boarding,

- flexibility in case of reservation changes,
- participation in loyalty partner programs.

Airlines	Main hubs	Staff	Fleet	Number of passengers	Destinations	Countries served	Daily departures
Air Canada	Toronto, Montreal, Vancouver	25800	246	19.2 million	120	26	1200
Air New Zealand	Auckland, L.A., Sydney	9560	79	6.4 million	48	15	470
All Nippon Airways	Tokyo, Osaka	14639	142	43.2 million	62	13	571
Ansett Australia	Melbourne, Sydney	14876	126	13.4 million	142	5	530
Austrian Airlines	Vienna, Salzburg, Innsbruck	7162	90	8 million	125	67	410
British Midland	London Heathrow	6309	60	6 million	32	12	306
Lufthansa	Frankfurt, Munich	31305	287	43.8 million	340	91	1349
Mexicana Airlines	Mexico City, Cancun, Los Angeles	6369	54	7.1 million	50	9	235
SAS	Copenhagen, Oslo, Stockholm	25754	190	22.2 million	105	31	1050
Singapore Airlines	Singapore	28000	91	12.8 million	99	42	222
Thai Airways International	Bangkok, Chiang Mai, Phuket, Hat Yai	24148	78	16.3 million	76	35	286
United Airlines	Chicago, Denver, S. Francisco, L. A., Washington	100414	600	87 million	255	26	2475
Varig	Rio de Janeiro, Sao Paulo	17740	87	11 million	120	20	453
STAR ALLIANCE		312076	2130	296.4 million	815	> 130	9557

Table 15-15. Some key figures about the alliance

The interest of this type of alliance lies in commercial efficiency in terms of cost, service quality, and commercial coverage. For example, United Airlines estimates that it has gained 250 million euros in additional revenue thanks to this commercial partnership.

SKYTEAM: A NEW WORLDWIDE AIR NETWORK[7]

With the first long-term strategic agreement signed by Air France and Delta on June 22, 1999, the foundations of a new world alliance were laid. The goal was to bring other partners on board, creating an alliance based on the provision of specialized customer benefits offered by all the partners. Right from the start, the main objective was the customer rather than the signing of a traditional inter-company agreement based on mere overlapping interests.

In fact a new alliance was of capital importance given that North America, Europe and the North Atlantic represented 54% of the world's traffic. Thus on June 22, 2000 in New York, a new alliance was signed bringing together Air France, Delta, Aeromexico, and Korean Airlines. The name SkyTeam was chosen to best symbolize the spirit of the new alliance.

Figure 15-16. SkyTeam partners and the customer focused slogan

Building a reputation and image

An effective name had to be chosen for the new alliance to project a certain image and build a reputation. To this end, travelers from all over the world were consulted to determine their preference for a name out of a half a dozen pre-selected. The name SkyTeam was retained, as best expressing the idea of a team of airlines, sharing the same vision of an alliance exclusively at the service of air travelers. A logo was created to allow customers to easily recognize the alliance partners in airports, sales outlets, and the communication campaign. The logo uses the name of the alliance with graphics evoking comfort and quality of service. It is rounded, protective and friendly, with swirls resembling a written signature. This logo progressively appeared on the documents of the allied partners.

Figure 15-17. SkyTeam partners

15. Alliance Strategies

For the signing of the agreement, the partners presented the alliance's first advertising campaign, launched simultaneously in Europe, the United States, Latin America and Asia. SkyTeam alliance aimed to affirm its commitment to situating the customer at the heart of its strategy through an ad campaign centered on customer service. In fact, customer surveys revealed considerable passenger dissatisfaction with other alliances in terms of individual attention. Designed by Euro RSCG BETC, the campaign slogan was "you first". Each ad is designed as a puzzle with the main part missing. The missing part, which SkyTeam places, symbolizes the element that all the others have forgotten: the passenger. "What is missing from other airline alliances ? You". Two 15 second TV ads, completed the creative campaign which began June 23rd, and lasted over five days. The media included the big daily national papers, the economic and financial press as well as American, European, South American and Asian editions of international magazines.

The alliance

The alliance satisfied an ever-greater need: the possibility of getting from one place in the world to another, simply and comfortably, in accordance with consistently high standards of quality. The worldwide SkyTeam network, based on connecting hubs strategically situated in Latin America, Europe, the United States and Asia, offers passengers 6,402 daily flights to 451 destinations in 98 countries[8]. With this aim in mind, the four airlines in SkyTeam have coordinated their hours and connection times. The SkyTeam network serves all major destinations in the Northern hemisphere towards which 80% of the world's traffic is concentrated. SkyTeam offers more destinations without stopovers between the United States and Europe than any other alliance. SkyTeam partners and their staff of 151,000 provide 174 million passengers a year with harmonized, consistent, quality service:

- standards in terms of punctuality, waiting time at the airport, and luggage arrival time,
- airport service norms,
- customer assistance procedures in case of operating irregularities (information, reservation changes, etc.).

This alliance offers a wide range of advantages to customers.

- **More destinations, more easily**

Thanks to one of the best hub systems in the world, SkyTeam has multiplied the possibilities for connections and flight times, making travelling considerably easier for their customers. SkyTeam offers a worldwide network to companies, which in the age of globalization, expect airlines to offer genuinely global service: thus the main connections of the alliance cover a territory concentrating over half of the world's air traffic. Moreover, SkyTeam serves all the major cities in the Northern hemisphere. In addition to its vast network, SkyTeam has other assets to ensure development. Roissy-Charles-de-Gaulle, the hub of Air France, and Seoul-Inchon, the future hub of Korean Air, are among the few airports with growth capacity. Opportunities for opening additional runways and creating new terminals, in turn allowing more frequent flights, are an invaluable advantage for SkyTeam customers.

In this way, SkyTeam heads one of the biggest networks in the world, organized around hubs situated in strategic zones in Asia, Europe, Latin America and the United States:

- Aeromexico's hub at Benito Juarez de Mexico, gives SkyTeam access to the number one Latin American hub, with more than 6,000 passengers per hour.

- In Paris-Charles-de-Gaulle airport, Air France has developed one of the most outstanding hubs in Europe: more than 14,000 weekly connections possible under two hours, more than Frankfurt, Amsterdam, Heathrow, Brussels or Zurich.

- At Atlanta-Hartsfield, Delta offer passengers the largest number of possible connections in the world, with 1,316 daily flights to 126 destinations.

- One of the largest companies in Asia, Korean Air, serves 78 airports in 29 countries including Japan, China, Korea and south east Asia. From their Seoul hub, they offer rapid connections to the world's main cities.

- **Better sales and check-in services**

SkyTeam, via its agencies or by using the most up-to-date technology, offers customers different ways of obtaining information. It has teams in towns or airports which are in touch with passengers all over the world. By using Inter/Intranet links between the airline partners, passengers can interrogate sales points or actually go to one of the 659 allied agencies in order to check prices and times, find out the services available (type of plane, seating plan, meals, etc.) and make or modify their reservations. A new website, www.skyteam.com gives customers practical information about the group, with dedicated links to each company member's site.

SkyTeam agents are trained to give customers the best possible service and to answer their travel needs. They can contact a 24 hour assistance service where specially trained teams will help them deal with customer requests. Moreover, the customers benefit from numerous service upgrades such as back-to-back check-in for 451 world destinations (luggage is labeled to the final destination).

- **New quality norms for on-board service**

- On board SkyTeam flights, passengers benefit from common quality norms covering the four airlines, while at the same time enjoying the specific styles of each one,

- On international flights, there is always a member of the cabin crew who speaks English plus someone for the language of departure and destination.

- **New advantages for frequent fliers**

Passengers can accumulate miles on the four partner companies and obtain bonuses (air tickets for the whole SkyTeam network, etc.). In addition, SkyTeam customers who are members of the frequent flier club, automatically benefit from SkyTeam Elite and Elite Plus status which gives them extra advantages with the group companies. Elite and Elite Plus are in addition to the four company's customer loyalty programs.

SkyTeam Elite (Aeromexico Club Premier Oro and Club Premier Oro Corporativo, Air France Fréquence Plus Bleu, Delta SkyMiles Silver and Gold Medallion, Korean Air Morning Calm card holders):

- Access to separate airport lounges for first and business class passengers,
- Choice of seat on reservation,
- Waiting list priority (reservation and airport),
- Priority check-in desks,
- Boarding priority.

15. Alliance Strategies

SkyTeam Elite Plus (Aeromexico Club Premier Platino, Air France Fréquence Plus Rouge, Delta SkyMiles Platinum Medallion, Korean Air Morning Calm Premium and Million Miler card holders). Customers benefit from all the Elite advantages, plus:
- Access to 265 separate airport lounges for international flights, regardless of class,
- Luggage priority on international flights,
- Guarantee of economy class reservation on long haul flights, even if fully booked (24 hours notice for purchase of full price economy class ticket).

- **Towards better cooperation for cargo**

The partners intend to extend their cooperation to freight transport, providing quality service and a world network. Already, customers benefit from the alliance thanks to reserved freight areas and sales agreements passed between company members. This network will be progressively enlarged and the products harmonized.

	DELTA	AIR FRANCE	AEROMEXICO	KOREAN AIR	SKYTEAM
Founded	1924	1933	1934	1969	2000
Main hubs	Atlanta Salt Lake City Cincinnati Dallas-Ft. Worth New York JFK	Paris-CDG Paris-Orly Lyon	Mexico City Monterrey Guadalajara Hermosillo	Seoul Pusan Cheju	Mexico City Paris-CDG Atlanta Seoul
Fleet	584	220 subsonic aircraft 6 Concorde	68	107	979 subsonic aircraft 6 supersonic aircraft Total: 985 A/C
Average age of fleet	11.6 years	9 years (less Concorde)	12 years	7.2 years	10.6 years (less Concorde)
Destinations *	227	198	50	78	451
Countries served *	29	83	7	29	98
Departures Daily *	4 442	1 231	309	420	6 402
Passengers per year	106 million	39 million	8.8 million	20.5 million	174.3 million
Internet Address	www.delta-air.com	www.airfrance.net	www.aeromexico.com	www.koreanair.com	www.skyteam.com
Airport lounges	87	128	12	53	265
Town and airport agencies	282	192	62	123	659
Loyalty program members	24 million	3 million	0,9 million	7 million	34,9 million
Frequent Flyer Loyalty Programs	SkyMiles Silver / Gold / Platinum Medallion	Fréquence Plus Bleu / Rouge	Club Premier Clásico / Oro / Platino	Skypass Morning Calm, Premium, Million Miler	
Staff	74,000	54,000	7,000	16,000	151,000
• Hostesses and stewards	• 19.117	• 10.305	• 1.450	• 3.274	• 34.146
• Pilots	• 8.674	• 3.610	• 808	• 1,687	• 14.779
• Ground	• 46.209	• 39.876	• 4.592	• 11,073	• 101.750

(*non duplicated, own flights, franchised or chartered)

Figure 15-18. Figures for the SkyTeam alliance[9]

3.3 Strategic alliances with creation of a specific structure

This is usually two or more companies who create a specific structure, while still remaining independent or even in competition. The structure is created with a long-term aim and a contribution of capital, technical know-how, staff and industrial assets. The creation of CFM international in 1974 for aircraft engines, comes under this category. The two equal partners in the joint venture, Snecma Moteurs and General Electric, have maintained their own activities sometimes in direct competition. This alliance created more than 26 years ago still exists, without any further integration between the two founding companies. Other examples of strategic alliances are Eurocopter in the helicopter sector and Sea Launch, Starsem for launchers.

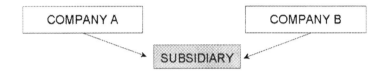

Figure 15-19. Alliance with the creation of a dedicated structure

THE CFM INTERNATIONAL ALLIANCE

CFM International is the joint company (50% each) of Snecma Moteurs and General Electric. Studies, marketing, production and product support are divided 50-50 between the two partners. For more than 20 years this partnership has been successful, notably in the development of engines used on 27 different types of aircraft. If in retrospect, the alliance of these two groups seems perfectly logical, it was much less so when the agreement was signed in the 1960's.

The nationalized, 90% military output, French Snecma company, was very experienced in international cooperation and wanted to get into the civil market with a 20 to 25 thousand pound thrust, dual flow engine for short and medium haul aircraft. They were thus looking for a partner who was highly competent in dual flow technology, used up till then only on larger engines, where it gave reduced fuel consumption and noise levels.

The undisputed leader Pratt & Whitney was not in favor of a 50-50 alliance, preferring a majority share, and wanting to retain its leadership in particular thanks to its JT8D engine. Quite apart from this, it did not believe in the viability of an engine with highly dilution rate for this application. Neither was Rolls-Royce a really potential partner for Snecma, in the light of the complicated past relations between the two companies and the commercial failure of two previous collaborations; the M45H and the Olympus of the Concorde.

15. Alliance Strategies

The private, American GE group present in the civil and military sectors had, up to then, very limited experience in international cooperation. They were in competition with Rolls-Royce and Pratt & Whitney on large engines, and wanted to develop their activities in Europe, even if that entailed defying Pratt & Whitney. In addition, the group had already signed a production agreement with Snecma for the CF6.

From this point of view, it seems only natural that the two groups should want to become allies. However, within the framework of alliances and in particular within the aeronautical and space sector, rational criteria are not the only things to be taken into consideration. One of the difficulties involved were the political stakes of such an alliance which, for the production of the proposed CFM56 engine[10], entailed:

- General Electric supplying the heart of the engine (high pressure), derived from the F101 which was fitted to the American strategic B1 bomber;
- Snecma supplying the low pressure modules.

Quite apart from American concerns about exporting sensitive technology, France itself also possessed a large military, aeronautical industry.

Nor did the French position in the eyes of NATO facilitate the signature of such an alliance. It needed all the determination of the two CEO's concerned, Mr. Gerhard Neumann and Mr. René Ravaud, as well as actions taken at the highest level of state between the American and French presidents, Richard Nixon and Georges Pompidou, to overcome these doubts. CFM International was created in 1974 with equal participation from the two groups, Snecma and General Electric. Traditionally run by a French person, CFM International's mission is to coordinate actions undertaken by the partners.

Figure 15-20. The CFM engine and the modules developed by the partners

After five years existence, CFM International still had no customers (which would have certainly doomed it by present day business criteria). But in 1979, the US Airways re-engining program for 110 DC8's (four engines per plane), gave the CFM56 project a commercial footing. This was followed by re-engining the US Air Force, KC-135 in-flight refueling tankers and above all the launch of the Boeing 737-300 in 1982. Sources at CFM International say that one of the reasons for the success of the alliance derives from the realistic attitude of the two groups, which have been able to share not just a business plan, but rather a strategic vision of the market. It must be said however that initially, in terms of operational practice, General Electric sealed the technically sensitive parts of its engines to deny its partner access. The alliance once established had to control transfer of technology and know-how between the two partner companies, and it was only after many years of effort that wider relations of confidence were achieved.

The alliance: facts and figures

Today, CFM International relies on the 45,000 employees of Snecma Moteurs and General Electric Aircraft Engines in more than 86 countries. CFM engines are found in aircraft from more than 320 customers/operators.

In terms of production

Each of the partners is entirely responsible for the design and production of its own modules. Two assembly plants also manufacture the engine parts and produce about 1,000 CFM engines per year over the last three years. These two plants are situated close to their main customers (Boeing, Airbus) to geographically rationalize supply: Villaroche in France and Cincinnati in the United States (Ohio). In this way, the engines are tested on the customers' test beds before delivery.

General Electric	Snecma
Core engine (high pressure body, combustion chamber) + Integration system design Control systems Main engine controls	Fan Low pressure turbine Drive control and transmission elements: operation mechanisms + Installation Fuel system and accessories

Figure 15-21. How the 50-50 alliance between Snecma and General Electric is split up for the CFM56

In financial terms

The two partners split the revenues but not the profit margin. They do not share internal costs, each one purchases independently. This ensures that they try to minimize their costs and maximize profits. For each program, the division of income is determined as a function of the effective value contributed by each of the partners.

In terms of sales

CFM International has a sales organization divided between two sites, which allows optimization of customer relations:

15. Alliance Strategies

- the General Electric's sales branch in the United States grouping together product support and logistics in Cincinnati. Here they are in charge of the Asian and North and South American markets.
- Snecma sales' division in France likewise groups together product support and logistics in Villaroche. Here they are in charge of sales for Europe, Africa, the Middle East, the Indian sub-continent and the former USSR region.

The sales force is CFM International's, with each of the partners managing its own prospecting and sales staff costs.

In terms of marketing

CFM International has been able to develop a range of CFM56 engines with thrusts ranging from 18,000 pounds (18Klb) to 34 Klb and fan diameters of 60 to 72 inches. The advantage for the customer is to be able to have engines designed around the same basic structure. This commonality means greatly reduced costs (maintenance, stock management, staff training, etc.).

CFM56-2 (1979)	CFM56-3 (1984)	CFM56-5A (1987)	CFM56-5B (1993)	CFM56-7 (1996)	CFM56-5C (1991)
Original Engine	60" Fan Dia	68" Fan Dia		61" Fan Dia	72" Fan Dia
DC8 C-135 FR E-3 (Awacs) KE-3 (Tanker) E-6	B 737-300/-400/-500	A319/A320	A318/A319/A320/A321	B 737-600/-700/-800/-900	A340-300
22 / 24 Klb	18,5 / 20 / 22/ 23,5 Klb	22 / 23,5 / 25 / 26,5 Klb	21.6 / 22 / 23,5 / 27 / 30 / 31 / 33 Klb	19,5 / 20,6 / 22,7 / 24,2 / 26,3 / 27,3 Klb	31,2 / 32,5 / 34 Klb

Figure 15-22. The CFM56 family of engines

CFM has succeeded in increasing its market share in the over 100 seat civil aircraft segment, to the point where today it accounts for 45,6% of all engines sold in the last 15 years all over the world (a market totaling almost 11,000 units).

CFM International's main competitor for the A320, B737 and MD90 is IAE, a company bringing together Pratt & Whitney (32,5%), Rolls-Royce (32,5%), MTU (15%), Japanese groups (20%) and Fiat. Confronting CFM for the A340-300 and the B777, there are Rolls-Royce (Trent 800), Pratt & Whitney (PW4084) and General Electric (GE90). Thus the three manufacturers are direct competitors for the B777 market.

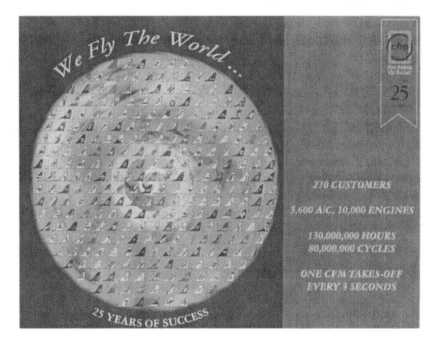

Figure 15-23. CFM International: 25 successful years

Another example of an alliance with the creation of a dedicated structure is Starsem.

STARSEM: AN ALLIANCE FOR A RELIABLE AND COMPETITIVE SPACE TRANSPORT SYSTEM

Figure 15-24. The Starsem alliance's symbolic logo

• **A rapidly evolving market**

The world launcher market underwent a profound evolution in the mid 1990's, with a large increase in the number of satellites launched and the development of " multimedia " satellite constellations in low and medium orbits. Requirements have thus been transformed over a few years, especially in view of the constantly increasing satellite masses. Whereas

15. Alliance Strategies

today satellites weighing less than 3 tons which account for more than half of the market, by 2005 the majority will be geostationary satellites of over 4 tons.

The launcher market, dominated up to now by Ariane 4 (60% of world launches), has had to adapt. The Americans, Boeing (Delta 4) and Lockheed-Martin, have invested in a new generation of Atlas 5 and Titan launchers (Evolutive Expandable Launch Vehicle) and these will soon be capable of carrying a payload of more than 8 tons (possibly 13 tons). The Ariane 5 launcher will also have to double its capacity to achieve 12 tons while reducing production costs at the same time. This is because the Russians, Krunichev (Ankara rocket) and the Ukrainians, Yuzhnoye as well as Chinese manufacturers, are offering cut price launches thus forcing the others to lower their prices. In this way, a kilo in orbit will soon only cost around 15,000 dollars. Consequently, there is a vital need for reliable and competitive space transport systems.

- **An original alliance: the Starsem philosophy**

Starsem was created in August 1996 in this competitive context, highlighted by an explosion in the demand for new capacity launchers. Their vocation is to ensure commercial launches using the Soyuz family of rockets. The alliance is based on a balanced partnership between different members:

- RKA (Rosaviacosmos / Russian Aviation and Space Agency) with 25%, the Russian Space Agency has overall charge of all the Russian space system and thus ensures a long term commitment. The Baikonur cosmodrome has some of the most modern technical (SPPF) installations in the world (1158m², class 100000). The Payload Preparation Facility (PPF) covers 286m², the Hazardous Processing Facility (HPF) 285m² and the Upper Composite Integration Facility (UCIF) is 587m²;

- Samara Space Center (TsSKB Progress), 25%. This space center gives access to the whole range of Soyuz launchers and allows development of new market-driven resources (Ikar re-ignitable stage for the GlobalStar program, improved Soyuz versions). This center produces launchers backed by its experience from the Soviet era. One of the pioneers of the Russian space program, it employs 5,000 people between the design office (TsSKB) and the production plant (Progress). It is also responsible for producing the supply launchers for the International Space Station (ISS);

- Aerospatiale-Matra 35% (EADS-European Aeronautic, Defence and Space Company), the architects behind Ariane, who have the know-how and mastery of complex systems;

- The Arianespace consortium 15%, world leader in launch services and the commercialization and exploitation of launchers.

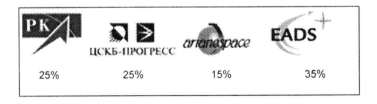

Figure 15-25. The European and Russian partners

Figure 15-26. Samara technical installations

The alliance is based on common interests: the Russian partners wish to develop their commercial activities internationally, and develop technologies with European players. The European partners want to develop and complete their commercial launch program in particular on the Russian market. All the partners were searching for an effective way of developing their activities, using their own know-how and the competence of the other members. At the heart of Starsem is a desire for cooperation between European and Russian manufacturers. The word Starsem symbolizes this alliance because it is based on the initials of European and Russian words: Space Technology Alliance with the R-7 launch vehicle (the Russian word for R-7 is Semyorka).

By joining together associated, the partners have been able to reliable and competitive launch solutions based on more than 40 years experience; Yuri Gagarin's first manned flight in the Soyuz launched sputnik in 1961, the Apollo mission in 1975 and the high frequency of launches from 1979 to 1982 (60 launches annually).

- **The Starsem offer**

The Soyuz launcher family is made up of:
- a basic, three stage version for unmanned missions,
- a Soyuz-Ikar version with greater performance and maneuverability,
- a Molniya, four stage version.

Other versions are being developed, e.g. the Fairing S and Fairing L.

15. Alliance Strategies

The idea is to offer access to space for scientific and commercial operators throughout the world by offering a variety of space applications from satellites right up to experimental manned flights. The main point is the ability to offer solutions using low, medium, high or very high elliptical terrestrial orbits.

- **The Starsem mission**

It consists of commercialization of the Soyuz launcher on the world market and its exploitation at the launch sites. The international team functions in on-going interaction with the shareholders, suppliers and customers.

Figure 15-27. Examples of launches n°s 1 and 10 by the Starsem Soyuz rocket

Starsem has its own marketing and sales structure in charge of the commercial applications of launches made by the family of Soyuz rockets. This division carries out market prospecting, sales and customer contract follow-up. The company finances the production of the commercial launchers and takes care of launch operations at the Baikonur (Kazakhstan) and Plesetsk (Russia) cosmodromes. The launcher division is thus in charge of overseeing the manufacture at Samara, the adaptation of the launchers to diverse customer missions, and the technical management of new space developments and programs between Russians and Europeans. The operations division prepares and conducts the launch campaigns, adapting and managing the necessary infrastructures. The financial division looks after financial, legal and human resources for the group. Starsem thus contributes to the commercial development and evolution of the Soyuz rockets as well as to the launch systems.

Orbit / launcher	Soyuz-Ikar	Soyuz-Fregat	Soyuz / ST	Soyuz / ST-Fregat
Circular 450 km Orbit	4 100 kg	5 000 kg	4 900 kg (almost circular orbit 450 x 495 km)	5 500 kg
Circular 1400 km Orbit	3 300 kg	4 200 kg	-	4 600 kg
Heliosynchronous 800 km Orbit	-	2 700 kg	-	2 900 kg
Geostationary Transfer Orbit (GTO) equivalent Cap Canaveral	-	1 350 kg	-	1 450 kg
GTO equivalent Kourou	-	1 100 kg	-	1 200 kg

Figure 15-28. Reference performances for Soyuz launchers from Baikonur

Starsem's main competitive advantages are its accumulated experience, proven know-how, very high launcher availability (manufacturing cycle less than one year) and launcher reliability (originally designed for manned flights) as well as the very high number of launches already made (more than 1600). These assets allow Starsem to make a highly competitive commercial offer which also completes and enlarges the Ariane range, allowing the European partner to have greater market cover on commercial launches.

- **Putting satellites into orbit**

The GlobalStar consortium, a vast satellite constellation, has called on Starsem's services to launch 24 satellites using 6 Soyuz-Ikar launchers. GlobalStar is a company founded jointly by Loral Space & Communications and Qualcomm, grouping together the world leaders in telecommunications: AirTouch Communications, China Telecom, Dacom/Hyundai, France Telecom/Alcatel, DaimlerChrysler, Vodafone, Alenia, Elsag Bailey and Finmeccanica. The first launches were on February 9, 1999 (ST01), then March 15, 1999 (ST02), April 15, 1999 (ST03), etc. The Soyuz-Ikar launcher has a re-ignitable stage, perfectly adapted to the GlobalStar program. Starsem developed this solution using the Kometa platform, to permit altitude changes in orbit for reconnaissance satellites.

Figure 15-29. The Soyuz rocket: horizontal integration of the stages

15. Alliance Strategies

In addition, the European Space Agency has asked Starsem to launch 4 Cluster II satellites using 2 Soyuz-Fregat rockets. Once again the reignitable stage system is used for the program, with propulsion derived from that of the interplanetary probes.

Starsem has therefore become the new opportunity on the market for low cost missions, offering up to 6 satellites per launch (150kg).

Starsem must:
- continue the rhythm of launches started with the GlobalStar (Soyuz) ST01, ST02 and ST03 as well as for ST07, ST08 and ST09 for Cluster II (Soyuz-Fregat) and carry out other planned launches;
- build on the value of these first launches in order to win new contracts;
- take the decision to expand their range in order to be more in phase with market needs, in particular concerning the development of the Soyuz/ST launcher destined for the Kourou pad.

Another alliance of the same type is the Sea Launch project.

SEA LAUNCH: FROM THE SEA INTO SPACE

Sea Launch was created in answer to the growing demand for commercial satellite launchers which are less costly and more reliable. It is a system using a sea platform, offering commercial customers from around the world a heavy payload launch service. This innovative launcher combines aerospace and maritime high technology to develop high value products and services.

Figure 15-30. The Odyssey platform and the Sea Launch Commander

First discussions between partners took place during the summer of 1993, followed by a preliminary study in spring 1994. The Sea Launch alliance was formed in April 1995. December 1995 marked the signing of the first order from the Hughes company, and the construction of the infrastructure and launch system began. Additional orders followed from Hughes and supplementary ones came from customers like Space Systems/Loral. From 1996 to 1998 the main structures and then the platform as well as the launchers were built and transported to Home Port, Long Beach. California (headquarters of the Sea Launch limited company) for assembly. The launch point is on the equator in the Pacific Ocean (longitude 154° west). In March 1999 there took place a demonstration launch and in October of the same year, with 19 firm orders, the first commercial launch (Directv 1-R satellite).

The Sea Launch partners are each involved in a part of the activities:

- the American company Boeing Commercial Space (40%): payload fairing, analytical and physical integration of launcher, mission operations;
- the Russian RSC Energia group (25%): last stage DM-SL module, launcher integration, mission operations;
- the Ukrainian SDO Yuzhnoye/PO Yuzhmash group (15%): the two first stages of the Zenit-3SL rocket, launcher integration support, mission operations;
- the Anglo-Norwegian, Kvaerner group (20%): operational services for the Odyssey platform, launcher assembly and guidance, Sea Launch Commander.

Mission operations group all tasks within the launch process from the loading, launcher integration and firing, right up to placing in orbit.

3.4 From alliance to merger

In North America as in Europe, the aeronautics and space industry has become much more concentrated over the past 30 years. Certain mergers have been created directly between ex-competitors as was the case with Boeing and McDonnell Douglas. Conversely, other mergers have come after a long alliance period which was the case with AIC, Airbus Integrated Company, result of the merger between the French, German and Spanish partners. From the moment that the merger takes place, the term alliance is meaningless since there is just one entity which integrates the previous partners. The objectives are unified and determined by a single decision making unit.

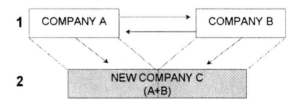

Figure 15-31. From alliance to merger

When there are mergers, the Federal Trade Commission (FTC) as well as the European Community, check to see if there are any potential effects in terms of competition. For example, these two institutions have given their agreement under certain conditions, for Boeing to buy Hughes Space and Communications (HSC) for the sum of 3.75 billion dollars. Actually it came down to guaranteeing loyal competition on the satellite and launcher market. An inspector was named to guarantee respect of conditions imposed on Boeing which:
- must split up its satellite and launcher activities, to the point where it cannot gather information about competitor's launchers via its satellite division;
- must in addition, supply information concerning its satellite interface to other launch operators.

Another example of an alliance giving rise to a merger is Airbus.

FROM GIE AIRBUS INDUSTRIE TO THE INTEGRATED COMPANY EADS

Airbus Industrie was initially created when the main European aeronautical constructors realized that cooperation was the only way of halting the progressive decline in their industry. It was therefore a pragmatic industrial decision which lead to the alliance, on December 18, 1970, of the different European aeronautical companies within a Grouping of Mutual Economic Interest. (GMEI). The GMEI system, introduced into France in 1967 and then into Europe in 1985[11], has turned out to be the best adapted structure for criss-cross international cooperation. The profits of the GMEI benefit all the partners as a function of the number of shares they hold.

This organization, much simpler than the cooperative programs used up till then, was set up in the following manner:

- a supervisory board with the partners grouped around a president: the German DASA[12] (37,9%), the French Aerospatiale[13] (37,9%), the British, British Aerospace (20%) and the Spanish CASA (4,2%).

- an executive committee made up of a President together with the Airbus directors from the partners, a financial director and an operational director.

The Airbus Industrie Consortium coordinates the industrial activities and is in charge of marketing, sales and technical support.

This GMEI developed relations with the Dutch Fokker and the Belgian Belairbus. The first flight of an aircraft built by Airbus Industrie took place in October 1972. This was the test flight of the twin-engined A300 which entered service in 1974 in order to meet American Airlines specifications. Then a range was produced, with another twin-engined plane, the A310 launched in 1978, the A320 in 1985, the A321 in 1989, the A319 in 1992 and the A330 and A340 in 1993. In 1993 the 1,000th Airbus plane was delivered and in 1998 the 3,000th order was signed. The Airbus alliance is without a doubt one of the most exemplary that has ever been made, whether from a technological, investment, intercultural, working methods or commercial success viewpoint.

Figure 15-32. The Airbus A320, the flagship product of the range, built by the Airbus Industrie consortium

On January 13, 1997, the four Airbus partners announced the project to transform the GMEI into an industrial company, designed to group together all the assets held by the partners. During this same period, the restructuring of the aeronautical and defense industries, started in 1994 -1995, spawned two new giants: Boeing McDonnell, Lockheed Martin and Raytheon. On the October14, 1999, the French and German governments announced the regrouping of their aeronautical, space and defense industries under the aegis of a new group called EADS. The Spanish group CASA joined Aerospatiale-Matra and DASA within EADS on December 2, 1999. EADS became the third space and military group in the world, the second for civil aviation (80% of Airbus) and the number 1 in the helicopter and space launcher (Ariane) sector. It is one of the world leaders for satellites (Astrium), military aviation (A400M, Eurofighter, Mirage, Rafale) and defense systems (second missile constructor).

On June 23, 2000, at the same time as the industrial launch of the A380 program, Airbus became a private firm in the form of a simplified joint stock company: Airbus Integrated Company (AIC).The company is controlled by two shareholders:

- EADS (European Aeronautics Defense and Space Company), company resulting from the merger between DASA, Aerospatiale-Matra and CASA (80%);
- BAe Systems (20%).

The merger allows EADS to hold 80% of Airbus Integrated Company and to continue the alliance with BAe Systems.

The company has two directing bodies:
- a shareholders committee of seven members, five named by EADS,
- an executive committee of ten members, eight chosen by EADS.

To avoid any blocked situations, most of the decisions (management, launching of new programs, commercial policy) are by simple majority vote. Transfer or acquisitions of more than 500 million euros need a unanimous vote.

A company with multiple origins

• Aerospatiale was created in 1970, in the wake of the merger of Sud-Aviation, Nord-Aviation, and Société d'Etudes et de Réalisation d'Engins Balistiques (SEREB). In June

15. Alliance Strategies

1999, Aerospatiale was privatized and merged with Matra Hautes Technologies (Groupe Lagardère), forming Aerospatiale-Matra.

• Created in 1923, CASA (Construcciones Aeronáuticas SA) is the Spanish leader in the aeronautics and space industry and for many years has participated in the main European programs in the sector. A subsidiary of EADS, CASA has maintained its name.

• DaimlerChrysler Aerospace (DASA) was created in 1989 under the name of Deutsche Aerospace to regroup the aerospatiale activities of the Daimler-Benz group. The merger was carried out with MBB (Messerschmitt-Bölkow-Blohm), Dornier, MTU (Motoren-und Turbinen-Union München/Friedrichshafen) and Telefunken Systemtechnik. Following the merger of Daimler-Benz with the American automotive constructor, Chrysler, in 1998, the company was renamed DaimlerChrysler Aerospace.

Main stages leading to the creation of EADS:

May 15, 1969	Signing of Airbus contract
December 18, 1970	Founding of Airbus Industrie
August 9, 1972	Creation of Euromissile
March 26, 1980	Creation of Arianespace
January 1, 1992	Creation of Eurocopter
May 14, 1999	Aerospatiale-Matra goes public
October 14, 1999	Announcement of the EADS merger in Strasbourg
October 20, 1999	Creation of Matra BAe Dynamics
December 2, 1999	CASA becomes a founding member of EADS
May 17, 2000	Start up of Astrium
June 23, 2000	Decision to create Airbus Integrated Company (AIC)
July 10, 2000	Start up of EADS

EUROCOPTER: ONCE A JOINT-VENTURE, TODAY A SUBSIDIARY

In 1992, the French Aerospatiale[14] (70%) and the German DaimlerChrysler Aerospace (30%) decided to create a common subsidiary in the helicopter sector: Eurocopter. Today this first totally European company is the world leader for the construction of helicopters, employing almost 10,000 people throughout the world (13 subsidiaries and 100 support centers), for a turnover of around 2 billion dollars. The activities of this company are split into 56% in the civil and parapublic sector, and 44% in the military sector. More than 1,740 customers in 132 countries have chosen Eurocopter helicopters giving a total of 8,414 aircraft in service throughout the world.

Out of this alliance has been born a complete range of helicopters from light craft (such as the BO105, AS 350/355, BK117, EC135 or EC120 "Colibri") through special combat craft (Tiger) to medium and heavy craft such as the AS365, EC 155, AS332 and NH90.

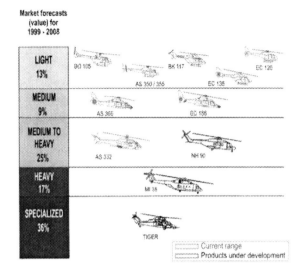

Figure 15-33. A very wide range of products

Always in tune with the present expectations of its customers and trying to anticipate their future needs, Eurocopter has gone down the road of technological innovation by developing helicopters with high performance in technical terms (speed, range, etc.) but also in economic terms (running costs, etc.) and environmental ones (protection of the environment).

Figure 15-34. Specific helicopters for different missions

Eurocopter works in collaboration with numerous countries. It has established commercial cooperation agreements creating subsidiaries, participation of other members

15. Alliance Strategies

within the Eurocopter programs being developed or by granting of manufacturing licenses. One of the most recent collaborations concerns the NH90 tactical helicopter which involves Agusta and Fokker in addition to Eurocopter. Five prototypes have initially been made in order to qualify the basic machine, and then the terrestrial version followed by the marine one. Agusta's experience with the EH-101 and Eurocopter's with the Tiger have facilitated the design of the basic avionics (displays, navigation, radios, surveillance and diagnosis, engine management, etc.). The systems from the Tiger have been adapted for the terrestrial version (helmet displays, digital map generators, electronic countermeasures, etc.). In addition the NH-90, the first in this category of craft, uses electrical control systems designed to a large extent thanks to research carried out by Aerospatiale-Matra for Airbus. This is one of the strong points of the EADS alliance strategy: allowing the group to transpose or exploit technological solutions from one aeronautical domain to another.

NOTES

1. Thomson-CSF was re-named Thales in 2000.
2. In-house sources: Turbomeca / Snecma.
3. In-house sources: Turbomeca / Snecma.
4. For further information on alliance strategies see Garrette, B. and Dussauge, P., (1995), *Les stratégies d'alliance*, Paris, Les Éditions d'Organisation.
5. See Chapter 2, The Individual and Organizational Purchase.
6. See Chapter 10, Project Marketing.
7. Source: www.skyteam.com and *Air France Magazine*, n°39, July 2000.
8. Destinations and countries not duplicated.
9. Source: www.skyteam.com (July 2000)
10. *The designation CFM 56 is a combination of the name "Commercial Fan", usually given to civil GE engines, and "M-56", abbreviation of the 56th engine project launched by Snecma.*
11. GMEI: Grouping of Mutual Economic Interest.
12. Which became DaimlerChrysler Aerospace Airbus
13. Which became Aerospatiale-Matra
14. Which became Aerospatiale-Matra and then merged with DASA to form EADS (including CASA).

REFERENCES

Dussauge, P., (1992), Alliances et coopérations dans les industries aérospatiale et de l'armement: une étude empirique des comportements stratégiques et des choix d'organisation, *Thèse de Doctorat en Sciences de Gestion*, Université de Dauphine, Paris IX.
Garrette, B. and Dussauge, P., (1995), *Les stratégies d'alliance*, Paris, Les Éditions d'Organisation.
Hochmuth, M.S., (1974), *Organizing the Transnational*, Sijthoff.
Lawrence, P.K. and Braddon, D.L., (1998), *Strategic Issues in European Aerospace*, Ashgate.
Lorenzoni, G. and Ornati, O., (1988), Constellations of Firms and New Ventures, *Journal of Business Venturing*, n°3.
Revue Aerospatiale, (2000), Finmeccanica opts to partner EADS, n°168, May, 2-5.
Rouach, D., (2000), Airbus-Boeing en Israël: un combat symbolique, *Les Echos*, 11th of May.

BIBLIOGRAPHY

BOOKS

Aaker, D.A. and Lendrevie, J., (1994), *Le management du capital de marque*, Paris, Dalloz.
Airbus, un succès industriel européen, (1995), Éditions Rive Droite.
Alba, J.W., Hutchinson, J.W. and Lynch, J.G., (1991), *Handbook of Consumer Behavior*, Englewood Cliffs, N.J., Prentice-Hall.
Altfeld, H.H., (2001), *Government Funding for Aerospace, a comparative analysis of government expenditure in the EU and the US*, AECMA.
Andrews, D.C. and Andrews, W.D, (1993), *Business Communication*, Macmillan.
Andrieu, O. and Lafont, D., (1995), *Internet et l'entreprise*, Paris, Eyrolles.
Benoun, M., Héliès-Hassid, M.-L. and Alphadeve, M., (1995), *Distribution, acteurs et stratégies*, Paris, Economica.
Bergadaà, M., (1997), *Révolution vente*, Paris, Village Mondial.
Bloch, A., (1996), *L'intelligence économique*, Paris, Economica.
Bonoma, T.V, Zaltman, G. and Johnston, W.J, (1997), *Industrial Buying Behavior*, Cambridge, Marketing Science Institute.
Brun, M. and Rasquinet, Ph., (1996), *L'identité visuelle de l'entreprise au-delà du logo*, Paris, Éditions d'Organisation.
Carre, A. D., (2001), *Aéroports et stratégie d'entreprise : Les aéroports, des entreprises à part entière*, Les Presses de l'Institut du Transport Aérien.
Chambelin, D., (1998), *Navigation Générale, JAR-FCL*, Jean Mermoz.
Chernatony, L. (de) and McDonald, M., (1992), *Creating Powerful Brands*, London, Butterworth-Heinemann.
Cova, B. and Salle, R., (1999), *Le marketing d'affaires*, Paris, Dunod.
Cowell, D., (1984), The Marketing of Services, William Heinemann.
Cross, R., (1997), *Revenue Management*, Broadway Books.
Cunningham, M., (2001), *eB2B*, Paris, Village Mondial.
Doganis, R., (1991), *Flying off course*, Harper Collins.
Doyle, P., (1994), *Marketing Management and Strategy*, Prentice Hall.
Dubois, P.-L. and Jolibert, A., (1992), *Le marketing, fondements et pratique*, Paris, Economica.

Dusclaud, M. and Soubeyrol, J., (1998), *Enjeux technologiques et relations internationales*, Paris, Economica.
Eiglier, P. and Langeard, E., (1987), *Servuction, le marketing des services*, Paris, McGraw-Hill.
England, W.B. and Leenders, M.R., (1975), *Purchasing and Material Management*, Homewood, Richard D. Irwin.
Farnel, F., (1994), *Le lobbying : stratégies et techniques d'intervention*, Paris, Les Éditions d'Organisation.
Filser, M., (1989), *Canaux de distribution*, Paris, Vuibert.
Ford, D., (1997), *Understanding Business Markets : Interactions, Relationships and Networks*, The International Marketing and Purchasing Group (IMP), The Dryden Press.
Garrette, B. and Dussauge, P., (1995), *Les stratégies d'alliance*, Paris, Les Éditions d'Organisation.
Gauchet, Y., (1996), *Achat industriel et marketing*, Paris, Publi-Union.
Ghauri, P.N. and Usunier, J.-C., (1996), *International Business Negotiations*, Oxford, Pergamon/Elsevier.
Grönroos, C., (1990), *Service Management and Marketing : Managing the moments of truth in service competition*, New York, Lexington Books.
Hakansson, H., (1982), *International Marketing and Purchasing of Industrial Goods*, New York, Wiley.
Hart, N.A, (1995), *Strategic Public Relations*, Macmillan.
Hartley, K., (1998), *Economic Evaluation of Offsets*, The University of York.
Hayward, K., (1986), *International Collaboration in Civil Aerospace*, Pinter.
Hochmuth, M.S., (1974), *Organizing the Transnational*, Sijthoff.
Howard, J. A. and Sheth, J. N., (1977), « Théorie du comportement de l'acheteur », *Encyclopédie du marketing*, Paris, Éditions Techniques.
Keegan, W.J., (1995), *Global Marketing Management*, Englewoods Cliffs, N.J., Prentice-Hall.
Kotler, P. and Dubois, B., (2000), *Marketing Management*, Paris, Publi-Union.
Kotler, P., Armstrong, G., Saunders, J. and Wong, V., (1999), *Principles of Marketing*, Prentice-Hall.
Lawrence, P. and Braddon, L., (1999), *Strategic Issues in European Aerospace*, Ashgate.
Le Picard, O., Adler, J.-C. and Bouvier, N., (2000), *Lobbying, les règles du jeu*, Paris, Les Éditions d'Organisation.
Lehu, J.-M., (1996), *Praximarket*, Paris, Éditions Jean-Pierre de Monza.
Levitt, Th., (1965), *Industrial Purchasing Behaviour, a Study of Communication Effects*, Harvard University.
Macquin, A., (1998), *Vendre*, Paris, Publi-Union.
Malaval, P., (2001), *Marketing Business to Business*, 2nd Ed., Paris, Pearson Education France.
Malaval, P., (1998), *Stratégie et Gestion de la Marque Industrielle*, Publi-Union.
Malaval, P., (1999), *L'Essentiel du Marketing Business to Business*, Publi-Union.
Malaval, P., (2000), *Strategy and Management of Industrial Brands*, Kluwer Academic Publishers.
Malaval, P. and Bénaroya, C., (2001), *Marketing aéronautique et spatial*, Paris, Pearson Education France.
March, J., (1991), *Décisions et Organisations*, Paris, Les Éditions d'Organisation.
McGuire, S., (1997), *Airbus Industrie : Co-operation and Conflict in EC-US Trade Relations*, Macmillan.
Minichiello, R.J., (1990), *Retail Merchandising and Control : Concepts and Problems*, Irwin, Burr Ridge.
Morrell, P.S., (2001), *Airline Finance*, Ashgate.
Moulinier, R., (2000), *Le cœur de la vente*, Paris, Les Echos Éditions.
Mueller, P., (1988), *L'Airbus. L'ambition européenne. Logiques de marché*, Paris, L'Harmattan.
Pache, G., (1994), *La logistique : enjeux stratégiques*, Dunod.
Parkinson, S.T. and Baker, M.J., (1986), *Organizational Buying Behavior*, Londres, The Macmillan Press.
Peppers, D. and Rogers, M., (1998), *Le One-to-One*, Les Éditions d'Organisation.
Picq, J., (1990), *Les ailes de l'Europe*, Paris, Livre de Poche.

Porter, M., (1980), *Competitive Strategy: Techniques for analysing Industries and Competitors*, New York, The Free Press.
Reeder, R.R., Brierty, E.G. and Reeder, B.H., (1991), *Industrial Marketing*, Englewood Cliffs, N.J., Prentice-Hall.
Ries, A. and Trout, J., (1986), *Le positionnement : la conquête de l'esprit*, Paris, McGraw-Hill.
Robinson, P., Faris, C.W. and Wind, Y., (1967), *Industrial Buying and Creative Marketing*, Allyn and Bacon.
Rodgers, E., (1996), *Flying High : The Story of Boeing and the Rise of the Jetliner Industry*, Atlantic Monthly Press.
Salle, R. and Sylvestre, H., (1992), *Vendre à l'industrie*, Paris, Éditions Liaison.
Sabbagh, K., (2001), *Twenty first century jet, the making and marketing of the Boeing 777*, Scribner.
Schiffman, L. and Lazar Kanuk, L., (1997), *Consumer behavior*, Prentice-Hall.
Shaw, S., (1990), *Airline Marketing & Management*, Pitman.
Szapiro, G., (1998), *Communication Business to Business, Les 7 pyramides de la réussite*, Paris, Les Éditions d'Organisation.
Treacy, M. and Wiersema, F., (1995), *L'Exigence du choix*, Paris, Village Mondial / Pearson.
Trout, J., (1996), *Les nouvelles lois du positionnement*, Paris, Village Mondial.
Von Hippel, E., (1988), *The sources of Innovation*, New York, Oxford University Press.
Webster, F.E. and Wind, Y., (1972), *Organizational Buying Behavior*, Englewood Cliffs, N.J., Prentice-Hall.
Williamson, O.E., (1975), *Markets and Hierarchies*, New York, The Free Press.
Williamson, O.E., (1985), *The Economic Institutions of Capitalism-Firms, Markets, Relational Contracting*, New York, The Free Press.
Woodside, A. and Vyas, N., (1984), *Industrial Purchasing Strategies*, Reading Mass., Addison-Wesley.
Zeyl, A. and Dayan, A., (1996), *Force de vente : direction, organisation, gestion*, Paris, Les Éditions d'Organisation.

SPECIALIZED JOURNALS AND TRADE PRESS

A/C Flyer
Advent Composite
Aero International
Aéronautique Business
Aéroports Magazine
Aerospace Daily
Air & Cosmos
Air & Space Europe
Air Cargo News
Air Cargo World
Air France Magazine
Air International
Air Transport Intelligence
Air Transport World
Aircraft Economics
Airfinance Journal
Airline Business
Airlines
American Economic Review (The)
Armada
Armed Force Journal
Armement
Aviation Daily
Aviation International News
Aviation Security International
Aviation Week & Space Technology
Avionics Magazine
Business & Commercial Aviation
Business Air News
Business Traveller
Business Weeks
Décisions Marketing
Defense Briefing
Defense Daily
Defense Helicopter
Defense News
DGA
Echos (Les)
Echos-Aéronautique Business (Les)
Economist (The)

European Journal of Marketing
Figaro Économie (Le)
Flight International
Gestion 2000
Guardian (The)
Harvard L'Expansion
Helicopter International
Helicopter News
Helicopter World
Helidata
Industrial Marketing Management
Industry Week
International Business Review
International Journal of Advertising
International Space Industry Report
Jane's
Journal of Business Venturing
Journal of Consumer Marketing
Journal of Electronic Defense
Journal of Interactive Marketing
Journal of Marketing
Journal of Personal Selling and Sales Management
Journal of Professional Services Marketing
Logistiques Magazine

Management International Review
Marketing Science Institute
Military Technology
Overhaul & Maintenance
Passager Aérien
Performance Materials
RAF
Recherche et Applications en Marketing
Research in International Marketing
Revue Aérospatiale
Revue des Marques (La)
Revue Française de Gestion
Revue Française du Marketing
Rotor and Wing
Satellite Finance
Sloan Management Review
Space & Communication
Space News
Space Technology
Time
Usine Nouvelle (L')
Wall Street Journal (The)
World Aircraft Sales
Yield & Revenue Management

SUBJECT INDEX

Ad hoc studies, 401
Advertising, 328
Aeronautics and space sector (specific characteristics), 1-2
Aeronautics marketing, 18-19
After sales, 469-471
Airline companies, 2-3, 134-139, 230-238, 432-435
Airport, 56, 3307-309, 314-315
Alliance (objectives), 492-498
Alliance strategy, 491-527
Alliance (different forms), 498-527
Audience panel, 103-104
Audience, 345

Barter, 485
BCG, 169-172
Benchmarking, 86
Bid, 17
Bidding, 47-48, 262-263, 311, 321
Bilateral Agreement, 2
Billboards, 376-377
Bottom-up segmentation, 122-128
Brand (awareness), 400-402
Brand (functions for the company), 404-406
Brand (functions for the customer), 406-413
Brand (image), 402-403
Brand (international policy), 424-425

Brand (loyalty), 403-404
Brand (performance facilitating), 410-413
Brand (positioning), 404-405
Brand printability, 415
Brand, 399-453
Brand/product communication, 334-336
Building demand, 313-317
Business jet market, 124-128
Business jet, 274-275
Business plan, 251-252
Business to Business, 15-16
Business to Consumer, 15-16
Bus mailing, 372-373
Buy back, 485
Buyclass, 44
Buyer, 35, 52, 62-63
Buygrid, 44
Buying (process), 30-34
Buying center, 34-36
Buying phases, 37

Catering, 119-121
Chicago Convention, 2
Co-branding, 409
Collective communication, 336
Communication (budget), 339-340
Communication (objectives), 328, 330

Communication (targets), 337-338
Communication plan, 337-345
Communication policy, 327-345
Communication, 291
Company news magazine, 365-366
Copy strategy, 341-343
Corporate brand, 425-426
Corporate communication, 331-334
Corporate philanthropy, 390
Cost-based pricing, 254-256
Counter-trade, 485
Criteria (rational, non-rational), 45-46
Cross functional teams, 194
Curative maintenance, 462
Customer intimacy, 317-319
Customer oriented, 15
Customer Relationship Management (CRM), 77, 231
Customer segmentation, 117-118, 121-124
Customers training, 472-484

Datamining, 231
Decision maker, 35, 52
Delphi method, 97
Department of Defense (DoD), 2
Deregulation Act, 3
Derived demand, 18, 247
Design to cost, 193
Desk research, 83
Differentiation, 141, 467
Direct channel, 280
Direct marketing, 371-373
Direct offset, 486-487
Distribution economy, 10
Distribution system, 280-286
Distribution, 277-306
Distributor panel, 104-105

Early adopters, 193
e-business, 20
EDI (Electronic Data Interchange), 20
Editorial style advertising, 364

ELLV (Evolutive Expandable Launch Vehicle), 2
Empowerment, 60
Environment economy, 10
E-procurement, 49-51
e-ticketing, 284-286
Experience effect, 162-163
External analysis, 337
External diagnosis, 149
External marketing, 233

Face to face interview, 96
Field research, 83
Flexibility strategy, 266
Focus group, 96-97
Fractional ownership, 274-275
Freight forwarders, 240-241

General press, 362-364
"Gray market", 270

Heavy maintenance, 458
Horizontal publications, 360
Hubs, 504-505

In suppliers, 39
Indirect channel, 280-282
Indirect offset, 487
Individual purchase, 23-34
Industrial brand ("purchaseability"), 414-415
Industrial brand (classification), 424-436
Industrial brand (visibility), 415-417
Industrial brand, 413-423
Interactive marketing, 233
Influence-oriented Marketing, 12
Influencer, 35-36, 53
Information sources, 82-95
Information system, 75-78
In-house communication, 330-331
In-house information, 82
Initial training, 482
Innovation, 182-210
Inseparability, 214
In-service training, 482

Subject Index

Intangibility, 211-213
Integrated, 241-242
Internal analysis, 337
Internal diagnosis, 149
Internal marketing, 233
Internet studies, 99-100
Internet, 268-269, 284-286, 366-371
Jingle, 441

Knowledge data, 231

Learning curve, 161-163
Leasing, 272-274
Letter of Intent (LoI), 455-456
Life cycle, 163-167
Light maintenance, 458
Little (model), 175-176
Lobbying, 377-388
Logistics, 277-280
Logotype, 436-438
Loyalty policy, 455-491

"Mirror" marketing, 61
Mailing, 372
Maintenance costs, 261
Maintenance, 455-471
Make or buy, 40-41, 44
Market economy, 10
Market survey, 88-94
Marketing (Function), 11-12
Marketing of services, 211-244
Marketing plan, 147
Marketing-mix, 14-15
Marketplace, 50-51, 222-227
Maslow, 28-29
McKinsey (model), 174-175
Media planning, 344-345
Memorandum of Understanding (MoU), 455-456
Merchandising, 286-291
Merger, 522-523
Modified rebuy, 42-43

Needs, 28-29
New task, 43-44
Niche market, 177

Offset, 484-490, 486-487
One-to-one, 20-21
Operating costs, 260
Operational costs, 248
Organizational purchase, 34-51
Out suppliers, 39

Passengers, 56-57, 121
Penetration strategy, 264-266
Perceived risk, 46, 322-323
Perceived value approach, 258-259
Perishability, 213-214
Pilots, 55, 475-481
Positioning (objectives), 133-139
Positioning, 133-144
Predictive maintenance, 463
Prescriber, 35-36
Preventive maintenance, 462
Pricing (approaches), 254-261
Pricing, 245-276
Primes, 393-394
Product brand, 426-430
Product oriented, 15
Product portfolio, 168-177, 252-253, 273
Product range, 178-182, 198
Product segmentation, 117-118
Production economy, 9-10
Products generation, 184-186
Professional directories, 365
Professional services, 220-230
Project marketing, 17, 307-326
Promotional discounts, 271
Prospecting, 292
Public relations, 388-390
Pull innovation, 189-191
Purchase (Influencing factors), 23-30
Purchase marketing, 60-65
Push innovation, 187-189

Qualitative study, 96-97
Quality, 456-457
Quantitative study, 97
Quantity discount, 270-271

Radio, 375-376
Rebates, 271
Reciprocity, 485
Regulatory organizations, 3-9
Related industries, 18-20
Research & development, 186-189
Retrofit, 259
Reverse engineering, 86
Reverse marketing, 61
Reverse shows, 61

Sales action plan (PAC), 156-160
Sales force (management), 296-306
Sales force (organization), 298
Sales force (remuneration), 301-304
Sales force (size), 299
Sales force (training), 304-305
Sales promotion, 328, 391-397
Sales representative, 291-296
Screening, 192
Segmentation (Bonoma et Shapiro), 129-130
Segmentation (objectives), 115-117
Segmentation (Wind & Cardozo), 128-129
Segmentation, 115-132
Selling process, 293-294
Senior marketing, 26-27
Service Bulletin, 3
Service Marketing, 26-27
Share of voice, 339
Skimming strategy, 263-264
Slogans, 439-441

Space marketing, 19
Specifications, 38-39, 320-326
Sponsoring, 390-391
STC (Supplemental Trade Certificate), 3
Straight rebuy, 42
Strategy (diversification), 492
Strategy (internationalization), 492
Strategy (vertical integration), 491
Supplier conventions, 63
Supply chain, 18-20
Surveillance, 76, 78-82
Survey-oriented Marketing, 12
SWOT, 150
Systematic maintenance, 462

Telephone interview, 96, 101
Television, 373-375
Top-down segmentation, 119-121
Trade press, 358-366
Trade shows, 347-358

Unique Selling Proposition, 135
User, 36-37, 52
Users panel, 102-103

Value-added pricing, 256-261
Variability, 214-216
Vertical publications, 360-366
Visual identity code, 436-442, 441-442

Yield management, 213, 266-269

Printed in the United States
79878LV00001B/83